Mathematics

for Caribbean Schools

Students' Book 3

Publishing for the Caribbean

Althea A Foster   Terry Tomlinson   Third Edition

Orders: please contact Hachette UK Distribution, Hely Hutchinson Centre, Milton Road, Didcot, Oxfordshire, OX11 7HH. Telephone: +44 (0)1235 827827. Email education@hachette.co.uk. Lines are open from 9 a.m. to 5 p.m., Monday to Friday. You can also order through our website: www.hoddereducation.com

The right of Althea Foster and Terry Tomlinson to be identified as authors of this work has been asserted by them in accordance with the Copyright, Designs and Patents Act 1988.

First published in 1988 by Longman Group Limited
Second edition published 2000
This edition published 2007

Published from 2015 by Hodder Education,
An Hachette UK Company
Carmelite House
50 Victoria Embankment
London EC4Y 0DZ
www.hoddereducation.com
First impression 2016

ISBN: 978–1–4058–4781–0

Set in 9.5/12 pt Stone Serif

Printed and bound by CPI Group (UK) Ltd, Croydon, CR0 4YY

Acknowledgements

The Publishers wish to acknowledge the work of J B Channon, A McLeish Smith, H C Head and M F Macrae which laid the foundation for this series.

The Publishers are grateful to the Caribbean Examinations Council for permission to reproduce examination questions as follows.

June 1986	Basic proficiency question 6 (17c question 15)
June 1987	General proficiency question 1b (17c question 9) question 1a (16c question 6)
June 1987	Basic proficiency question 4 (20r question 5) question 7a (22e question 3) question 8 (17c question 11)
June 1988	General proficiency question 1c (Revision exercise 3 question 9) question 7 (17a question 10) question 8b (21c question 11)
June 1988	Basic proficiency question 4 (21f question 4) question 6 (20r question 8) question 9a (23c question 2)
January 1989	General proficiency question 4c (20r question 7) question 5 (16c question 17) question 7 (24b question 10)
June 1989	General proficiency question 2c (20o question 7) question 6 (17a question 12) question 7b (Revision exercise 3 question 3)
June 1989	Basic proficiency question 7b (17b question 8)
January 1990	General proficiency question 1b (Revision test 3 question 7)
June 1990	General proficiency question 3 (General revision test A question 16)
June 1990	Basic proficiency question 3 (5d question 6) question 6b (21g question 4) question 8 (Revision test 6 question 7)
January 1991	General proficiency question 4 (21e question 4)
June 1991	Basic proficiency question 7 (5d question 1) question 10a (Revision exercise 6 question 3)
June 1992	General proficiency question 1c (17a question 11) question 9 (17b question 8)
June 1992	Basic proficiency question 3a (21b question 7) question 4b (Revision exercise 4 question 10) question 5b (21a question 8) question 6 (5c question 5)
June 1999	(Practice exercise P12.1 question 3)

Preface

This series of four volumes, of which this is the third, is intended for use primarily by students who are preparing to sit the certificate examinations held by the Caribbean Examinations Council and by the individual countries in the Caribbean.

Each volume represents material which may be covered by the average student in one year approximately (although some students may need more time) so that there is ample time for the series to be completed over a four to five year period. Emphasis has been placed on detailed explanation of concepts, principles and methods of working out problems. In addition, many problems have been included, both as worked examples to illustrate particular approaches to solving problems in the teaching text, and also as exercises for practice and reinforcement of the concepts. This has been done in a deliberate attempt to provide a stimulus to teachers in developing their strategies for teaching different topics; but, especially, to provide guidance to students as they work or revise on their own.

At the end of each chapter a summary of the main points developed in the chapter has been included. Practice exercises enable students to consolidate what they have learnt. Key reference words are printed in bold type throughout the text. The text has been arranged sequentially so that each chapter may use material covered in the previous chapters. However, in order to use relevant examples at some points in the text, ideas/concepts, which may need a short explanation or reminder by the teacher, may be introduced; these concepts will be fully explained in a later chapter. A few chapters are independent of the previous chapters and so may be omitted without loss of continuity at a first working. However, it is believed that the text will be used most efficiently by working through the chapters in the given order.

This revised edition seeks to incorporate changes in keeping with amendments to the syllabuses in the various Caribbean countries and that of the Caribbean Examinations Council (CXC). All topics in the core syllabus which are tested at the Basic proficiency level of the CXC examination have been completed in this volume. Attention has also been given to the suggestions of Caribbean teachers whose positive reaction and responses to the series have been most encouraging.

We are indebted to both the teachers and the students whose questions and responses over the years have undoubtedly influenced our thinking and our approach to the teaching of the subject as exemplified in this series.

We also wish to acknowledge the work of J B Channon, A McLeish Smith, H C Head and M F Macrae which laid the foundation for this series.

To the teacher

This edition of the series of four texts, revised with respect to content and its sequencing and to pedagogy (to a lesser extent), will be found useful in providing help and guidance in how the topics are taught and the order in which they are taken. The texts do not attempt to prescribe specific approaches to the teaching of any topic. Teachers are free to adopt or modify the suggested approaches. It is the teacher who must decide on the methodology to be used in creating the most suitable learning conditions in the classroom, in providing challenging activities which motivate the students to think and yet give them a chance to succeed in finding solutions.

It is vitally important that teachers use the 'Oral' exercises to initiate class discussion in the careful development of concepts. Teachers are urged to use, and thus reinforce, concepts taught earlier, so that, for example, the use of estimation and approximation in the calculation of numerical values is practised throughout the course.

The approach in the examples and exercises in Geometry is used as a means of developing the capacity to reason logically and to make valid deductions from stated data. The ability to select relevant information from a word problem is a

related faculty to which teachers must devote attention.

In addition, it is widely accepted that learning is aided by doing. Thus concrete/practical examples and real-life applications must be provided, whenever possible, as well as the use of pictures, flow-charts and other diagrammatic representations to deepen the understanding of abstract/theoretical ideas. Some problems in the exercises require the use of diagrams which are tedious and/or time-consuming to produce. In order to keep down costs to the student/school and reduce the tedium and time wastage, it is suggested that teachers should use a copying machine for producing the necessary material.

The number of 'Oral' exercises and the identification of 'Group Work' projects have been reduced in this volume. However, it is suggested that teachers continue to provide some opportunity for such activities where they consider it possible and relevant. The benefits to be derived both from the discussion of alternative strategies for solving a problem as well as the development of positive attitudes of friendly competition, of cooperation and team work among the students make it worthwhile.

In response to the positive feedback from students, teachers and adult learners throughout the Caribbean, the 'Preliminary Chapter' at the beginning of this volume, which is a review of the main concepts dealt with in Books 1 and 2, has been expanded. In addition, the concepts have been grouped under the topic headings as listed in the CXC syllabus as well as referenced to the specific chapters in which they are found in Books 1 and 2.

In an attempt to assist the teacher and the student in quickly identifying and reviewing necessary background knowledge, we have included information to be referenced under the heading 'Pre-requisites' at the beginning of the chapters. It must be remembered that the main new ideas of each chapter are highlighted in the 'Summary' at the end of the chapter. The point at which the 'Revision Exercises and tests' are used is a matter for the individual teacher's choice. If the series of related chapters has been taught as sequenced in the printed text, then the revision material may be used as that time, or may be omitted until the entire book has been completed. This material may also be used as supplemental problems for the quicker students.

Also included in this edition of Book 3 is a 'Revision Course' in the same format as that used in Book 4 (1st Edn) – text, worked examples, exercises for practice and a multiple-choice test in each chapter. This course revises all the material covered in the first three volumes – Books 1, 2 and 3 of the series.

In order to comply with the requests of teachers. 'Practice Examinations (Papers 1 and 2)' have also been included at the end of the book.

Finally, it is unfortunate that Mathematics is perceived by a large majority of students as a 'necessary evil', a subject that they have to 'get at CXC' in order to become employable. Teachers have a significant responsibility in helping to change this attitude, and in having students appreciate that, by acquiring the skills and techniques to solve problems in mathematics, they also acquire the tools and the ability to solve problems in the real world.

To the student

Before attempting the problems in the exercises, study and discuss the worked examples until you understand the concept. The oral exercises are intended to encourage discussion. This will help to clarify lingering misunderstandings. In particular, in solving word problems you first have to get thoroughly familiar with the problem. The next step involves translating the problem into mathematical symbols and language, for example, into an equation or an inequality, or into a graph. The next steps are applying the required mathematical operations, and finally checking the original word problem. Remember to check that the variables are in the same units. Another useful hint is to look for patterns in similar problems and in the methods of solution. Concrete materials such as cans, coins, balls, stones and boxes are very useful aids for clearing up doubts – not for wasting time!

A calculator is an excellent machine when used wisely, but you must bear in mind that it needs to be used by a clear-thinking human who fully comprehends the mathematical concepts. Dividing 4.0 by 3 and giving your answer as 1.3333333 indicates, among other things, a lack of appreciation of the concept of accuracy.

Nothing can replace the neat appearance of an answer that is well set out – the date, the page and exercise from which the problem has been taken, a statement of the facts given, the necessary calculations performed and the conclusions drawn which result in a final answer. This whole process helps you to think clearly.

Students are urged to make full use of the 'Revision Exercises and Tests', and the 'Revision Course', – these will assist you in identifying the concepts to which you may need to give extra attention and time.

Finally, the more a concept is applied. the clearer it becomes. Thus, PRACTICE and more practice in working out examples is an essential ingredient to success.

Althea A. Foster
E. M. Tomlinson

Contents

Contents

Before beginning Book 3, readers should be familiar with the contents of Books 1 and 2. The following summary contains those parts of Books 1 and 2 which appear in the Mathematics syllabus for regional examinations. References to earlier books are given at the end of each section.

Number theory and computation

(a) The Number System includes all the representations of number ideas used for counting, measurement and calculation. These numbers can be represented on the **number line**, shown in Fig. Pr1.

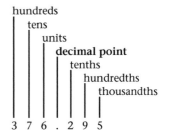

Fig. Pr1

N is the set of **natural counting numbers**. These are all **positive** numbers. When the number zero [0] is included, the set becomes the set of **whole numbers**, **W**. The set of positive and **negative** whole numbers is called **Z**, the set of **integers**. When fractions, that is, exact parts of a whole, are included, the set of numbers is called the set of **rational numbers**, **Q**. When the numbers for approximate values, for example, π, are included, the set is called the set of **real numbers**, **R**. The relation between these sets of numbers is seen as:

$$N \subset W \subset Z \subset Q \subset R$$

(b) The order in which the basic arithmetic operations is done is

first: *brackets*
next: *of*
then: *multiplication and division*
last: *addition and subtraction*

Addition and multiplication obey the **commutative** and **associative** laws. Multiplication is **distributive** over addition and subtraction.

(c) Numbers are normally written in the decimal **place value** system (Fig. Pr2):

```
hundreds
  tens
    units
      decimal point
        tenths
          hundredths
            thousandths

3 7 6 . 2 9 5
```

Fig. Pr2

The symbols 0, 1, 2, 3, 4, 5, 6, 7, 8, 9 are called **digits**.

(d) $28 \div 7 = 4$. 7 is a whole number which divides exactly into another whole number, 28.

7 is a **factor** of 28. 28 is a **multiple** of 7.

(e) A **prime number** has only two factors, itself and 1. 1 is *not* a prime number. 2, 3, 5, 7, 11, 13, 17, ..., are prime numbers. Numbers which have more than two factors are **composite numbers**. The **prime factors** of a number are those factors which are prime. For example, 2 and 5 are the prime factors of 40. 40 can be written as a **product of prime factors**; either $2 \times 2 \times 2 \times 5 = 40$, or, in **index form**, $2^3 \times 5 = 40$.

(f) The numbers 18, 24 and 30 all have 3 as a factor. 3 is a **common factor** of all the numbers. The **highest common factor** (HCF) is the largest of the common factors of a given set of numbers. For example, 2, 3 and 6 are the common factors of 18, 24 and 30; 6 is the HCF.

The number 48 is a multiple of 4 and a multiple of 6. 48 is a **common multiple** of 4 and 6. The **lowest common multiple** (LCM) is the smallest of the common multiples of a given set of numbers. For example, 12 is the LCM of 4 and 6.

(g) A **fraction** is the number obtained when one number (the **numerator**) is divided by another number (the **denominator**). The fraction $\frac{5}{8}$ means $5 \div 8$ (Fig. Pr3).

$$\frac{5}{8}$$

— numerator
— dividing line
— denominator

Fig. Pr3

Fractions are used to describe parts of quantities (Fig. Pr4 (a) and (b)).

(a) $\frac{5}{8}$ *of the circle is shaded.*

(b) $\frac{2}{5}$ *of the quadrilaterals are squares.*

Fig. Pr4

The fractions $\frac{5}{8}, \frac{10}{16}, \frac{15}{24}$ all represent the same amount; they are **equivalent fractions**. $\frac{5}{8}$ is the **simplest form** of $\frac{15}{24}$, i.e. $\frac{15}{24}$ in its **lowest terms** is $\frac{5}{8}$.

To add or subtract fractions, change them to equivalent fractions with a **common denominator**. For example:

$$\frac{5}{8} + \frac{2}{3} = \frac{15}{24} + \frac{16}{24} = \frac{15 + 16}{24} = \frac{31}{24} \left(= 1\frac{7}{24}\right)$$

$$\frac{13}{16} - \frac{5}{8} = \frac{13}{16} - \frac{10}{16} = \frac{13 - 10}{16} = \frac{3}{16}$$

To multiply fractions, multiply numerator by numerator and denominator by denominator. For example:

$$\frac{5}{8} \times \frac{2}{3} = \frac{5 \times 2}{8 \times 3} = \frac{10}{14} \left(= \frac{5}{12} \text{ in simplest form}\right)$$

$$12 \times \frac{5}{8} = \frac{12}{1} \times \frac{5}{8} = \frac{12 \times 5}{1 \times 8} = \frac{60}{8} \left(= \frac{15}{2} = 7\frac{1}{2}\right)$$

To divide by a fraction, multiply by the **reciprocal** of the fraction. For example:

$$35 \div \frac{5}{8} = \frac{35}{1} \times \frac{8}{5} = \frac{35 \times 8}{1 \times 5} = \frac{7 \times 8}{1} = 56$$

$$\frac{5}{8} \div 3\frac{3}{4} = \frac{5}{8} \div \frac{15}{4} = \frac{5}{8} \times \frac{4}{15} = \frac{5 \times 4}{8 \times 15}$$

$$= \frac{20}{120} \left(= \frac{1}{6}\right)$$

(h) $x\%$ is short for $\frac{x}{100}$. 64% means $\frac{64}{100}$.

Hence, since a **percentage** is the numerator of a fraction, when the denominator is 100, to write $\frac{5}{8}$ as $P\%$

then $$\frac{5}{8} = \frac{P}{100}$$

and $$\frac{P}{100} \times 100 = \frac{5}{8} \times 100$$

so that $$P = \frac{5}{8} \times 100$$

and $$= \frac{5}{8} \times 100$$

$$= \frac{500}{8}\% = \frac{125}{2}\% = 62\frac{1}{2}\%$$

To change a fraction to an equivalent percentage, multiply the fraction by 100 and write the % symbol.

(i) To change a fraction to a **decimal fraction**, divide the numerator by the denominator. For example:

$$\frac{5}{8} = 0.625$$

$$\begin{array}{r} 0.625 \\ 8\overline{)5.00} \\ \underline{4\ 8} \\ 20 \\ \underline{16} \\ 40 \\ \underline{40} \end{array}$$

When adding or subtracting decimals, write the numbers in a column with the decimal points exactly under each other. For example: *Add 2.29, 0.084 and 4.3, then subtract the result from* 11.06

$$\begin{array}{r} 2.29 \\ 0.084 \\ + \ 4.3 \\ \hline 6.674 \end{array} \qquad \begin{array}{r} 11.06 \\ - \ 6.674 \\ \hline 4.386 \end{array}$$

To multiply decimals, ignore the decimal points and multiply the given numbers as if they are whole numbers. Then place the decimal point so that the answer has as many digits after the point as there are in the given numbers together. For example:

$$0.08 \times 0.3$$
$$8 \times 3 = 24$$

There are 3 digits after the decimal point in the given numbers, so $0.08 \times 0.3 = 0.024$ as shown below.

$$0.08 \times 0.3 = \frac{8}{100} \times \frac{3}{10}$$
$$= \frac{24}{1000}$$
$$= 0.024$$

To divide by decimals, make an equivalent division such that the divisor is a whole number. For example $5.6 \div 0.07$:

$$5.6 \div 0.07 = \frac{5.6}{0.07} = \frac{5.6 \times 100}{0.07 \times 100} = \frac{560}{7} = 80$$

(j) Numbers may be positive or negative. Positive and negative numbers are called **directed numbers**. Directed numbers can be shown on a **number line** (Fig. Pr5).

Fig. Pr5

The following examples show how directed numbers are added, subtracted, multiplied and divided.

addition	subtraction
$(+8) + (+3) = +11$	$(+9) - (+4) = \ +5$
$(+8) + (-3) = \ +5$	$(+9) - (-4) = +13$
$(-8) + (+3) = \ -5$	$(-9) - (+4) = -13$
$(-8) + (-3) = -11$	$(-9) - (-4) = \ -5$

multiplication	division
$(+2) \times (+7) = +14$	$(+6) \div (+3) = \ +2$
$(+2) \times (-7) = -14$	$(+6) \div (-3) = \ -2$
$(-2) \times (+7) = -14$	$(-6) \div (+3) = \ -2$
$(-2) \times (-7) = -14$	$(-6) \div (-3) = \ +2$

Remember that an **integer** is any positive or negative *whole* number.

(k) The number 3.7×10^4 is in **standard form**. The first part of the product is between 1 and 10 and the second is a power of 10.

(l) When **rounding off** numbers, the digits 1, 2, 3, 4 are rounded down and the digits 5, 6, 7, 8, 9 are rounded up. For example,

$$3425 = 3430 \text{ to 3 } \textbf{significant figures}$$
$$= 3400 \text{ to the } \textbf{nearest hundred}$$
$$7.283 = 7.28 \text{ to 2 } \textbf{decimal places}$$
$$= 7.3 \text{ to 1 decimal place}$$
$$= 7 \text{ to the } \textbf{nearest whole number}$$

(m) Numbers may be written in **bases** other than ten. For example,

in base ten (*decimal system*)
$$5279 = 5 \times 10^3 + 2 \times 10^2 + 7 \times 10^1 + 9$$

in base two (*binary system*)
$$110\,101_{two} = 1 \times 2^5 + 1 \times 2^4 + 0 \times 2^3 +$$
$$1 \times 2^2 + 0 \times 2^1 + 1$$
$$(= 32 + 16 + 0 + 4 + 0 + 1 = 53_{ten})$$

in base eight (*octal system*)
$$352_{eight} = 3 \times 8^2 + 5 \times 8^1 + 2$$
$$(= 192 + 40 + 2 = 234_{ten})$$

To change a base ten number to base n, divide repeatedly by n until the quotient is zero and further division is impossible. The remainders in each line of working, reading upwards, give the required number.

For example:
Change 29_{ten} to a binary number

2	29	rem.
2	14	1
2	7	0
2	3	1
2	1	1
	0	1

$$29_{ten} = 11\,101_{two}$$

(n) In order to compare quantities of the same kind, **ratios** are used. The quantities must be given in the *same units*. The ratio may then be expressed as a fraction. For example, if there are two pieces of wire, 10 cm an 30 cm long, the ratio of the lengths is $10:30$ or $1:3$, and may be written as $\frac{1}{3}$.

Quantities may also be divided or shared in a given ratio. For example, the sum of money, \$120, may be shared in the ratio $5:7$. Then the two amounts are $\frac{5}{(5+7)}$ and $\frac{7}{(5+7)}$, that is $\frac{5}{12}$ and $\frac{7}{12}$ of the total amount. The amounts are $\$(\frac{5}{12} \times 120)$ and $\$(\frac{7}{12} \times 120)$, that is \$50 and \$70.

If quantities are of different kinds, the quantities are compared in the form of a **rate**. For example, when distance is compared with time, the rate is speed, that is, $\frac{distance}{time} = speed.$

Detailed coverage of number theory and computation is given in Book 1, Chapters 1, 3, 5, 8, 9, 13, 20, and in Book 2, Chapters 1, 3, 7.

Consumer arithmetic

(a) When a sum of money $P called the **principal** is invested or borrowed, extra money, $I called the **simple interest** is paid/charged on the principal according to the formula

$$I = \frac{P \times R \times T}{100},$$ where R = rate % per annum and T = time in years.

Compound interest is calculated using the same formula but the interest is added to the principal at the end of a time interval which has been agreed beforehand to give a new value for the principal for the following time interval.

(b) The money of different countries is exchanged according to agreed **currency exchange rates** to enable the transfer of money and the payment for goods and services. These rates are not constant but may vary daily.

(c) The price paid by a buyer is called the **cost price** (C.P.) and the price charged by the person selling the item is called the **selling price** (S.P.). The difference (S.P. − C.P.) is a **profit** when this difference is greater than zero and a **loss** when this difference is less than zero. The profit or loss may be expressed as a percentage of the C.P., or of the S.P. A **discount** is a reduction in price and is usually given as a percentage of the price, or as a rate.

(d) Items may be bought on **hire purchase**. An **instalment** or part payment is made and the remainder is paid over an agreed period of time while the buyer hires the use of the item. **Commission** is payment to a salesman for selling an item.

(e) A **tax** is money charged by government to pay for services provided by the government. **Income tax** is calculated on the income of the citizen. **Sales tax** is charged when items are bought. Tax systems vary from country to country; the tax rates charged and the non-taxable allowances granted vary.

Detailed coverage of consumer arithmetic is given in Book 1, Chapter 22, and in Book 2, Chapter 21.

Set theory

(a) A **set** is a collection of objects. The **members** or **elements** of a set may be defined in a number of ways.

by description:
A = {first five counting numbers}

by listing elements:
A = {1, 2, 3, 4, 5}

in set-builder notation:
A = {x: $1 \leqslant x \leqslant 5$, $x \in \mathbf{Z}$}

The last statement may be read as: A is the set of values x such that x lies between 1 and 5 inclusively, where x is an integer. Note the use of curly brackets to contain sets, the use of commas to separate the elements of a set, the use of capital letters for naming a set and the use of **Z** as an abbreviation for the set of integers. A complete list of symbols and language of sets is given in Chapter 6 Table 6.1 on page 64.

The **empty set** or **null set** is the set with no elements and is represented as { } or ∅.

The elements in the set **N** = {1, 2, 3, 4, ...} never end. This is an **infinite set**.

(b) A set can be represented on a **Venn diagram** (Fig. Pr6).

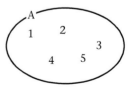

Fig. Pr6

(c) Given a **universal set** U = {a, b, c, d, e} and sets X = {a, b, e} and Y = {d, e}:

X is a **subset** of U, written $X \subset U$. Also $Y \subset U$. The **union** of X and Y, written $X \cup Y$, is the set whose elements are members of X or Y or both X and Y.

$$X \cup Y = \{a, b, d, e\}$$

The **intersection** of X and Y, written $X \cap Y$, is the set whose elements are members of both X and Y.

$$X \cap Y = \{e\}$$

The **complement** of X, written X' is the set whose elements are in the universal set but not in X.

$$X' = \{c, d\}$$
$$\text{and} \quad Y' = \{a, b, c\}$$

(d) The number of elements in X, written n(X), is 3.

$$n(X) = 3$$

Detailed coverage of set theory is given in Book 1, Chapter 2, and in Book 2, Chapter 8.

Algebraic processes

(a) $3y^2 + 2x - 7x$ is an example of an **algebraic expression**. The letters y and x stand for numbers. $3y^2$, $2x$ and $7x$ are the **terms** of the expression. $3y^2$ is short for $3 \times y \times y$. 3 is the **coefficient** of y^2. Algebraic terms may be **simplified** by combining **like terms**. Thus

$$3y^2 + 2x - 7x = 3y^2 - 5x$$

$2x$ and $7x$ are like terms (i.e. both terms in x).

(b) $3(5x - 2) = 11x$ is an **algebraic sentence** containing an equal sign; it is an **equation** in x. x is the **unknown** of the equation. To **solve an equation** means to find the value of the unknown which makes the equation true. We use the **balance method** to solve equations.

$$3(5x - 2) = 11x$$

clear brackets

$$15x - 6 = 11x$$

subtract $11x$ from both sides

$$15x - 11x - 6 = 11x - 11x$$
$$4x - 6 = 0$$

add 6 to both sides

$$4x - 6 + 6 = 0 + 6$$
$$4x = 6$$

divide both sides by 4

$$\frac{4x}{4} = \frac{6}{4}$$
$$x = 1\tfrac{1}{2}$$

The **solution set** of an equation is the set of values of the unknown which make the equation true. $\{1\tfrac{1}{2}\}$ is the solution set of the equation $3(5x - 2) = 11x$.

In general, when solving an equation, (i) first clear brackets and fractions, (ii) using equal additions and/or subtractions on both sides of the equation, collect unknown terms on one side of the equals sign and known terms on the other, (iii) where necessary, divide or multiply both sides of the equation by the same number to find the unknown.

(c) An **inequality** is an algebraic sentence which contains an inequality sign:

$<$ is less than
$\leq$ is less than or equal to
$>$ is greater than
$\geq$ is greater than or equal to

Inequalities are solved in much the same way as equations. However, when both sides of an inequality are multiplied or divided by a negative number, the inequality sign is *reversed*.

For example,

if $\qquad -3a \leq 12$

divide both sides by -3 and reverse the inequality.

Then $\qquad a \geq -4$

(d) Algebraic expressions may be factorised or expanded in accordance with the basic rules of arithmetic. Some examples are given next.

expansion

$$3(a - 2b) = 3a - 6b$$
$$(5 + 8x)x = 5x + 8x^2$$
$$(a + b)(c + d) = c(a + b) + d(a + b)$$
$$= ac + bc + ad + bd$$
$$(3x + 2)(x - 4) = 3x^2 + 2x - 12x - 8$$
$$= 3x^2 - 10x - 8$$
$$(a - 5b)^2 = a^2 - 10ab + 25b^2$$

factorisation

common factor

$$5y - 10y^2 = 5y(1 - 2y)$$
$$4x - 8 + 3bx - 6b = 4(x - 2) + 3b(x - 2)$$
$$= (x - 2)(4 + 3b)$$

fractions

$$\frac{2m}{3} - \frac{3m}{5} = \frac{5(2m) - 3(3m)}{15}$$

$$= \frac{m}{15}$$

$$\frac{7}{y} + \frac{4}{3y} - \frac{1}{2y} = \frac{6(7) + 2(4) - 3(1)}{6y}$$

$$= \frac{47}{6y}$$

$$\frac{2x - 3}{4} + \frac{3x + 1}{3} = \frac{3(2x - 3) + 4(3x + 1)}{12}$$

$$= \frac{18x - 5}{12}$$

$$\frac{3}{y - 1} + \frac{2}{y} = \frac{3y + 2(y - 1)}{y(y - 1)}$$

$$= \frac{5y - 2}{y(y - 1)}$$

(e) The following laws of indices are true for all values of a, b and x.

$$x^a \times x^b = x^{a + b} \qquad x^a \div x^b = x^{a - b}$$
$$x^0 = 1 \qquad x^{-a} = \frac{1}{x^a}$$

Detailed coverage of algebraic processes is given in Book 1, Chapters 6, 11, 15, and Book 2, Chapters 5, 10, 16, 20.

Relations, functions and graphs

(a) **Relations** and **mappings** show a connection between the members of two sets. These may be shown as arrow diagrams, tables, ordered pairs or graphs.

(b) A **function** is defined as a relation between members of two sets, set A and set B, in which each member of set A relates to one and only one member of set B.

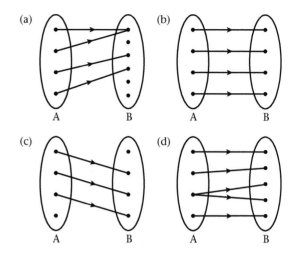

Fig. Pr7

In the above arrow diagrams, (a) and (b) represent functions; (c) and (d) do not represent functions. In (c) not all the elements in set A are related to elements in set B; in (d) one element in set A is related to two elements in set B.

(c) A **graph** of an algebraic sentence is a picture representing the meaning of the sentence, that is representing the values of the variables which satisfy the equation or the inequality.

Graphs of equations and inequalities in one variable can be shown on the number line (Fig. Pr8).

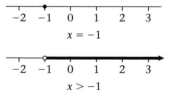

Fig. Pr8

For graphs, connecting two variables, two **axes** are drawn at right angles to each other to give a **Cartesian plane**. The horizontal *x*-axis and the vertical *y*-axis cross at their zero-point, the **origin** of the plane. The corresponding values of these two variables which satisfy the equation are defined by the **coordinates** of the points which lie on a straight line or on a curve. The *order* of the coordinates is important: the **x-coordinate** is given first, the **y-coordinate** second.

Straight-line graphs can be drawn to represent two connected variables, for example cost and quantity, distance and time, temperature and time. Straight-line graphs can also be drawn to show conversions between currencies.

To draw a straight-line graph, plot at least three points which satisfy the given equation. See the table of values in Fig. Pr9. At point A in Fig. Pr9, $x = 2$ and $y = 1$, written A(2, 1).

Fig. Pr9 is the graph of the equation
$$y = 2x - 3.$$

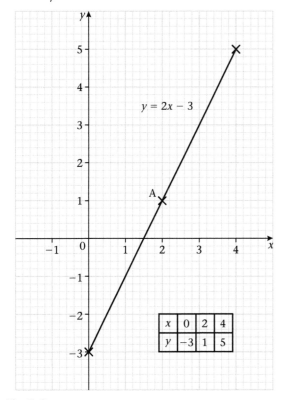

x	0	2	4
y	-3	1	5

Fig. Pr9

Detailed coverage of relations, functions and graphs is given in Book 1, Chapter 19, and Book 2, Chapters 4, 8, 9, 13.

Geometry and measurement

(a) Fig. Pr10 gives sketches and names of some **common solids**.

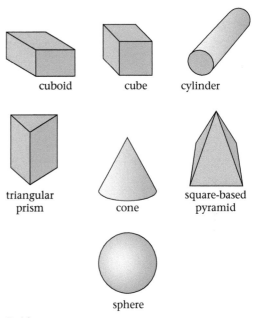

cuboid   cube   cylinder

triangular prism   cone   square-based pyramid

sphere

Fig. Pr10

All solids have **faces**; most solids have **edges** and **vertices** (Fig. Pr11).

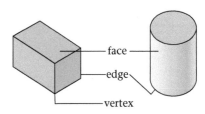

face — edge — vertex

Fig. Pr11

Formulae for the **surface area** and **volume** of common solids are given in the tables on page 264.

(b) **Angle** is a measure of rotation or turning.

1 **revolution** = 360 **degrees** (1 rev = 360°)
1 **degree** = 60 **minutes** (1° = 60′)

The names of angles change with their size. See Fig. Pr12.

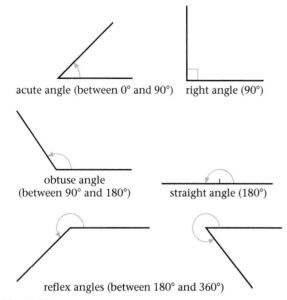

acute angle (between 0° and 90°)   right angle (90°)

obtuse angle
(between 90° and 180°)   straight angle (180°)

reflex angles (between 180° and 360°)

Fig. Pr12

Angles are measured and constructed using a **protractor**.

Fig. Pr13 shows some properties of angles formed when straight lines meet.

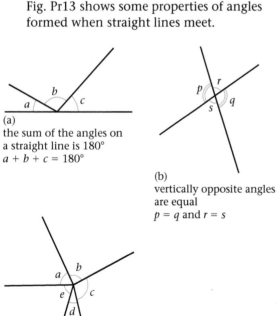

(a)
the sum of the angles on a straight line is 180°
$a + b + c = 180°$

(b)
vertically opposite angles are equal
$p = q$ and $r = s$

(c)
the sum of the angles at a point is 360°
$a + b + c + d + e = 360°$

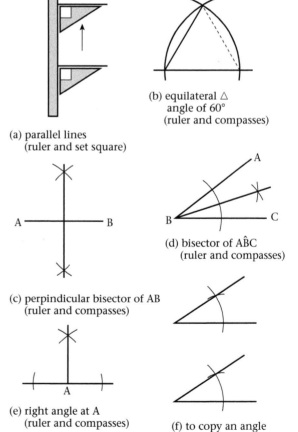

(d)
alternate angles on parallel lines are equal
$x = y$ and $m = n$

(e)
corresponding angles on parallel lines are equal
$a = b$ and $p = q$

Fig. Pr13

(c) The sketches in Fig. Pr14 (a)–(f) show the main features of the common **geometrical constructions**.

To construct angles of 45° and 30°, bisect angles of 90° and 60° respectively.

(a) parallel lines
(ruler and set square)

(b) equilateral △
angle of 60°
(ruler and compasses)

(c) perpindicular bisector of AB
(ruler and compasses)

(d) bisector of AB̂C
(ruler and compasses)

(e) right angle at A
(ruler and compasses)

(f) to copy an angle
(ruler and compasses)

Fig. Pr14

A **polygon** is a plane shape bounded by three or more straight sides.

(d) Fig. Pr15 shows the names and properties of some **common triangles**.

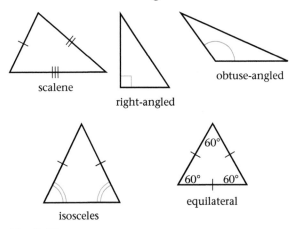

Fig. Pr15

Fig. Pr16 shows the names and properties of some **common quadrilaterals**.

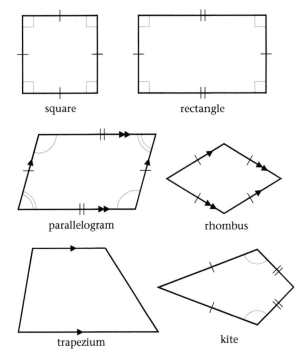

Fig. Pr16

A **regular polygon** has all its sides of equal length and all its angles of equal size.

Fig. Pr17 gives the names of some regular polygons.

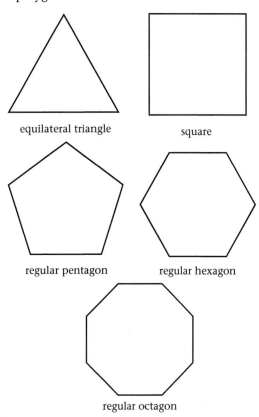

regular octagon

Fig. Pr17

The **sum of the angles of an *n*-sided polygon** is $(n - 2) \times 180°$. In particular, the sum of the angles of a triangle is 180° and the sum of the angles of a quadrilateral is 360°.

Fig. Pr18 gives the names of lines and regions in a circle.

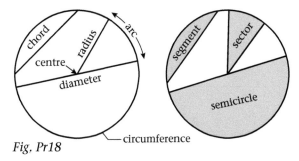

Fig. Pr18

(e) Formulae for the **perimeter and area of plane shapes** are given in the table on page 264.

The **SI system of units** is given in the tables on page 263 and 264.

(f) To **solve a triangle** means to calculate the sizes of its sides and angles. Right-angled triangles can be solved using **Pythagoras' rule** and the **tangent**, **sine** and **cosine** ratios.

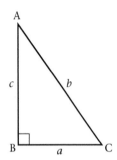

Fig. Pr19

In Fig. Pr19 △ABC is right-angled at B; side AC is the **hypotenuse**.

Pythagoras' rule: $b^2 = a^2 + c^2$

tangent ratio: $\tan \widehat{C} = \frac{c}{a}$ and $\tan \widehat{A} = \frac{a}{c}$

sine ratio: $\sin \widehat{C} = \frac{c}{b}$ and $\sin \widehat{A} = \frac{a}{b}$

cosine ratio: $\cos \widehat{C} = \frac{a}{b}$ and $\cos \widehat{A} = \frac{c}{b}$

(g) In Fig. Pr20, α is the **angle of elevation** of the top of the flag-pole from the girl and β is the **angle of depression** of the girl from the boy.

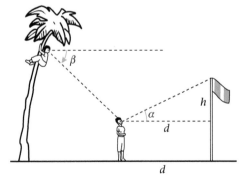

Fig. Pr20

The trigonometric ratios are often used to find a distance or an angle in problems involving the angle of elevation (or depression). For example, in Fig. Pr20,

where d is the distance from the foot of the pole, and h is the height of the pole above the girl's eyes, when two of the values of d, h or α are given, the unknown quantity can be calculated, since $\tan \alpha = \frac{h}{d}$.

(h) In a **scale drawing** of a plane figure, the shape of the actual figure remains the same but the size is different. The **scale** is the ratio of the length of the lines in the drawing to the actual lengths in the shape. Maps may be drawn to large scales such as 1 cm to 5 km, or 1 : 500 000.

(i) A shape is **transformed** when its position or dimensions (or both) change. The **image** of a shape is the figure that results after a transformation. If the image has the same dimensions as the original shape, the transformation is called a **congruency**. Fig. Pr21 shows the three basic congruencies, (a) **translation**, (b) **reflection**, (c) **rotation**.

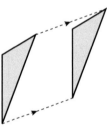 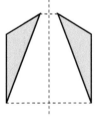

(a) translation   (b) reflection

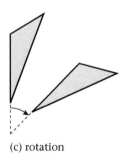

(c) rotation

Fig. Pr21

(j) A shape may fit upon itself when it is rotated about a point in the shape, or it may be folded about a line so that one section fits exactly over the other section. The shape is then said to have **symmetry** about the point (**rotational symmetry**) or about the line (**line symmetry**).

Detailed coverage of geometry and measurement is given in Book 1, Chapters 4, 7, 10, 12, 14, 16, 17, 21, and Book 2, Chapters 2, 6, 11, 12, 14, 15, 17, 19, 23.

Statistics

(a) Information in numerical form is called **statistics**. Statistical **data** may be given in **rank order** (i.e. in order of increasing size), such as in the following marks obtained in a test out of 5:

0, 1, 1, 2, 2, 2, 3, 3, 5

Data may also be given in a **frequency table** (Table Pr1).

Table Pr1

Mark	0	1	2	3	4	5
Frequency	1	2	3	2	0	1

The **frequency** is the number of times each piece of data occurs.

Statistics can also be presented in graphical form. Fig. Pr22 shows the above data in a **pictogram**, a **bar chart** and a **pie chart**.

(b) The **average** of a set of statistics is a number which is representative of the whole set. The three most common averages are the **mean**, the **median** and the **mode**. For the 9 numbers given in rank order in the previous paragraph (a),

$$mean = \frac{0 + 1 + 1 + 2 + 2 + 2 + 3 + 3 + 5}{9}$$

$$= 2\frac{1}{9}$$

the *median* is the middle number when the data is arranged in order of size (2);
the *mode* is the number with the greatest frequency (also 2 in this case).

Detailed coverage of statistics is given in Book 1, Chapter 18, and Book 2, Chapter 18.

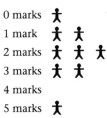

pictogram

0 marks
1 mark
2 marks
3 marks
4 marks
5 marks

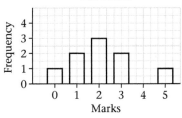

bar chart

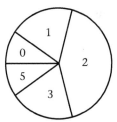

pie chart

Fig. Pr22

Chapter 1

Geometrical proofs (1)
Triangles and other polygons

Geometry is the study of the properties of shapes. In Books 1 and 2 many geometrical properties of lines, angles, plane shapes and solids were discovered experimentally, often by drawing and measuring particular figures. However, these methods do not provide sufficient proof that the properties and facts discovered will always be true for all such figures. In theoretical or **formal geometry** the facts are shown for general cases by a method of argument or reasoning rather than by measurement.

Some geometrical facts are more important than others. These basic facts are called **theorems**. Theorems form the foundations upon which formal geometry is built. While we no longer need to carry out rigid proofs of these theorems, we still need to use these basic facts to develop other ideas in geometry. Some of the basic facts of the angle properties of polygons have already been shown by measurement, e.g. the sum of the angles of a polygon is $(2n - 4)$ right angles.

Example 1

Show that the sum of the angles of triangle ABC, Fig. 1.1, is 180°.

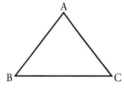

Fig. 1.1

Note that in Fig. 1.2 BC is produced to X and CP is drawn parallel to BA.

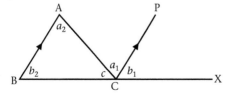

Fig. 1.2

With the lettering of Fig. 1.2,
$$a_1 = a_2 \text{ (alternate angles)}$$
$$b_1 = b_2 \text{ (corresponding angles)}$$
$$c + a_1 + b_1 = 180° \text{ (}B\widehat{C}X \text{ is a straight angle)}$$
$$\therefore c + a_2 + b_2 = 180°$$
$$\therefore A\widehat{C}B + \widehat{A} + \widehat{B} = 180°$$
$$\therefore \widehat{A} + \widehat{B} + \widehat{C} = 180°$$

Note that the reasons are given for the statements above.

In the following section this basic fact (theorem) will be used to develop other angle properties of triangles and other polygons.

Interior and exterior angles of triangles and other polygons

Example 2

Show that the exterior angle of a triangle is equal to the sum of the interior opposite angles.

In Fig. 1.3 $A\widehat{C}D + A\widehat{C}D = 180°$ (*angles on a straight line*)
$$A\widehat{B}C + B\widehat{A}C + A\widehat{C}B = 180° \text{ (}angle \text{ } sum \text{ } of \text{ } \triangle\text{)}$$
$$\therefore A\widehat{C}D = A\widehat{B}C + B\widehat{A}C$$

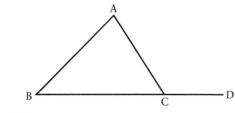

Fig. 1.3

Example 3

Show, without measuring, that the sum of the angles of a quadrilateral is 360°.

The quadrilateral PQRS, Fig. 1.4, is divided into two triangles.

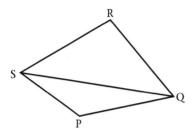

Fig. 1.4

Sum of angles of △PQS = 180°
Sum of angles of △QRS = 180° (*angle sum of* △)
Sum of angles of △PQS and △QRS
= 180° + 180° + 360°
∴ Sum of angles of quadrilaterals PQRS = 360°

Would this be true for all quadrilaterals?

Example 4

Find the sum of the angles of a pentagon. Do not measure the angles.

In Fig. 1.5, the pentagon ABCDE is divided into three triangles.

The sum of the angles of the pentagon equals the sum of the angles of the three triangles.

Sum of angles of △ABE = 180°
Sum of angles of △BCE = 180°
Sum of angles of △CDE = 180°
∴ Sum of angles of △ABCDE = 540°

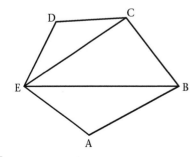

Fig. 1.5

Group work

Working in groups of four, work through the following activities.

① Draw polygons having 5, 6, 7, 8 sides.

② Divide each polygon into either triangles only or triangles and quadrilaterals.

③ Find, without measuring, the sum of the angles of each polygon.

④ Copy and complete Table 1.1

Table 1.1

Polygon	No. of sides	Sum of angles	No. of right angles
triangle	3	180°	2
quadrilateral	4	360°	4
pentagon	5	540°	6
	6		
	7		
	8		

⑤ What patterns do you notice?

⑥ Using the patterns noticed, continue the table to include polygons having 9, 10, 12, *n* sides.

You can now state another basic fact (theorem). The sum of the interior angles of any *n*-sided convex polygon is $(2n - 4)$ right angles.

Note: A **convex polygon** does not contain any reflex angles. A polygon which contains reflex angles is called a **re-entrant polygon** (Fig. 1.6).

convex polygon   re-entrant polygon

Fig. 1.6

Group work

Working in small groups, draw a convex polygon having n sides and produce each side as shown in Fig. 1.7.

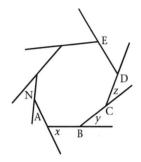

Fig. 1.7

What can you say about the angles at each of the points A, B, C, D, E, ...

What can you say about the sum of all the angles above?

What can you say about the sum of the interior angles of the polygon?

What can you now about the sum of the angles $x + y + z + ...$?

Remember to use the facts already known.

Do you think this would hold for all convex polygons?

You can now state another basic fact. The sum of the exterior angles of a convex polygon is 4 right angles.

These basic facts can now be applied to new situations. Notice that reasons must always be given for your statements.

Example 5

In the polygon ABCDE, Fig. 1.8, AB is parallel to DC and AB = BC. If angle CAB = 57° and the angles marked x are equal, calculate the size of the angles of the polygon.

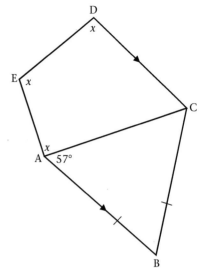

Fig. 1.8

In $\triangle$BAC

$\qquad$ $\widehat{CAB} = \widehat{BCA} = 57°$ $\qquad$ (*base angles of isosceles triangle*)

$\therefore$ $\widehat{ABC} = 180° - 57° - 57°$ (*sum of angles of triangle*)

$\qquad\qquad$ $= 66°$

Also $\quad$ $\widehat{BAC} = \widehat{ACD} = 57°$ $\qquad$ (*alternate angles*)

$\qquad$ $\therefore$ $\widehat{BCD} = 114°$

Now

$x + x + x + 57° + 66° + 114° = 6 \times 90°$

$\qquad\qquad\qquad\qquad$ (*sum of interior angles of polygon*)

$\qquad\qquad$ $3x + 237° = 540°$

$\qquad\qquad\qquad$ $3x = 303°$

$\qquad\qquad\qquad$ $x = 101°$

$\qquad\qquad\qquad$ $= \widehat{CDE}$

$\qquad\qquad\qquad$ $= \widehat{DEA}$

and $\qquad\qquad$ $\widehat{BAE} = 158°$

Example 6

ABCDE is a regular pentagon. The sides AB and DC are produced to meet at X. Show that △BDX has two equal angles.

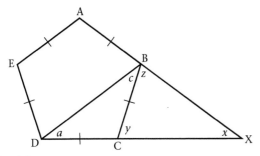

Fig. 1.9

With the lettering of Fig. 1.9:

In △BXC,

$$y = z \qquad \text{(regular polygon.}$$
$$\therefore \text{ equal exterior angles)}$$

$$y = z = \frac{360°}{5} \qquad \text{(5 ext. angles add to } 360°\text{)}$$

$$\therefore y = z = 72°$$
$$\therefore x + y + z = 180° \quad \text{(angle sum of △)}$$
$$x = 180° - y - z$$
$$= 180° - 72° - 72°$$
$$= 36°$$

In △BCD,

$$y = a + c \qquad \text{(ext. angle of △)}$$

But $a = c$

$$\therefore y = 2a$$
$$\therefore a = \tfrac{1}{2}y$$
$$= \tfrac{1}{2} \times 72°$$
$$= 36°$$

In △BDX, $a = x = 36°$
∴ △BDX has two equal angles.

Exercise 1a

1. The angles of a triangle are x, $2x$ and $3x$. Find the value of x in degrees.

2. An isosceles triangle is such that each of the base angles is twice the vertical angle. Find the angles of the triangle.

3. In a right-angled triangle one of the acute angles is 20° greater than the other. Find the angles of the triangle.

4. The angles of a quadrilateral, taken in order, are x, $5x$, $4x$ and $2x$. Find these angles. Draw a rough sketch of the quadrilateral. What kind of quadrilateral is it?

5. Find the size of each interior angle of a regular polygon which has (a) 6, (b) 10, (c) 20 sides.

6. Find, to the nearest degree, the size of the angles of a regular heptagon (7 sides).

7. A regular polygon has angles of size 150°. How many sides has the polygon?

8. ABCDE is a regular pentagon. Find the angles of △ADC.

9. Four angles of a pentagon are equal and the fifth is 60°. Find the equal angles and show that two sides of the pentagon are parallel.

10. In △ABC, the side BC is produced to D. If the bisector of $A\widehat{C}D$ is parallel to AB, show that the two angles of △ABC are equal.

11. In Fig. 1.10, what is the value of the angle marked m?

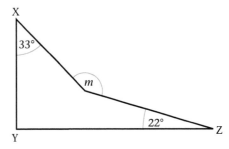

Fig. 1.10

⑫ In Fig. 1.11, BX is the bisector of $A\widehat{B}C$ and CX is the bisector of $A\widehat{C}B$. If $\widehat{A} = 68°$, find the size of $\widehat{X}$.

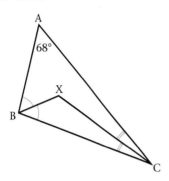

Fig. 1.11

⑬ In △PQR, X is a point on QR such that $R\widehat{P}X = \widehat{Q}$. Prove that $P\widehat{X}R = Q\widehat{P}R$.

⑭ In △ABC, the bisectors of $\widehat{B}$ and $\widehat{C}$ meet at I. Prove that $B\widehat{I}C = 90° + \frac{1}{2}\widehat{A}$.

Congruent triangles

Exercise 1b (Group work)

① Draw the following triangles.
  (a) ABC in which AB = 3 cm, BC = 5 cm, $A\widehat{B}C = 60°$
  (b) PQR in which PQ = 4 cm, QR = 6 cm, PR = 7 cm
  (c) XYZ in which $\widehat{X} = 57°$, $\widehat{Y} = 63°$, XY = 6 cm
  (d) DEF in which $\widehat{D} = 72°$, DE = 5 cm, EF = 5 cm

② Cut out and compare your triangles from question 1 with those drawn by the other members of your group. What do you notice?

Notice that your triangles ABC, PQR and XYZ have exactly the same shape and size as those of your classmates. The triangles DEF may, however, be different in shape and size.

Two figures are **congruent** if they have exactly the same shape and size. Congruent shapes fit exactly one over the other.

In Fig. 1.12, the two triangles are congruent, although at first sight they may appear different.

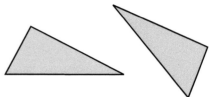

Fig. 1.12

If one of the triangles were cut out and turned over it could be arranged to fit exactly over the other.

Exercise 1c (Group work)

① Copy and cut out each pair of triangles in Fig. 1.13.

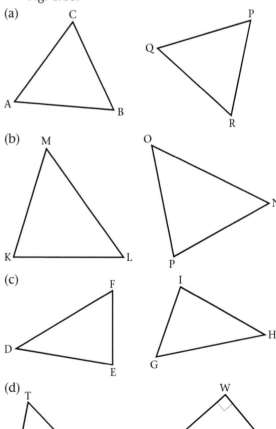

(a)
(b)
(c)
(d)

Fig. 1.13

② State whether each pair of triangles are congruent.

③ Name any two pairs of equal sides and angles.

Naming congruent triangles

When naming congruent triangles, give the letters in the correct order so that it is clear which letters of the triangles correspond to each other. For example, in Fig. 1.14, △FGE is congruent to △LMK, *not* △KLM or △LKM, etc. When congruent triangles are properly named, it is possible to find pairs of equal sides or equal angles without looking at the figure.

The symbol ≡ means 'is identically equal to', or 'is congruent to'. Thus △EFG ≡ △KLM is short for 'triangle EFG is congruent to triangle KLM'.

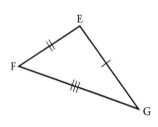

Fig. 1.14

Exercise 1d

In each part of Fig. 1.15, pairs of triangles have equal sides or equal angles shown with marks. Working in pairs, state whether, in each case, the triangles are congruent or not congruent.

(a)

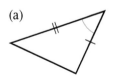

(b)

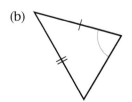

(c)

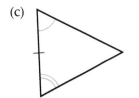

(d)

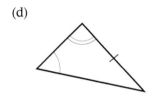

(e)

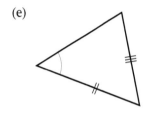

(f)

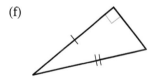
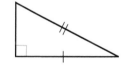

(g)

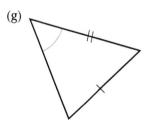
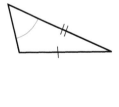

Fig. 1.15

Isosceles and equilateral triangles

An **isosceles triangle** has two equal sides. The third side is called the **base**. The equal sides meet at the **vertex**. The angle at the vertex is called the **vertical angle**.

Triangle ABC, Fig. 1.16, is an isosceles triangle with AB = AC.

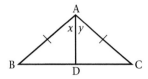

Fig. 1.16

If triangle ABC is folded along AD so that AB fits exactly over AC, then triangles ABD and ADC are congruent. Which angles are equal? Another basic fact about an isosceles triangle is:

the base angles of an isosceles triangle are equal.

Because AD divides △ABC into two equal parts then

AD bisects BC so that BD = DC
AD bisects angle BAC so that $x = y$
∴ the other properties of isosceles triangles which follow from the fact that
△ABD ≡ △ACD in Fig. 1.16 are as follows.

1 The bisector of the vertical angle bisects the base (BD = CD).
2 The bisector of the vertical angle meets the base at right angles.
(AD̂B = AD̂C and AD̂B + AD̂C = 180°,
∴ AD̂B = AD̂C = 90°)

An **equilateral triangle** is a special isosceles triangle in which the three sides are of equal length. Each angle in an equilateral triangle is 60°.

Example 7

Isosceles triangles ABC and ABD are drawn on opposite sides of a common base AB. If AB̂C = 70° and AD̂B = 118°, calculate AĈB and CB̂D.

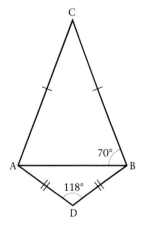

Fig. 1.17

Fig. 1.17 contains the data of the question.

In △ABC,

$$AB̂C = 70° \qquad \text{(given)}$$
$$∴ BÂC = 70° \qquad \text{(base angles of isos. △)}$$
$$∴ AĈB = 180° − 70° − 70° \qquad \text{(angle sum of △)}$$
$$= 40°$$

In △ABD,

$$AD̂B = 118° \qquad \text{(given)}$$
$$∴ AB̂D + BÂD = 180° − 118° \qquad \text{(angle sum of △)}$$
$$= 62°$$
$$∴ 2 × AB̂D = 62° \qquad \text{(base angles of isos, △)}$$
$$∴ AB̂D = 31°$$
$$CB̂D = CB̂A + AB̂D$$
$$= 70° + 31°$$
$$= 101°$$
$$AĈB = 40° \text{ and } CB̂D = 101°$$

Exercise 1e

1 In Fig. 1.18, calculate the angles marked with letters.

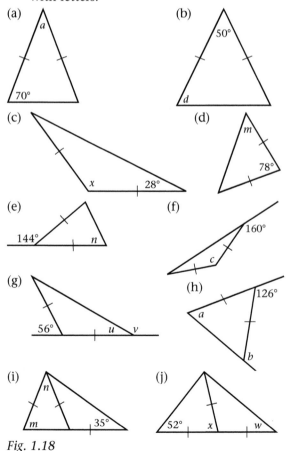

Fig. 1.18

❷ In isosceles △ABC, AB = AC. If $\widehat{B}$ = 55°, calculate $\widehat{A}$.

❸ PQR is an isosceles △ in which PQ = PR and $\widehat{P}$ = 58°. Calculate $\widehat{Q}$.

❹ In isosceles △XYZ, $\widehat{X}$ = 117°, Calculate $\widehat{Z}$.

❺ The base JK of isosceles △HJK is produced to L. If $\widehat{J}$ = 69°, calculate $H\widehat{K}L$.

❻ ABC is an isosceles triangle with its base BC produced to D. If $\widehat{A}$ = 75°, calculate $A\widehat{C}D$.

❼ The base QR of isosceles △PQR is produced to S. If PRS = 102°, calculate $\widehat{Q}$.

❽ The base VW of isosceles △UVW is produced to X. If $U\widehat{W}X$ = 121°, calculate U.

❾ In Fig. 1.19, ABC is an equilateral triangle. P is a point on AC such that $P\widehat{B}C$ = 46°. Calculate $A\widehat{P}B$.

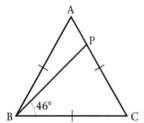

Fig. 1.19

❿ PQR is an equilateral triangle and QP is produced to S so that PS = QP. Calculate $Q\widehat{R}S$.

⓫ In △ABC, $\widehat{B}$ = 67° and $\widehat{C}$ = 46°. Show that CA = CB.

⓬ Given the data of Fig. 1.20, show that △PQR is isosceles.

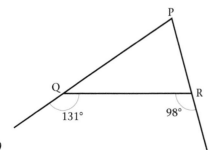

Fig. 1.20

⓭ An isosceles triangle is such that the vertical angle is 4 times the size of a base angle. What is the size of a base angle?

⓮ Calculate the angles of an isosceles triangle in which each base angle is 4 times the vertical angle.

⓯ In Fig. 1.21, △XYZ is equilateral. Use the data in the figure to calculate $N\widehat{M}X$.

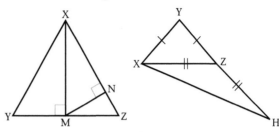

Fig. 1.21   Fig. 1.22

⓰ In Fig. 1.22, XY = YZ, XZ = ZH and $X\widehat{Y}Z$ = 52°. Calculate $Z\widehat{H}X$.

⓱ In Fig. 1.23, PR = PQ, QS = QR and $R\widehat{P}Q$ = 40°. Calculate $P\widehat{Q}S$.

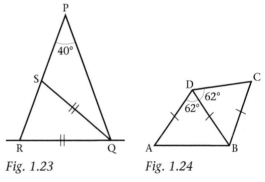

Fig. 1.23   Fig. 1.24

⓲ In Fig. 1.24, AD = DB = BC and $A\widehat{D}B$ = $B\widehat{D}C$ = 62°. Calculate $A\widehat{B}C$.

⓳ In Fig. 1.25, ABC is an isosceles triangle with AB = AC. BC is produced to D such that AC = CD. If $A\widehat{B}C$ = 2x°, $B\widehat{A}C$ = x° and $A\widehat{D}C$ = y°, show that x = y.

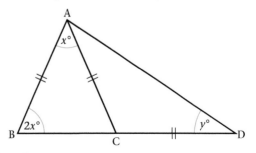

Fig. 1.25

20 In Fig. 1.26, DEF is a triangle with EDF = 2x°. DE and DF are produced to G and H respectively so that EF = EG = FH. EH and FG intersect at K. Show that $E\hat{K}G = 90° = x°$.

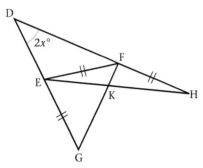

Fig. 1.26

Parallelograms

Definitions

A **parallelogram** is a quadrilateral which has both pairs of opposite sides parallel (Fig. 1.27(a)).

A **rhombus** is a parallelogram with sides of equal length (Fig. 1.27(b)).

A **rectangle** is a quadrilateral which has all its angles right angles (Fig. 1.27(c)).

A **square** is a rectangle with sides of equal length (Fig. 1.27(d)).

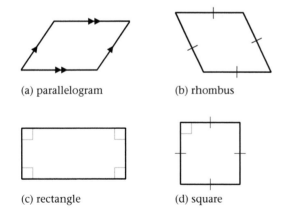

(a) parallelogram

(b) rhombus

(c) rectangle

(d) square

Fig. 1.27 Parallelograms

From the definitions it can be seen that the rhombus, rectangle and square are all special examples of parallelograms. Any properties which can be proved for a parallelogram will also be true of a rhombus, rectangle or square.

Group work

Work in pairs.

1 Copy and cut out the parallelogram in Fig. 1.28.

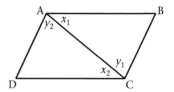

Fig. 1.28

Cut along the diagonal AC and arrange triangles ADC, ABC until they fit exactly over each other. You should now be able to say that the diagonal of a parallelogram divides the parallelogram into two triangles of equal area (congruent triangles).

You have also shown that $x_1 = x_2$, $y_1 = y_2$
$$AB = DC, AD = BC$$
∴ the opposite sides and the opposite angles of a parallelogram are equal. Would this be true for all parallelograms?

2 Copy and cut out the parallelogram ABCD, Fig. 1.29.

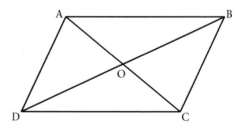

Fig. 1.29

Show by folding that DO = OB and AO = OC.
What can you say about the diagonals of the parallelogram?

The diagonals of a parallelogram bisect one another.

We can now look at all the properties of parallelograms.

Summary of properties

In a **parallelogram**:
1 the opposite sides are parallel,
2 the opposite sides are equal,
3 the opposite angles are equal,
4 the diagonals bisect one another.

In a **rhombus**:
1 all four sides are equal,
2 the opposite sides are parallel,
3 the opposite angles are equal,
4 the diagonals bisect one another at right angles,
5 the diagonals bisect the angles.

In a **rectangle**:
1 all of the properties of a parallelogram are found,
2 all four angles are right angles.

In a **square**:
1 all of the properties of a rhombus are found,
2 all four angles are right angles.

Exercise 1f

1 In Fig. 1.30 how many parallelograms are there altogether?

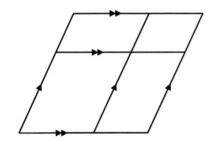

Fig. 1.30

2 In Fig. 1.31, AB = DC, AB ∥ DC, CQ ∥ PA and QAB and DCP are straight lines.
(a) Name all the parallelograms in the figure.
(b) Name all the pairs of congruent triangles in the figure.

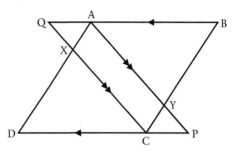

Fig. 1.31

3 In Fig. 1.32, ABCD and CDEF are parallelograms and ABEF is a straight line. If BE = 2 cm and DC = 3 cm, find AF.

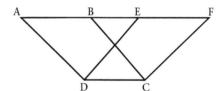

Fig. 1.32

4 Given the data of Fig. 1.32, if AF = 11 cm and BE = 3 cm, find DC.

5 In Fig. 1.33, ABCD is a rhombus and APCQ is a square. If $P\widehat{A}B = 21°$, calculate the four angles of ABCD.

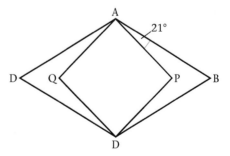

Fig. 1.33

6 ABCDEF is a regular hexagon with O at its centre. What kind of quadrilaterals are ABCO and ACDF?

7 ABCD is a square with centre O. P is a point on AB such that AP = AO. Calculate $P\widehat{O}B$.

8. ABC is a triangle and M is the mid-point of BC. A line through C parallel to AB cuts AM produced at X. Show that MX = MA.

9. PQRS is a parallelogram. SP is produced to a point X so that PX = PS. XR cuts PQ at Y. Show that Y is the mid-point of PQ.

10. In Fig. 1.34 ABCDE is such that ABCD is a parallelogram and ABDE is a square. By measurement, MD = 3 cm and AD = 8.5 cm. Calculate what the perimeter of ABCDE will measure.

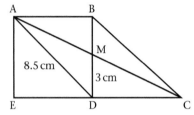

Fig. 1.34

Summary

The sum of the angles of a triangle is 180°.

The exterior angle of a triangle is equal to the sum of the opposite interior angles.

The sum of the interior angles of any n-sided convex polygon is $(2n - 4)$ right angles.

The sum of the exterior angles of any convex polygon is 4 right angles.

The base angles of an isosceles triangle are equal.

The opposite sides and angles of a parallelogram are equal.

The diagonals of a parallelogram bisect one another.

1. In △ABC, $B\widehat{A}C$ = 78°. If BC is produced to D and $A\widehat{C}D$ = 129°, show that △ABC is an isosceles triangle.

2. PQR is an equilateral triangle. PRS is a triangle such that RS ∥ QP and PS ∥ QR. Show that △PRS is an equilateral triangle.

3. The angles of a pentagon are $2x$, x, $3x$, $4x$, $5x$. Find the size of the smallest angle.

4. If the exterior angle of a regular polygon is 40°, calculate the number of sides in the polygon.

5. In a triangle, one angle is obtuse, the second is one third the size of the obtuse angle, and the third is half the size of the second angle.
   (a) Write an equation representing the information above.
   (b) Calculate the size of the angles of the triangle.

6. The sum of the angles of a regular polygon is twice the sum of the angles of a quadrilateral.
   (a) Calculate the number of sides of the regular polygon.
   (b) Calculate the size of the interior angles of the regular polygon.

7. △PQR is an equilateral triangle. If S is a point such that $R\widehat{S}P$ is 60°, show that PQRS is a parallelogram.

8. The angles of an octagon are 152°, 128°, 144°, 126°, 140°, 147°, $2x$, x.
   (a) Write an equation to show the sum of the angles of the octagon.
   (b) Using this equation, find the size of the angle measuring x.

Simplifying algebraic expressions (3)
Binary operations

Pre-requisites
- expansion of algebraic terms; algebraic fractions

Adding and subtracting algebraic fractions

Just as in arithmetic, fractions in algebraic form must have a common denominator before they can be added or subtracted.

Example 1
Simplify the following.

(a) $\dfrac{3x}{7} + \dfrac{2x}{7}$

(b) $\dfrac{5}{2a} - \dfrac{3}{2a}$

(c) $\dfrac{1}{4x} + \dfrac{2}{3x}$

(d) $\dfrac{7}{3a} - \dfrac{6}{5c}$

(a) $\dfrac{3x}{7} + \dfrac{2x}{7} = \dfrac{3x + 2x}{7} = \dfrac{5x}{7}$

(b) $\dfrac{5}{2a} - \dfrac{3}{2a} = \dfrac{5 - 3}{2a} = \dfrac{2}{2a} = \dfrac{2 \div 2}{2a \div 2} = \dfrac{1}{a}$

(c) The LCM of $4x$ and $3x$ is $12x$. $12x$ is the common denominator.

$$\dfrac{1}{4x} + \dfrac{2}{3x} = \dfrac{1 \times 3}{4x \times 3} + \dfrac{2 \times 4}{3x \times 4}$$
$$= \dfrac{3}{12x} + \dfrac{8}{12x}$$
$$= \dfrac{3 + 8}{12x}$$
$$= \dfrac{11}{12x}$$

(d) The LCM of $3a$ and $5c$ is $15ac$. $15ac$ is the common denominator.

$$\dfrac{7}{3a} - \dfrac{6}{5c} = \dfrac{7 \times 5c}{3a \times 5c} - \dfrac{6 \times 3a}{5c \times 3a}$$
$$= \dfrac{35c}{15ac} - \dfrac{18a}{15ac}$$
$$= \dfrac{35c - 18a}{15ac}$$

This does not simplify further.

$\dfrac{x - 2}{4}$ is a short way of writing $\dfrac{(x - 2)}{4}$ or $\frac{1}{4}(x - 2)$.

Similarly $\dfrac{3x}{x - y}$ is a short way of writing $\dfrac{3x}{(x - y)}$.

In each case, consider the terms inside the brackets to be a single term until the brackets can be properly removed.

Example 2
Simplify the following.

(a) $\dfrac{4a + 13}{5} - \dfrac{2a + 3}{3}$

(b) $2 + \dfrac{5(2a + 1)}{6a} - \dfrac{4b - 3}{2b}$

(a) The LCM of 5 and 3 is 15. 15 is the common denominator.

$$\dfrac{4a + 13}{5} - \dfrac{2a + 3}{3}$$
$$= \dfrac{3(4a + 13)}{5 \times 3} - \dfrac{5(2a + 3)}{3 \times 5}$$
$$= \dfrac{3(4a + 13) - 5(2a + 3)}{15}$$
$$= \dfrac{12a + 39 - 10a - 15}{15}$$
$$= \dfrac{2a + 24}{15} = \dfrac{2(a + 12)}{15}$$

(b) The LCM of $6a$ and $2b$ is $6ab$. Make equivalent fractions with denominators of $6ab$.

$$2 + \dfrac{5(2a + 1)}{6a} - \dfrac{4b - 3}{2b}$$
$$= \dfrac{6ab \times 2}{6ab} + \dfrac{b \times 5(2a + 1)}{6ab} - \dfrac{3a \times (4b - 3)}{6ab}$$
$$= \dfrac{12ab + 5b(2a + 1) - 3a(4b - 3)}{6ab}$$
$$= \dfrac{12ab + 10ab + 5b - 12ab + 9a}{6ab}$$
$$= \dfrac{10ab + 9a + 5b}{6ab}$$

This does not simplify further.

Example 3

Express each of the following as a single fraction in its simplest form.

(a) $\dfrac{3}{m + 2n} - \dfrac{2}{m - 3n}$ (b) $\dfrac{x}{x + 2} - \dfrac{x - 2}{x - 3}$

(a) The denominators are $(m + 2n)$ and $(m - 3n)$. Their LCM is $(m + 2n)(m - 3n)$. Make equivalent fractions with a common denominator of $(m + 2n)(m - 3n)$.

$$\dfrac{3}{m + 2n} - \dfrac{2}{m - 3n}$$

$$= \dfrac{3(m - 3n)}{(m + 2n)(m - 3n)} - \dfrac{2(m + 2n)}{(m + 2n)(m - 3n)}$$

$$= \dfrac{3(m - 3n) - 2(m + 2n)}{(m + 2n)(m - 3n)}$$

$$= \dfrac{3m - 9n - 2m - 4n}{(m + 2n)(m - 3n)}$$

$$= \dfrac{m - 13n}{(m + 2n)(m - 3n)}$$

This does not simplify further.

(b) $\dfrac{x}{x + 2} - \dfrac{x - 2}{x - 3}$

$$= \dfrac{x(x - 3) - (x + 2)(x - 2)}{(x + 2)(x - 3)}$$

$$= \dfrac{x^2 - 3x - (x^2 - 4)}{(x + 2)(x - 3)}$$

$$= \dfrac{x^2 - 3x - x^2 + 4}{(x + 2)(x - 3)}$$

$$= \dfrac{4 - 3x}{(x + 2)(x - 3)}$$

Exercise 2a

1. Express each of the following as a single fraction.

(a) $\dfrac{4x}{9} + \dfrac{x}{9}$ (b) $\dfrac{5y}{2} - \dfrac{2y}{3}$

(c) $\dfrac{3}{a} - \dfrac{2}{b}$ (d) $\dfrac{1}{2x} + \dfrac{1}{7x}$

(e) $\dfrac{3}{4a} - \dfrac{4}{3b}$ (f) $\dfrac{3}{2ab} + \dfrac{4}{3bc}$

2. Express $\dfrac{2}{a} + \dfrac{7}{b} - \dfrac{3}{c}$ as a single fraction.

3. Reduce $\dfrac{3}{x} - \dfrac{x}{2} + 5$ to a single fraction.

4. Simplify $\dfrac{5}{2cd} + \dfrac{4}{3de}$.

5. Simplify each of the following.

(a) $\dfrac{2x - 3}{3} + \dfrac{x - 1}{3}$

(b) $\dfrac{7x + 2}{5} - \dfrac{5x + 3}{5}$

(c) $\dfrac{a - 2}{3} - \dfrac{a - 4}{5}$

(d) $\frac{1}{2}(3x + 1) + \frac{1}{3}(x + 2)$

(e) $\dfrac{2(3x - 2)}{5} + \dfrac{x + 5}{6}$

(f) $\dfrac{3a - 4}{5} - \dfrac{2a + 19}{15}$

6. Simplify the following expression.

$\dfrac{3x + 2}{3} - \dfrac{x - 1}{4} - \dfrac{5}{12}$

7. Simplify the following.

(a) $\dfrac{6x + 1}{2a} - \dfrac{x - 2}{2a}$

(b) $\dfrac{x - 2}{6x} + \dfrac{2x + 1}{3x}$

(c) $\dfrac{4a + 1}{a} + \dfrac{3b - 2}{b}$

(d) $3 - \dfrac{6a - 5}{2a}$

(e) $\dfrac{2a - 1}{5a} - \dfrac{4b - 3}{10b}$

(f) $\dfrac{2x + 1}{x} + \dfrac{3y - 2}{y} - 5$

8. Express $\dfrac{a - 4}{2b} - \dfrac{b - 3}{6b} + 4$ as a single fraction in its lowest terms.

9. Simplify the following.

(a) $3 + \dfrac{2b}{a - b}$ (b) $2 - \dfrac{x}{x + 2y}$

(c) $\dfrac{5}{a + 4} - \dfrac{2}{a - 2}$ (d) $\dfrac{2}{t + 1} + \dfrac{3}{t + 2}$

(e) $\dfrac{3x}{x - 1} - \dfrac{x}{x + 2}$ (f) $\dfrac{x}{x - 3} - \dfrac{8}{x} - 2$

(g) $\dfrac{1}{n - 6} + \dfrac{1}{n - 4} - \dfrac{2}{n - 5}$

(h) $\dfrac{x - 2}{x + 2} - \dfrac{x - 1}{x + 3}$

10. If $X = \dfrac{2a + 3}{3a - 2}$, express $\dfrac{X - 1}{2X + 1}$ in terms of a.

Common factors

Example 4

Complete the bracket in the statement
$8p - 20q = 4(\quad)$.

4 is the HCF of $8p$ and $20q$. Divide $8p$ by 4 and
$20q$ by 4 to find the terms inside the bracket.
$8p \div 4 = 2p$ and $20q \div 4 = 5q$
$8p - 20q = 4(2p - 5q)$

Here, 4 and $(2p - 5q)$ are the **factors** of $8p - 20q$.
4 is the **common factor** of the given terms.

Example 5

Factorise the following.
(a) $2a^3 - 5a^2 - a$
(b) $15a^3b^4c - 6a^2b^5c^2$

(a) The common factor is a.
   $2a^3 - 5a^2 - a = a(2a^2 - 5a - 1)$
(b) The HCF of the two terms is $3a^2b^4c$.
   $15a^3b^4c - 6a^2b^5c^2 = 3a^2b^4c(5a - 2bc)$

Exercise 2b (Oral)

Factorise the following either by completing the
brackets or by finding the highest common
factors of the given terms.

① $2m + 8n = 2(\quad)$

② $3a - 15b = 3(\quad)$

③ $10x - 5 = 5(\quad)$

④ $-3h - 12k = -3(\quad)$

⑤ $-2x + 18 = -2(\quad)$

⑥ $5a - 8ab = a(\quad)$

⑦ $9x + 3xz = 3x(\quad)$

⑧ $8cm + 12dm - 16em = 4m(\quad)$

⑨ $3x^3 - 12x^2 - 9x = 3x(\quad)$

⑩ $10a^2b^2 - 15a^2b + 20ab^2 = 5ab(\quad)$

⑪ $4a - 8b$ ⑫ $9x + 12y$

⑬ $3ab - 6ac + 3ad$ ⑭ $8px - 4qx + 8rx$

⑮ $3m^3 - 2m^2 + m$ ⑯ $6n^4 - 2n^3 + 4n^2$

⑰ $5ab + 4a^2b + 6ab^2$ ⑱ $2abx + 7acx - 3a^2x$

⑲ $4a^4 + 2a^3b - 10a^2b^2$

⑳ $24a^2bc^2 + 30a^2c^2x - 18a^3cx^2$

Example 6

Factorise $12x^2 + 3x - 4x - 1$.

The terms $12x^2$ and $3x$ have $3x$ in common.
The terms $-4x$ and -1 have -1 in common.
$12x^2 + 3x - 4x - 1$
$\quad = 3x(4x + 1) - 1(4x + 1)$
$\quad = (4x + 1)(3x - 1)$
(since the expressions $3x(4x + 1)$ and
$- 1(4x + 1)$ both have $(4x + 1)$ in common).

Example 7

Factorise $3x - 2dy + 3y - 2dx$.

The terms $3x$ and $3y$ both have 3 in common;
and $-2dx$ and $-2dy$ both have $-2d$ in common.
Grouping the given terms in this order,
$3x - 2dy + 3y - 2dx$
$\quad = (3x + 3y) - (2dx + 2dy)$
$\quad = 3(x + y) - 2d(x + y)$
$\quad = (x + y)(3 - 2d)$

Exercise 2c

Factorise the following by grouping terms in pairs.

① $3x + 9b + 5ax + 15ab$

② $2ce - 2cf + de - df$

③ $4u + 4v + vx + ux$

④ $mn - 3my - 3nx + 9xy$

⑤ $6a^2 - 3a + 4a - 2$

⑥ $cd - ce - d^2 + de$

⑦ $12eg - 4eh - 6fg + 2fh$

⑧ $4ab + 6bn - 2a - 3n$

⑨ $3ax - 2a - 6bx + 4b$

⑩ $ax - a + x - 1$

⑪ $ac + ad - bc - bd$

⑫ $4a - 7b + 28bx - 16ax$

⑬ $p + q + 5ap + 5aq$

⑭ $2c^2 - 8cm - 3cm + 12m^2$

⑮ $x^2 + 2x + 5x - 10$

⑯ $2pr - sq + 2qr - ps$

⑰ $y^2 - 5y + 4y - 20$

⑱ $6ac + bd - 3bc - 2ad$

⑲ $2ab - 10cd - 5bc + 4ad$

⑳ $6xy - 2z - 4y + 3xz$

Substitution

Example 8

Find the value of $4(3d - e) - 2f$ when $d = 2$, $e = 4$ and $f = 3$.

Either:

$$4(3d - e) - 2f = 4(3 \times 2 - 4) - 2 \times 3$$
$$= 4(6 - 4) - 6 \quad \text{(multiply before subtracting)}$$
$$= 4 \times 2 - 6$$
$$= 8 - 6$$
$$= 2$$

or:

$$4(3d - e) - 2f = 12d - 4e - 2f$$
$$= 12 \times 2 - 4 \times 4 - 2 \times 3$$
$$= 24 - 16 - 6$$
$$= 2$$

Example 9

Evaluate $(m - n)(u + v)$ if $m = 5$, $n = 3$, $u = 1$ and $v = 2$.

$$(m - n)(u + v) = (5 - 3)(1 + 2)$$
$$= 2 \times 3 \quad \text{(brackets first)}$$
$$= 6$$

Notice the order of operations in Examples 8 and 9. First evaluate the contents of brackets, then do multiplication (or division) before addition or subtraction.

Exercise 2d

If $a = 1$, $b = 2$, $c = 3$, $m = 4$ and $n = 5$, find the value of the following.

1. bc
2. $bc + n$
3. $4n - 2b$
4. $4(n - 2b)$
5. $3(a + 2m)$
6. $3a + 2m$
7. $2(a + 3c) - 4n$
8. $3(a - b + c)$
9. $(a + c)(m - b)$
10. $(2m - n)(c + 2a)$
11. $b(3m - n)$
12. $(m - 1)(m + 1)$
13. $\dfrac{bcn}{b + c}$
14. $\dfrac{mn + b}{cm - a}$
15. $a + \dfrac{m}{b}$
16. $\dfrac{a + m}{b}$
17. $\dfrac{c^2 + m}{n^2 + a}$
18. $cm^2 - c^2m$
19. $\sqrt{(c^2 + m^2)}$
20. $\sqrt{n(n + b) + a}$

Example 10

Evaluate $\dfrac{2a^2bc}{2b - c}$ when $a = 3$, $b = -4$, $c = -5$.

$$\frac{2a^2bc}{2b - c} = \frac{2 \times (3)^2 \times (-4) \times (-5)}{2 \times (-4) - (-5)}$$
$$= \frac{2 \times 9 \times 20}{-8 + 5}$$
$$= \frac{2 \times 9 \times 20}{-3}$$
$$= -120$$

Notice in Example 10 that the denominator, $(2b - c)$, must be reduced to a single number before division is possible.

Exercise 2e

1. Find the value of the following when $a = 1$, $b = 0$ and $c = -3$.

   (a) ac
   (b) abc
   (c) $a - c$
   (d) $ab - ac$
   (e) $a + b + c$
   (f) $b - (a + c)$
   (g) $\dfrac{c - a}{2a}$
   (h) $(a + 2c)^b$
   (i) $c(a + c)$
   (j) $\sqrt{(a - 5c)}$

2. Evaluate the following when $x = 4$, $y = -5$ and $z = -2$.

   (a) $x + y + z$
   (b) $x - y + z$
   (c) $x - (y + z)$
   (d) $x + y - z$
   (e) xyz
   (f) $z(x + y)$
   (g) $y^2 - 2x$
   (h) $3xy - y^3$
   (i) $\dfrac{x + yz}{y + z}$
   (j) $\dfrac{y^2 - 1}{x - z}$

3. What is the value of $p^2q - q^2p$ if $p = 3$ and $q = -1$?

4. Evaluate $x^2 + 3x + 2$ when (a) $x = 1$, (b) $x = 0$, (c) $x = -1$, (d) $x = -2$.

5. Evaluate $p^2 - 2p - 3$ when (a) $p = 4$, (b) $p = 3$, (c) $p = 1$, (d) $p = 0$ (e) $p = -1$ (f) $p = -2$.

6. Evaluate $3a^2 - 2a + 5$ if (a) $a = 2$, (b) $a = 0$, (c) $a = -2$, (d) $a = -4$.

7. Evaluate $z^3 - z$ if (a) $z = 2$, (b) $z = 1$, (c) $z = 0$, (d) $z = -1$, (e) $z = -2$.

⑧ Evaluate $ab\sqrt{(c^2 + b^2)}$, given that $a = 2$, $b = -3$, $c = 4$.

⑨ If $a = -3$, $b = 2$, $c = 1$ and $d = -4$, find the values of (a) $5b - ad$, (b) $(ba)^2 - bd^2$, and (c) $\left(\dfrac{c - a}{b - d}\right)^2$.

⑩ If $x = -7$ and $y = 3$, calculate the values of

(a) $\left(\dfrac{x + y}{x - y}\right)^2$ (b) $2x^2y + y^2x$,

(c) $27x^2 \div 12y^2$

Laws of indices

The following laws of indices are true for all values of a, b and x.

1   $x^a \times x^b = x^{a + b}$

2   $x^a \div x^b = x^{a - b}$

3   $x^0 = 1$

4   $x^{-a} = \dfrac{1}{x^a}$

Example 11

Simplify (a) $10^2 \times 10^3$ (b) $22n^7 \div 2n^3$ (c) 19^0 (d) 5^{-2} (e) $2^3 \times \left(\frac{1}{6}\right)^{-1}$ (f) $r \times r^0 \times r^{-5}$

(a) $10^2 \times 10^3 = 10^{2+3} = 10^5$ (Law 1)
   Check:
   $10^2 \times 10^3 = (10 \times 10) \times (10 \times 10 \times 10)$
   $= 10 \times 10 \times 10 \times 10 \times 10$
   $= 10^5$

(b) $22n^7 \div 2n^3 = \frac{22}{2} \times n^{7-3} = 11n^4$ (Law 2)
   Check:
   $22n^7 \div 2n^3 = \dfrac{22 \times n \times n \times n \times n \times n \times n \times n}{2 \times n \times n \times n}$
   $= 11 \times n \times n \times n \times n = 11 \times n^4$
   $= 11n^4$

(c) $19^0 = 1$ (Law 3)
   Check:
   $19^0 = 19^{a - a}$ (since $a - a = 0$)
   $= \dfrac{19^a}{19^a}$ (Law 2)
   $= 1$ (*numerator and denominator are equal*)

(d) $5^{-2} = \dfrac{1}{5^2} = \dfrac{1}{25}$ (Law 4)
   Check:
   $5^{-2} = 5^{0-2} = \dfrac{5^0}{5^2}$ (Law 2)
   $= \dfrac{1}{5^2}$ (Law 3)
   $= \frac{1}{25}$

(e) $2^3 \times \left(\frac{1}{6}\right)^{-1} = 8 \times \dfrac{1}{\frac{1}{6}}$ (Law 4)
   $= 8 \times 6 = 48$

(f) $r \times r^0 \times r^{-5} = r^{1 + 0 + (-5)}$ (Law 1)
   $= r^{-4} = \dfrac{1}{r^4}$
   or
   $r \times r^0 \times r^{-5} = r^1 \times 1 \times \dfrac{1}{r^5}$ (Laws 3 and 4)
   $= \dfrac{1}{r^4}$

Exercise 2f (Revision)

Simplify the following.

① $10^5 \times 10^4$ ② $5y \times 4y^4$

③ $m^8 \div m^5$ ④ $\dfrac{24x^6}{8x^4}$

⑤ 2^0 ⑥ 4^{-3}

⑦ $\left(\frac{1}{4}\right)^{-2}$ ⑧ $x^3 \div x^{-5}$

⑨ $(3x)^{-3}$ ⑩ $5x^2 \times 4x^0 \times 2x^{-6}$

Products of indices $(x^a)^b$

Example 12

Simplify (a) $(x^2)^3$ (b) $(y^4)^2$ (c) $(z^3)^5$.

(a) $(x^2)^3 = x^2 \times x^2 \times x^2$
   $= x^{2 + 2 + 2} = x^{2 \times 3}$
   $= x^6$

(b) $(y^4)^2 = y^4 \times y^4$
   $= y^{4 \times 2}$
   $= y^8$

(c) $(z^3)^5 = z^3 \times z^3 \times z^3 \times z^3 \times z^3$
   $= z^{3 \times 5}$
   $= z^{15}$

Notice that in each part of Example 12 the index in the result is the product of the given indices, e.g. $(x^2)^3 = x^{2 \times 3} = x^6$.

In general, $(x^{ab}) = x^{a \times b} = x^{ab}$.

Exercise 2g (Oral)

Simplify the following.

1. $(a^3)^2$
2. $(b^2)^4$
3. $(c^5)^3$
4. $(d^4)^3$
5. $(e^3)^3$
6. $(f^0)^8$
7. $(g^{-2})^5$
8. $(h^4)^{-5}$
9. $(5^2)^{-1}$
10. $(3^{-2})^{-2}$
11. $(10^2)^7$
12. $(2^{-3})^2$

Example 13

Simplify the following.

(a) $(-3d^3)^2$ (b) $-3(d^3)^2$
(c) $(-4g^5)^3$ (d) $(a^3b)^4$

(a) $(-3d^3)^2 = (-3d^3) \times (-3d^3)$
$\quad\quad\quad\quad = -3 \times -3 \times d^3 \times d^3$
$\quad\quad\quad\quad = +9d^6$

(b) $-3(d^3)^2 = -3 \times d^3 \times d^3$
$\quad\quad\quad\quad = -3d^6$

(c) $(-4g^5)^3 = (-4g^5) \times (-4g^5) \times (-4g^5)$
$\quad\quad\quad\quad = -64 \times g^{5 \times 3}$
$\quad\quad\quad\quad = -64 \times g^{15}$

Or, more quickly:
$(-4g^5)^3 = (-4)^3 \times g^{5 \times 3} = -64g^{15}$

(c) $(a^3b)^4 \quad = (a^3b^1)^4 = a^{3 \times 4} \times b^{1 \times 4} = a^{12}b^4$

Notice the following in Example 13.

1. The power outside the bracket raises everything inside the bracket to that power.
2. A negative number raised to an odd power is negative. A negative number raised to an even power is positive.

Exercise 2h

Simplify the following.

1. $(3m^4)^2$
2. $(2n^5)^3$
3. $(4v^3)^2$
4. $-2(a^2)^3$
5. $(-c^3)^2$
6. $(-e^4)^3$
7. $-(c^5)^4$
8. $(-d^5)^4$
9. $(mn^2)^4$
10. $(a^2b)^3$
11. $(-u^3v^2)^4$
12. $(5mn^3)^3$

Binary operations

In a **binary operation**, two quantities are combined by some method to give a 'result'.

In $3 + 7 = 10$, 3 and 7 are added together. The binary operation is *addition*.

In $2b \times 5c = 10bc$, the terms $2b$ and $5c$ have been multiplied. The binary operation is *multiplication*.

The four basic arithmetic binary operations are *addition, subtraction, multiplication* and *division*. The symbols used to represent these operations are well known, namely, $+$, $-$, $\times$, and $\div$. These symbols are called **binary operators**.

Other operations to combine two quantities may be defined by using more than one of the four basic arithmetic operations at the same time. For example, finding the arithmetic mean of two numbers is a binary operation since it combines the two numbers in a new way. The method of finding the average involves adding the two numbers and dividing their sum by 2. If we use the symbol $\square$ to represent this operation of finding the average, we can say that

$$8 \square 6 = 7,$$

where the symbol $\square$ means add 8 and 6, and divide the result by 2;
that is,

$$8 \square 6 = \frac{8 + 6}{2}$$

Other binary operations which have been defined in this mathematics course include:

raising one number to a power which is a second number (called the index), for example, 2^3 where the number 2 is raised to the second number, the index 3;

and taking the logarithm of one number to the second number, called the base, for example, $\log_{10} 1000$, where we take the logarithm of the number 1000 to the second number, the base $_{10}$.

We may specific different special symbols in an algebraic statement to define different binary operations, for example, we may use $*$, $\boxplus$, $\square$, $[\,]$, $(\,)$, $\oslash$, $\boxtimes$, and so on.

In algebra, we use letters to represent general numbers, so that to represent the average by the symbol $\square$, we may write

$$a \square b = \frac{a + b}{2}$$

Example 14

Given that $x \square y = 3x + y$, find the value of $2 \square 5$.

$$2 \square 5 = 3 \times 2 + 5 = 11$$

The rule about the order of doing operations holds, so that the multiplication is done first.

Example 15

Given that $\quad a \oslash b = 4a - b^3$,
show that $7 \oslash (3 \oslash 2) = -36$.

$$\begin{aligned}
7 \oslash (3 \oslash 2) &= 7 \oslash (4(3) - (2)^3) \\
&= 7 \oslash (12 - 8) \\
&= 7 \oslash 4 \\
&= 4(7) - (4)^3 \\
&= 28 - 64 \\
&= -36
\end{aligned}$$

Example 16

Given that $m * 2 = 2m + n$, evaluate
(a) $3 * 1$, (b) $2 * (3 * 1)$.

(a) $3 * 1 = 2(3) + 1 = 7$
(b) $2 * (3 * 1) = 2 * 7 = 2(2) + 7 = 11$

Note that in (b) the operation in the bracket is done first.

Example 17

If $r \otimes s = \dfrac{3r - s}{3r + s}$ show that $5 \otimes 9 = \dfrac{1}{4}$.

$$5 \otimes 9 = \frac{3 \times 5 - 9}{3 \times 5 + 9} = \frac{6}{24} = \frac{1}{4}$$

Exercise 2i

① If $p * q = p + \dfrac{q}{2}$. find the value of
(a) $1 * 4$　　　　　　(b) $4 * 1$
(c) $3 * (1 * 4)$　　　(d) $3 * (4 * 1)$

② If $x \boxplus y = \frac{1}{2}(3x + y)$,
(a) show that $3 \boxplus 1 = 5$,
(b) find the value of m if $m \boxplus 3 = 6$.

③ If $a \otimes b = \dfrac{2a - 1}{3b}$ find the value of x in the following.
(a) $x \otimes 2 = 3$
(b) $2x \otimes x = 1$
(c) $x \otimes (x \otimes 2) = \dfrac{x}{2}$

The binary operations defined may be commutative, or assosciative, or both, or neither. Remember that

$$5 + 3 = 3 + 5$$
also $2y + 8y = 8y + 2y$

This property is called the **commutative** property. Changing the order in which the addition of the two terms is performed does not alter the result. Multiplication is also commutative:

$$5 \times 3 = 3 \times 5$$
and $7a \times -2b = -2b \times 7a$

Also, notice that

$$(7 + 5) + 3 = 7 + (5 + 3)$$
and $(7 \times 5) \times 3 = 7 \times (5 \times 3)$

This property is called the **associative** property.

Subtraction and division are neither commutative nor associative:

$$8 - 3 \neq 3 - 8$$
and $(7 - 5) - 3 \neq 7 - (5 - 3)$
Similarly $6 \div 2 \neq 2 \div 6$
and $(10 \div 5) \div 2 \neq 10 \div (5 \div 2)$

Multiplication is **distributive** over addition so that

$$5(2 + 4) = 5(2) + 5(4)$$
and $(2 + 4)5 = 2(5) + 4(5)$

It is also true that

$$5(2 - 4) = 5(2 + (-4))$$

since subtraction is the same process as addition of a negative number. So multiplication is also distributive over subtraction.

Example 18

If $a \square b = \dfrac{2a + b}{3}$, is $7 \square 4 = 4 \square 7$? Hence, state whether the operation is commutative.

$$7 \square 4 = \frac{2 \times 7 + 4}{3} = \frac{14 + 4}{3} = 6$$

$$4 \square 7 = \frac{2 \times 4 + 7}{3} = \frac{8 + 7}{3} = 5$$

Hence $7 \square 4 \neq 4 \square 7$, that is, the operation is not commutative.

Note that if there is *only one* example where an operation is neither commutative nor associative, we can say that the operation does not, in general, have that property. However, the reverse is *not* necessarily true. If a property (or law) holds for one (or more) example(s) of an operation, it does not prove that it is true for all examples.

Example 19

If $a \odot b = \dfrac{3ab}{2(a + b)}$, by finding the value of $4 \odot 5$ and $5 \odot 4$, show that this operation is commutative for 5 and 4.

$$4 \odot 5 = \frac{3(4 \times 5)}{3(4 + 5)}$$

and

$$5 \odot 4 = \frac{3(5 \times 4)}{2(5 + 4)}$$

that is, $4 \odot 5 = 5 \odot 4$.

Note that this is *not* a general proof since we have used only one example. However, since we know that both the numerator and the denominator contain the commutative operations of multiplication and addition, we can deduce that the operation is generally commutative.

Example 20

If $h \boxtimes k = 5h + \dfrac{k}{2}$, find the value of

(a) $4 \boxtimes 6$,   (b) $6 \boxtimes 4$,
(c) $2 \boxtimes (6 \boxtimes 4)$,   (d) $(2 \boxtimes 6) \boxtimes 4$.
(e) State whether the operation is commutative and/or associative.

(a) $4 \boxtimes 6 = 5 \times 4 + \dfrac{6}{2} = 20 + 3 = 23$

(b) $6 \boxtimes 4 = 5 \times 6 + \dfrac{4}{2} = 30 + 2 = 32$

(c) $2 \boxtimes (6 \boxtimes 4) = 2 \boxtimes 32$

$$= 5 \times 2 + \frac{32}{2}$$

$$= 10 + 16 = 26$$

(d) $(2 \boxtimes 6) \boxtimes 4 = \left(5 \times 2 + \dfrac{6}{2}\right) \boxtimes 4$

$$= 13 \boxtimes 4$$

$$= 5 \times 13 + \frac{4}{2}$$

$$= 65 + 2 = 67$$

(e) The operation is neither commutative (since $4 \boxtimes 6 \neq 6 \boxtimes 4$) nor associative (since $2 \boxtimes [6 \boxtimes 4] \neq [2 \boxtimes 6] \boxtimes 4$).

Example 21

If $b * d = \dfrac{3b + 2}{d}$, find the value of y in the following.

(a) $y * 2 = 7$ (b) $2y * (y * 3) = 2$.

(a) $y * 2 = 7$

$$\frac{3y + 2}{2} = 7$$

$$3y + 2 = 14$$

$$3y = 12$$

$$y = 4$$

(b) $2y * (y * 3) = 2$

$$2y * \frac{3y + 2}{3} = 2$$

$$\frac{6y + 2}{\left(\dfrac{3y + 2}{3}\right)} = 2$$

$$\frac{3(6y + 2)}{3y + 2} = 2$$

$$3(6y + 2) = 2(3y + 2)$$

$$18y + 6 = 6y + 4$$

$$18y - 6y = 4 - 6$$

$$12y = -2$$

$$y = -\tfrac{1}{6}$$

Exercise 2j

1. Given that $a \odot b = 2a - 3b$, find the value of
   (a) $5 \odot 2$,   (b) $2 \odot 5$.
   (c) Is $a \odot b$ commutative?

2. Given that $c \boxplus d = \dfrac{(2c + d)^2}{3}$, evaluate
   (a) $2 \boxplus 5$,   (b) $5 \boxplus 2$,
   (c) $2 \boxplus (5 \boxplus 2)$,   (d) $(2 \boxplus 5) \boxplus 2$.
   (e) State whether the operation is
        (i) commutative,   (ii) associative.

3 Given that $x * y = \dfrac{x + y}{2xy}$ show that the operation is commutative but not associative.

4 Use the commutative property and/or the associative property to evaluate the following.

(a) 102×47

(b) $89 + 17 + 211$

(c) $72 \times 27 + 28 \times 27$

Summary

The order of operations which applies in arithmetic also applies in simplifying algebraic expressions. Generally the first step is to collect **like terms**.

Simplification may involve **expanding** brackets or finding common **factors**. In simplifying expressions which include algebraic fractions, it is important to use common factors to find the simplest common denominator.

It is important to distinguish between **simplifying an algebraic expression** and **solving an algebraic equation**.

In simplifying an expression, the value of the given expression must not change. However, the form of the expression changes; for example, by collecting like terms; and, when there are fractions, by using a common denominator and/ or by dividing both numerator and denominator by common factors.

To solve an equation is to find the particular value(s) of the variable(s) which make the LHS and RHS of the equation equal. Note that in solving an equation, it is often necessary to first simplify algebraic expressions.

In a **binary operation** two elements are combined to produce a 'result' according to given combinations of the basic binary arithmetic operations of addition, subtraction, multiplication and division. Axioms which relate to these operations include the statements defining the **commutative**, **associative** and **distributive** properties.

Practice exercise P2.1

Simplify the following expressions.

1 $2a + 3 - 5a$

2 $\frac{1}{2}f - \frac{1}{4}f + 2$

3 $-\frac{1}{3}(3k) + 5k - 9$

4 $c + 7 + 2c - 3$

5 $\dfrac{y}{2} - \dfrac{y}{3} - \dfrac{2}{3}$

6 $\dfrac{3d}{2} - \dfrac{1}{2} + 3 - \dfrac{d}{4}$

7 $\dfrac{x + 1}{4} + \dfrac{2x - 3}{3}$

8 $\dfrac{2b}{3} - \dfrac{b - 1}{2}$

9 $\dfrac{h + 1}{h} + \dfrac{h - 1}{2h}$

10 $\dfrac{1 + n}{2} - \dfrac{3n - 2}{3}$

Practice exercise P2.2

Simplify the following expressions.

1 $-3k + 5 + 2(2k - 1)$

2 $4(m - 2) - 2(2m - 3) + m$

3 $4 - (1 + 2y)2 - (2 - 3y)3 + 11$

4 $\frac{2}{3}b + 1 + \frac{1}{2}(b - 1)$

5 $c(c - 5) + 6 - (c - 1)(c - 2)$

6 $3t - 2t(4 - 2t) - 5$

7 $\dfrac{x + 1}{4} - \dfrac{2x - 3}{3}$

8 $\dfrac{2x + 3}{2} + \dfrac{3x - 4}{3} - \dfrac{7x}{4}$

9 $\dfrac{1}{2} - \dfrac{x + 1}{2x + 1} - \dfrac{1}{3}$

Practice exercise P2.3

Factorise the following, first simplifying if necessary.

1 $3a - 6b$

2 $12cd - 100d^2$

3 $3a^2 - 3ab^2 + 9ab$

4 $-3(v - 2)(v + 2) + 3v^2 + 2v$

5 $\dfrac{x + 2}{2} + \dfrac{1 - 3x}{3} - \dfrac{1}{3}$

6 $\dfrac{x - 2}{x + 2} - \dfrac{3}{5}$

7 $c(c - 5) + 5 + (c - 1)(c - 2)$

8 $2x(3x + y) + y(x - 2y)$

Practice exercise P2.4

Factorise the following expressions by grouping.

1 $af + ag + bf + bg$

2 $ab - ac + gb - gc$

3 $2ak - ch - 2ah + ck$

④ $3a^2 - 2ac - 6ab + 4bc$

⑤ $2a^2 + axy - 2a - xy$

⑥ $2jm - 2kn + 2km - 2jn$

⑦ $hm - 2km - 2hn + 4kn$

⑧ $3sx + tx - 6sv - 2tv$

Practice exercise P2.5

If $a = 2$, $b = -1$, $c = -2$ and $d = 3$, find the values of the following.

① $3ad + bc$

② $ab(c - 2d)$

③ $(a - 3c)(d - 3b)$

④ $abd + 3c^2$

⑤ $a^3b - 2cd + 5$

⑥ $d(2a + b - 3c + 2)$

Practice exercise P2.6

① Find the values of the following.
   (a) 2^6 (b) 4^5 (c) $(-4)^3$ (d) 0.3^3 (e) 1.1^3

② Use index form to write the answers to the following.
   (a) $m^3 \times m^7$ (b) $k^4 \times k^3$
   (c) $(mp)^2 \times mp$ (d) $m^8 \div m^5$
   (e) $k \div k^7$ (f) $(mp)^6 \div (mp)^5$

③ For each calculation:
   (i) find the answer in index form;
   (ii) check that the answer is correct by converting to ordinary numbers.
   (a) $3^4 \times 3^3$ (b) $5^4 \times 5^2$
   (c) $2^5 \times 2^3$ (d) $0.4^3 \times 0.4^2$
   (e) $(-2)^5 \times (-2)^3$ (f) $10^5 \times 10^4$
   (g) $3^7 \div 3^2$ (h) $2^4 \div 2^6$
   (j) $2^8 \div 2^3$ (k) $0.9^4 \div 0.9^2$
   (m) $(-5)^5 \div (-5)^2$ (n) $10^3 \div 10^6$

④ Write each negative power as:
   (i) a fraction with an index;
   (ii) an ordinary fraction;
   (iii) a decimal, rounded to 3 sig. figs. where necessary [see first example].
   (a) 2^{-4} (i) $\frac{1}{2^4}$ (ii) $\frac{1}{16}$ (iii) 0.0625
   (b) 4^{-3} (c) 5^{-2} (d) 3^{-5} (e) 6^{-2}
   (f) 10^{-4} (g) 3^{-4} (h) 11^{-1}

⑤ Write each fraction as a negative power, using the smallest possible base.
   (a) $\frac{1}{5}$ (b) $\frac{1}{49}$ (c) $\frac{1}{8}$ (d) $\frac{1}{10}$
   (e) $\frac{1}{125}$ (f) $\frac{1}{27}$ (g) $\frac{1}{1000}$ (h) $\frac{1}{128}$

⑥ Simplify the following expression and write as an ordinary fraction.
   $(12^2 \times 10^{-3}) \div (6^3 \times 5^{-4})$

⑦ Simplify the following, writing your answer:
   (i) in index form;
   (ii) as an integer or fraction.
   (a) $(3^2)^3$ (b) $(5^2)^2$
   (c) $(2^2)^3 \times (2^3)^{-2}$ (d) $(0.4^3)^2$
   (e) $(3^7)^2 \div (3^2)^5$ (f) $(7^4)^3 \div (7^6)^2$
   (g) $(2^8)^2 \div (2^3)^6$ (h) $(0.9^2)^4 \div 0.9^7$
   (i) $(-5)^4 \div (-5^4)^2$ (j) $(10^3)^2 \div 10^2$

Practice exercise P2.7

① Given that $a \,§\, b = 3a \times \sqrt{b}$,
   (a) evaluate
       (i) $2 \,§\, 9$ (ii) $9 \,§\, 2$
       (iii) $2 \,§\, (9 \,§\, 4)$ (iv) $(2 \,§\, 9) \,§\, 4$
   (b) state, giving the reasons, whether $a \,§\, b$ obeys the commutative and the associative laws.

② The operation $\square$ is defined for any two numbers:
   $$h \,\square\, k = (h + k)(h - k).$$
   Find the value of
   (a) $3 \,\square\, 4$ (b) $4 \,\square\, 3$
   (c) $3 \,\square\, -1$ (d) $3 \,\square\, (3 \,\square\, -1)$
   (e) $(3 \,\square\, -1) \,\square\, 3$

③ If $d \,\square\, y = \frac{1}{d} + \frac{1}{y}$,
   (a) show that $2 \,\square\, 3 \,\square\, -1 = \frac{1}{5}$
   (b) evaluate $\frac{1}{2} \,\square\, \frac{1}{3}$

④ Given that $c \,\boxdot\, d = 2c - d$,
   (a) evaluate
       (i) $2 \,\boxdot\, 4$ (ii) $4 \,\boxdot\, 2$
       (iii) $1 \,\boxdot\, (2 \,\boxdot\, 4)$ (iv) $(1 \,\boxdot\, 2) \,\boxdot\, 4$
   (b) Is the operation (i) commutative?
                    (ii) associative?
   Give reasons.

⑤ If $m \,\oslash\, n = m^n$,
   (a) find the values of
       (i) $3 \,\oslash\, 2$ (ii) $5 \,\oslash\, 2$
       (iii) $2 \,\oslash\, 3$ (iv) $(3 \,\oslash\, 2) \,\oslash\, 2$.
   (b) Is the operation (i) commutative?
                    (ii) associative?
   Give reasons.

Algebraic equations (3)
Formulae

Equations

Solving linear equations (revision)

An **equation** is a statement that two algebraic expressions are equal in value. For example, $4 - 4x = 9 - 12x$ is a linear equation with an **unknown** x. This equation is only true when x has a particular numerical value. To **solve** an equation means to find the real number value of the unknown which makes the equation true.

Example 1

Solve $4 - 4x = 9 - 12x$.

$4 - 4x = 9 - 12x$

Add $12x$ to both sides of the equation.
$$4 - 4x + 12x = 9 - 12x + 12x$$
$$4 + 8x = 9$$
Subtract 4 from both sides of the equation.
$$4 + 8x - 4 = 9 - 4$$
$$8x = 5$$
Divide both sides of the equation by 8.
$$\frac{8x}{8} = \frac{5}{8}$$
$$x = \frac{5}{8}$$

$\frac{5}{8}$ is the solution or **root** of the equation.

Check: When $x = \frac{5}{8}$.

LHS $= 4 - 4 \times \frac{5}{8} = 4 - 2\frac{1}{2} = 1\frac{1}{2}$

RHS $= 9 - 12 \times \frac{5}{8} = 9 - 7\frac{1}{2} = 1\frac{1}{2} =$ LHS

The equation in Example 1 was solved by the **balance method**. Compare the equation with a pair of scales. If the expressions on opposite sides of the equals sign 'balance', they will continue to do so if the same amounts are added to or subtracted from both sides. They will also balance if both sides are multiplied or divided by the same amounts.

Exercise 3a (Revision)

Solve the following equations.

1. $4x = 3x + 5$
2. $7m = 8 + 5m$
3. $2a = 9 - a$
4. $\frac{1}{4}d = 3$
5. $5y + 6 = 21$
6. $2n - 3 = 6$
7. $4 + 3x = 17$
8. $15 = 4a + 3$
9. $3 = 3m - 4$
10. $\frac{2}{3}t = 8$
11. $6x + 5 = 13 + 4x$
12. $2a + 4 = 16 - 3a$
13. $1\frac{1}{2}y = 9$
14. $7b - 9 = 5 + 3b$
15. $2 - 5t = 20 - 8t$
16. $22 = 7 + 2\frac{1}{2}x$
17. $8a - 19 = 5 + 3a$
18. $3 + 2y - 24 = 14 - 3y$
19. $2e = 20 - 3e - 9$
20. $2x + 19 - 5x = x - 5$

Equations with brackets (revision)

If an equation contains brackets, remove the brackets by multiplying before collecting terms.

Example 2

Solve $3(4c - 7) - 4(4c - 1) = 0$.

$3(4c - 7) - 4(4c - 1) = 0$
Remove brackets by multiplying.
$12c - 21 - 16c + 4 = 0$
Collect like terms.
$-4c - 17 = 0$
Add 17 to both sides.
$-4c = 17$
Divide both sides by -4.
$c = -\frac{17}{4}$
$c = -4\frac{1}{4}$

Check: When $c = -4\frac{1}{4}$

LHS $= 3(-17 - 7) - 4(-17 - 1)$
$= 3(-24) - 4(-18)$
$= -72 + 72 = 0 =$ RHS

Exercise 3b (Revision)

Solve the following equations.

1. $4a - (3 - a) = 17$
2. $8b - (3b + 4) = 11$
3. $3 - (3m - 7) = 43$
4. $8n - (5n + 13) = 7$
5. $12t + (1 - 7t) = 31$
6. $0 = 5 - (2x - 17)$
7. $9 - (5 - 7y) = 13 + 4y$
8. $d = 12 - (11 + 4d)$
9. $2 - 3(a + 5) = -10$
10. $3x - 2(x + 3) = 0$
11. $3(5c - 1) = 4(3c + 2)$
12. $5(3m + 4) = 3(4m + 7)$
13. $2(2x - 5) = 3(x - 6)$
14. $5(a + 2) - 3(3a - 5) = 1$
15. $7(5n - 4) - 10(3n - 2) = 0$
16. $4(3x - 1) = 11x - 3(x - 4)$
17. $5(v + 2) + 3(v - 5) = 19$
18. $3(6 + 7y) + 2(1 - 5y) = 42$
19. $2 = 5(5z - 2) - 9(3z - 2)$
20. $3x - [3(1 - x) - 2x] = 3$

Equations with fractions (revision)

Always clear fractions before beginning to solve
an equation. To clear fractions, multiply each
term in the equation by the LCM of the
denominators of the fractions.

Example 3

Solve the equation $\frac{3}{4}x - 1\frac{2}{3} = \frac{2}{3}x$.

Express the given equation as follows.

$$\frac{3x}{4} - \frac{5}{3} = \frac{2x}{3}$$

The denominators are 4, 3 and 3. Their LCM is 12.

Multiply every term by 12.

$$12 \times \frac{3x}{4} - 12 \times \frac{5}{3} = 12 \times \frac{2x}{3}$$
$$3 \times 3x - 4 \times 5 = 4 \times 2x$$
$$9x - 20 = 8x$$

Subtract $8x$ from both sides.

$$x - 20 = 0$$

Add 20 to both sides.

$$x = 20$$

Check: When $x = 20$,
$$\text{LHS} = \frac{3}{4} \times 20 - 1\frac{2}{3} = 15 - 1\frac{2}{3} = 13\frac{1}{3}$$
$$\text{RHS} = \frac{2}{3} \times 20 = 13\frac{1}{3} = \text{LHS}$$

Example 4

Solve the equation $\frac{3x + 2}{6} - \frac{2x - 7}{9} = 0$.

The LCM of 6 and 9 is 18. Multiply every term by 18,
remembering that the dividing line of the given
fraction acts like a bracket on the numerators.

$$18 \times \frac{(3x + 2)}{6} - 18 \times \frac{(2x - 7)}{9} = 18 \times 0$$
$$3(3x + 2) - 2(2x - 7) = 0$$

Clear brackets.

$$9x + 6 - 4x + 14 = 0$$
$$5x + 20 = 0$$
$$5x = -20$$
$$x = -4$$

Check: When $x = -4$,
$$\text{LHS} = \frac{-12 + 2}{6} - \frac{-8 - 7}{9}$$
$$= \frac{-10}{6} - \frac{-15}{9}$$
$$= -\frac{5}{3} + \frac{5}{3} = 0 = \text{RHS}$$

Example 5

Solve the equation $\frac{2}{5}(6x - 1) = \frac{3}{4}(3x + 2) - 2$.

Express the given equation as follows.

$$\frac{2(6x - 1)}{5} = \frac{3(3x + 2)}{4} - 2$$

The LCM of 5 and 4 is 20. Multiply every term by 20.

$$20 \times \frac{2(6x - 1)}{5} = 20 \times \frac{3(3x + 2)}{4} - 20 \times 2$$
$$4 \times 2(6x - 1) = 5 \times 3(3x + 2) - 40$$
$$8(6x - 1) = 15(3x + 2) - 40$$

Clear brackets.

$$48x - 8 = 45x + 30 - 40$$

Collect terms.

$$48x - 45x = 30 - 40 + 8$$
$$3x = -2$$
$$x = -\frac{2}{3}$$

Check: When $x = -\frac{2}{3}$,
$$\text{LHS} = \frac{2}{5}(-4 - 1) = -2$$
$$\text{RHS} = \frac{3}{4}(-2 + 2) - 2$$
$$= 0 - 2 = -2 = \text{LHS}$$

Notice in each example that *every* term is multiplied by the LCM of the denominators, whether the term is a fraction or not. Also notice that the solution can be checked by substituting the value of the unknown into the *original* equation.

Exercise 3c

Solve the following equations.

1 $\dfrac{x}{2} = \dfrac{x}{3} + \dfrac{1}{2}$

2 $\dfrac{1}{2}a + 1\dfrac{1}{4} = \dfrac{1}{4}a$

3 $\dfrac{5d}{3} + 1 = \dfrac{d}{6} + 3$

4 $3y - \dfrac{2}{9} = 4\dfrac{7}{9}$

5 $\dfrac{1}{5}x + \dfrac{1}{7}x = \dfrac{7}{15}$

6 $\dfrac{9y}{10} + \dfrac{3}{5} = \dfrac{2y}{5} + \dfrac{7}{10}$

7 $\dfrac{2d + 7}{6} + \dfrac{d - 5}{3} = 0$

8 $\dfrac{6a + 3}{7} = \dfrac{2a - 1}{3}$

9 $\dfrac{7x + 2}{3} - \dfrac{9x - 2}{5} = 2$

10 $\dfrac{3x + 2}{5} - \dfrac{2x + 3}{3} = 3$

11 $\dfrac{2x - 1}{3} - \dfrac{3 - x}{2} = \dfrac{x}{4}$

12 $\dfrac{4n + 1}{3} - 1\dfrac{1}{2} = \dfrac{2n + 5}{6}$

13 $\dfrac{3}{8}(4a - 5) - \dfrac{5}{12}(3a - 4) = \dfrac{1}{6}$

14 $\dfrac{4}{5}(2y - 5) = \dfrac{2}{3}(2y - 7) + \dfrac{2}{15}$

15 $\dfrac{7}{6}(3x - 1) - 8\dfrac{1}{3} = \dfrac{3}{2}(2x - 5)$

16 $\dfrac{5x + 16}{4} + \dfrac{x}{2} = \dfrac{4x + 2}{3}$

17 $\dfrac{2(5z - 3)}{3} - \dfrac{3(5z + 2)}{5} = \dfrac{8}{15}$

18 $\dfrac{4m}{3} - \dfrac{17}{21} = \dfrac{6m - 1}{7}$

19 $\dfrac{x}{5} + 1\dfrac{1}{2}x + \dfrac{11}{20} = \dfrac{5x + 1}{4}$

20 $7(x + 4) - 2[x - 3(5 + x)] = \dfrac{2}{3}(x - 6)$

Equations from word problems

Example 6

A man walks for 2 hours at a certain speed. He then cycles at 3 times that speed for 4 hours. He goes 77 km altogether. Find the speed at which he walks.

Since we are to find walking speed, let this be v km/h.

From the 1st sentence,
distance walked = v km/h × 2 h = $2v$ km

From the 2nd sentence,
  cycling speed = $3v$ km/h
  distance cycled = $3v$ km/h × 4 h = $12v$ km

From the 3rd sentence,
  total distance = 77 km

Hence $2v + 12v = 77$
$$14v = 77$$
$$v = \dfrac{77}{14} = \dfrac{11}{2} = 5\dfrac{1}{2}$$

The man walks at $5\dfrac{1}{2}$ km/h.

Check:
Distance in 2 h at $5\dfrac{1}{2}$ km/h   = 11 km
Distance in 4 h at $16\dfrac{1}{2}$ km/h = 66 km
Total distance   = 77 km

Example 7

Joseph is 11 years older than Zena. In 5 years' time, Joseph will be twice as old as Zena. Find their present ages.

Let Zena's age now be x years.
Joseph's age is $(x + 11)$ years (*from the 1st sentence*)

In 5 years' time,
  Zena's age = $(x + 5)$ years
Joseph's age = $(x + 11) + 5$ years

Hence, from the 2nd sentence,
$$(x + 11) + 5 = 2(x + 5)$$
$$x + 16 = 2x + 10$$
$$6 = x$$
$$x = 6$$

Zena is 6 years old and Joseph is 17 years old.

Check: In 5 years' time,
Zena's age = 6 + 5 = 11 yr
Joseph's age = 17 + 5 = 22 yr = 2 × Zena's age.

Example 8

A trader buys n oranges at the rate of 5 oranges for $4. Eight of the oranges are bad. She sells the rest at the rate of 4 oranges for $4 and makes a profit of $18. Find n.

If 5 oranges cost $4
   1 orange cost $\frac{4}{5}$
$\therefore n$ oranges cost $\frac{4n}{5}$
8 oranges are bad,
$\therefore$ number of oranges sold $= n - 8$
If 4 oranges sell for $4
   1 orange sells for $\frac{4}{4} = \$1$
$\therefore (n - 8)$ oranges sell for $(n - 8)$
Profit = selling price − cost price
$18 = (n - 8) - \dfrac{4n}{5}$
Multiply each term by 5
  $90 = 5(n - 8) - 4n$
  $90 = 5n - 40 - 4n$
$130 = n$
  $n = 130$
Check: C.P. of 130 oranges $= \frac{130}{5} \times \$4 = \$104$
S.P. of 122 oranges $= \frac{122}{4} \times \$4 = \$122$
Profit $= \$122 - \$104 = \$18$

Where necessary, choose a letter to represent the unknown quantity. Express the data of the question in terms of the letter. Form an equation and solve the equation. Check that the solution agrees with the data of the question.

Exercise 3d

1. x represents a certain number. When the number is multiplied by 3, the result is the same as that of adding 34 to the number. Find x.

2. A rectangle is 3 times as long as it is wide. If its perimeter is 56 cm, find the width of the rectangle.

3. Indra and Nina share $147 between them so that Indra gets $19 more than Nina. Find how much money each gets.

4. A woman is 3 times as old as her daughter. 6 years ago the sum of their ages was 36. Find the age of the daughter.

5. A man walked for 2 hours at 6 km/h. He then cycled for a certain time at 16 km/h. If he travelled 36 km altogether, for how many hours did he cycle?

6. A sum of $1.24 is made up of 5-cent coins and 1-cent coins. There are 6 times as many 5-cent coins as there are 1-cent coins. Find the number of 5-cent coins.

7. Ruth has $30 and Tom has $186. If Ruth saves $5 a day and Tom spends $7 a day, after how many days will they have equal amounts?

8. Divide 59 ml into two parts so that one part is 7 ml less than 5 times the other part.

9. The result of taking 3 from x and multiplying the answer by 4 is the same as taking 3 from 5 times x.
   (a) Express this statement as an algebraic equation.
   (b) Hence find the value of x.

10. The sum of 6 and one third of n is one more than twice n.
    (a) Express this statement as an algebraic equation.
    (b) Hence find the value of n.

11. A boy is 10 years old and his father is 37 years old. In how many years' time will the father be twice as old as his son?

12. One farmer has 119 goats and another has 73. After they sell the same number of goats, one is left with 3 times as many goats as the other. How many goats did each sell?

13. A motorist travels regularly between two towns. She usually takes 5 hours when travelling at a certain speed. She finds that if she increases her average speed by 15 km/h the journey takes 1 hour less. Find her usual speed.

14. A water tank contains 5 times as much as another water tank. When 20 litres of water are poured from the first tank into the second, the first contains 3 times as much as the second. How much water did each tank contain originally?

15 A trader buys some eggs at 45c each. She finds that 6 of them are broken. She sells the rest at 72c each and makes a profit of $14.31. How many eggs did she buy?

16 The result of adding 15 to x and dividing the answer by 4 is the same as taking x from 80.
    (a) Express this statement as an algebraic equation.
    (b) Hence find the value of x.

17 One stick is 9 cm longer than another. $\frac{2}{5}$ of the longer stick is equal to $\frac{1}{2}$ of the shorter stick. Find the length of the longer stick.

18 A man cycles to a village at 18 km/h and returns at 12 km/h. If he takes $6\frac{1}{4}$ hours for the double journey, how far does he ride altogether?

19 A total of m matches is needed to fill 30 matchboxes with the same number of matches in each box.
    (a) How many matches are in each box?
    (b) If 3 fewer matches are put into each box, there are enough for 32 boxes. What is the value of m?

20 A train travels a certain journey and is supposed to arrive at midday When its average speed is 40 km/h, it arrives at 1 p.m. When its average speed is 48 km/h, it arrives at 11 a.m. What is the length of the journey?

Formulae

A **formula** is an equation with letters which stand for quantities. For example.

$$C = 2\pi r$$

is the formulae which gives the circumference, C, of a circle of radius r.

In science,

$$I = \frac{V}{R}$$

is the formula which shows the relation between the current, I amps, voltage, V volts, and resistance, R ohms, in an electrical circuit.

In arithmetic,

$$I = \frac{PRT}{100}$$

is the formula which gives the simple interest. I, gained on a principal, P, invested at $R\%$ per annum for T years.

Notice in the above formulae that sometimes the same letter can stand for different quantities in different formulae. For example, I stands for current in the science formula and I stands for interest in the arithmetic formula. (**Formulae** is the plural of formula.)

Substitution in formulae

The area, A, of a triangle of base length b and height h is given by the formula $A = \frac{1}{2}bh$, where b and h are in the same units. In this example, A is the **subject** of the formula. By substituting values of b and h into the formula, corresponding values of A can be found.

Notice also that units are not stated in formulae, although the quantities that the letters stand for must be in appropriate units.

Example 9

Use the formula $A = \frac{1}{2}bh$ to calculate the area of a triangle (a) of base 3.2 cm and height 5 cm, (b) of base 4 km and height 600 m.

(a) $A = \frac{1}{2}bh$
    When $b = 3.2$ and $h = 5$
        $A = \frac{1}{2} \times 3.2 \times 5$
          $= 1.6 \times 5$
          $= 8$
    The area is 8 cm²

(b) Working in km, 600 m = 0.6 km.
    When $b = 4$ and $h = 0.6$,
        $A = \frac{1}{2} \times 4 \times 0.6$
          $= 2 \times 0.6$
          $= 1.2$
    The area is 1.2 km²

The following examples show how a formula can be used to find the value of a quantity which is not the subject of the formula.

Example 10

The formula $F = \dfrac{9C}{5} + 32$ shows the relationship between temperature in degrees Fahrenheit (F) and degrees Centigrade (C).

Find (a) F when $C = 40$, (b) C when $F = 100$, (c) the temperature when $F = C$.

(a) $F = \dfrac{9C}{5} + 32$

When $C = 40$,
$$F = \dfrac{9 \times 40}{5} + 32$$
$$= 9 \times 8 + 32$$
$$= 72 + 32 = 104$$
The temperature is $104\,°F$

(b) When $F = 100$,
$$100 = \dfrac{9C}{5} + 32$$

Solve the equation for C.
Multiply each term by 5.
$$500 = 9C + 160$$
$$340 = 9C$$
$$C = \tfrac{340}{9} = 37\tfrac{7}{9}$$
The temperature is $37\tfrac{7}{9}\,°C$

(c) When $F = C$,
$$C = \dfrac{9C}{5} + 32$$

Multiply each term by 5.
$$5C = 9C + 160$$
$$-4C = 160$$
$$C = \dfrac{160}{-4} = -40$$
$$F = C = -40$$
The temperature is $-40\,°F$ (or $-40\,°C$).

Example 11

A gas at a temperature of $\theta\,°C$ has an absolute temperature of $T\,K$, where $T = \theta + 273$.

(a) Find the absolute temperature of a gas at a temperature of $68\,°C$.
(b) If the absolute temperature of a gas is $380\,K$, find its temperature in $°C$.

(a) $\qquad T = \theta + 273$
When $\theta = 68$,
$$T = 68 + 273 = 341$$
The absolute temperature is $341\,K$.

(b) $\qquad T = \theta + 273$
When $\quad T = 380$,
$$380 = \theta + 273$$
Subtract 273 from both sides
$$380 - 273 = \theta$$
$$107 = \theta$$
The temperature of the gas is $107\,°C$.

Example 12

The formula $W = VI$ gives the power, W watts, used by an electrical item when a current of I amps flows through a circuit of V volts.

(a) An air conditioner on maximum power needs a current of 25 amps in a 120 volt circuit. Find the power being used.
(b) An electric light bulb is marked 100 watts, 240 volts. Find the current required to light the bulb.

(a) $\qquad W = VI$
When $\quad V = 120$ and $I = 25$,
$$W = 120 \times 25$$
$$= 3000$$
The maximum power is 3000 watts.

(b) $\qquad W = VI$
When $\quad V = 100$ and $V = 240$,
$$100 = 240I$$
Divide both sides by 240
$$\dfrac{100}{240} = I$$
$$I = \dfrac{10}{24} = \dfrac{5}{12}$$
The current required is $\tfrac{5}{12}$ amp.

Example 13

The monthly cost, d dollars, of running a household of n people is given by the formula $d = 12n + 35$.

(a) Find the monthly cost for 5 people.
(b) How many people are there if the monthly cost is \$119?

(a) There are 5 people, thus $n = 5$.
$$d = 12n + 35$$
When $\quad n = 5$,
$$d = 12 \times 5 + 35 = 60 + 35$$
$$= 95$$
The monthly cost is \$95.

(b) The monthly cost is $119, thus $d = 119$.
$$d = 12n + 35$$
When $d = 119$,
$$119 = 12n + 35$$
Subtract 35 from both sides
$$84 = 12n$$
Divide both sides by 12
$$7 = n$$
There are 7 people.

Exercise 3e

1 A gas at a temperature of $\theta\,°C$ has an absolute temperature of $T\,K$, where $T = \theta + 273$.
 (a) Find the absolute temperature of a gas at a temperature of $36\,°C$.
 (b) If the absolute temperature of a gas is $400\,K$, find its temperature in $°C$.

2 The perimeter of a rhombus of side $d\,cm$ is $p\,cm$, where $p = 4d$.
 (a) Find the perimeter of a rhombus of side $3.2\,cm$.
 (b) Find the length of a side of a rhombus of perimeter $14\,cm$.

3 Two quantities x and y are connected by the formula, $y = 7 - 9x$.
 (a) Find the value of y when $x = 0$.
 (b) Find the value of x when $y = 0$.

4 A rectangle l units long and b units wide has an area of A square units, where $A = lb$.
 (a) Find the floor-area of a room $5\,m$ long and $3\,m$ wide.
 (b) Find the width of a postcard of area $112\,cm^2$, the length being $14\,cm$.
 (c) Find the length of a rectangular piece of plastic of area $171\,cm^2$ and width $9\frac{1}{2}\,cm$.
 (d) The area of a picture is $3.125\,m^2$ and its width is $1.25\,m$. Find its length.

5 The simple interest formula
$$I = \frac{PRT}{100}$$
gives the interest I on a principal P invested at a rate $R\%$ per annum for T years.

 (a) Find the interest when $1500 is invested at 5% per annum for 4 years.

 (b) Find the principal that gains an interest of $161 in 5 years at 7% per annum

6 A circuit of voltage V volts and resistance R ohms has a current of I amps, where
$$I = \frac{V}{R}$$
 (a) Find the current when the voltage is 240 volts and the resistance is 80 ohms.
 (b) Find the voltage when the current is 0.6 amps and the resistance is 5 ohms.
 (c) Find the resistance when the current is 0.1 amp and the voltage is 9 volts.

7 A rectangular room $l\,m$ long and $b\,m$ wide has a perimeter p, where $p = 2l + 2b$.
 (a) Find the perimeter of a room which is $3.5\,m$ long and $2\,m$ wide.
 (b) Find the length of a room of perimeter $20\,m$ wide and width $3\,m$.

8 The mass of water in a rectangular tank $l\,m$ long, $b\,m$ wide and $h\,m$ deep is $M\,kg$, where $m = 1000lbh$.
 (a) What is the mass of water in a tank $5\,m$ long, $4\,m$ wide and $3\,m$ deep?
 (b) How deep is the water in a tank $4\,m$ long and $3\,m$ wide if its mass is $24\,000\,kg$?
 (c) How wide is a tank $3\,m$ long and $0.5m$ deep if it holds exactly one tonne of water?

9 The circumference, C units, of a circle of radius r units is given by the formula $C = 2\pi r$ where $\pi = \frac{22}{7}$.
 (a) What is the circumference of a circle of radius $7\,cm$?
 (b) What is the radius of a circle whose circumference is $22\,m$?
 (c) What is the circumference of a circle of radius one metre?
 (d) What is the radius of a circle of circumference $2.75\,m$?

10 The speed $s\,km/h$ of a certain car t seconds after starting is given by the formula $s = 12t$.
 (a) Find the speed of the car 5 seconds after starting.
 (b) How long does it take the car to reach a speed of $75\,km/h$?

Example 14

If $y = 3 - x$. find the values of y when $x = 1, 2, 3, 4, 5$.

When $x = 1, y = 3 - 1 = \quad 2$
When $x = 2, y = 3 - 2 = \quad 1$
When $x = 3, y = 3 - 3 = \quad 0$
When $x = 4, y = 3 - 4 = -1$
When $x = 5, y = 3 - 5 = -2$

The working and results in Example 14 can be set out more neatly in a **table of values**, like Table 3.1.

Table 3.1 $y = 3 - x$

x	1	2	3	4	5
3	3	3	3	3	3
$-x$	-1	-2	-3	-4	-5
$y = 3 - x$	2	1	0	-1	-2

Example 15

If $y = 2x - 5$, make a table of values of y for $x = -1, 0, 1, 2, 3$.

Table 3.2 $y = 2x - 5$

x	-1	0	1	2	3
$2x$	-2	0	2	4	6
-5	-5	-5	-5	-5	-5
y	-7	-5	-3	-1	1

Exercise 3f

1. If $y = 5 - x$, find the values of y when $x = 1, 2, 3, 4, 5$.

2. If $d = c + 3$, find the values of d when $c = -2, -1, 0, 1, 2$.

3. If $y = 2x + 1$, find the values of y when $x = 0, 1, 2, 3, 4$.

4. Given that $y = 3x + 2$, copy and complete Table 3.3 to show values of y for $x = -1, 0, 1, 2, 3$.

   Table 3.3 $y = 3x - 2$

x	-1	0	1	2	3
$3x$					
$+2$	$+2$	$+2$	$+2$	$+2$	$+2$
y					

5. If $y = 17 - 6x$, make a table of values of y for $x = 0, 1, 2, 3, 4, 5$.

6. The cost, $\$c$, of hiring a car for a journey of d km is given by the formula
   $$c = 0.5d + 85$$
   (a) Find the cost of hiring a car for a journey of 420 km.
   (b) How long is a journey if the cost of car hire is $145?

7. The time, t min, to cook meat is given by the formula $t = 40m + 25$, where m is the mass of the meat in kg.
   (a) Find how long it takes to cook a piece of meat of mass 1.2 kg.
   (b) Find the mass of a piece of meat which takes 2 h 15 min to cook.

8. On a certain island the tax, $\$T$, paid on an income of $\$I$ is given by the formula $T = 0.2I - 50$.
   (a) How much tax is paid on an income of
       (i) $1000,
       (ii) $6225?
   (b) What income would have a tax of $450?

9. A car starts a journey, with a full petrol tank. The amount of petrol, p litres, left in the tank after travelling for t hours is given by the formula $p = 63 - 10t$.
   (a) Find the amount of petrol left after travelling for $2\frac{1}{2}$ hours.
   (b) If there are 18 litres of petrol left, how long has the car been travelling?
   (c) How long will it take the car to run out of petrol (i.e. find t when $p = 0$)?

10. A closed cylinder of height h cm and base radius r cm has a surface area, A cm², where $A = 2\pi r^2 + 2\pi rh$. Use the value $\frac{22}{7}$ for π to find
    (a) the surface area of a closed cylinder of height 9 cm and base radius 5 cm;
    (b) the height of a closed cylinder of base radius 7 cm and surface area 1012 cm².

Example 16

If $y = 5x^2 - 1$, find (a) the value of y when $x = -3$, (b) the values of x when $y = 79$.

(a) $\qquad y = 5x^2 - 1$

When $x = -3$
$$y = 5 \times (-3)^2 - 1$$
$$= 5 \times (+9) - 1$$
$$= 45 - 1$$
$$= 44$$

(b) $\qquad y = 5x^2 - 1$

When $y = 79$
$$79 = 5x^2 - 1$$
Add 1 to both sides
$$80 = 5x^2$$
Divide both sides by 5
$$16 = x^2$$
Take the square root of both sides
$$\sqrt{16} = x$$
$$x = +4 \text{ or } -4$$

Notice that there are two possible values for x. We can shorten this to $x = \pm 4$ where $\pm$ is short for '+ or −'.

Example 17

From a height of h metres above sea level it is possible to see a distance of approximately d kilometres, where d and h are connected by the formula $2d^2 = 25h$.
(a) From what height is it possible to see a distance of 10 km?
(b) What distance can be seen from a height of 18 m?

(a) $\qquad 2d^2 = 25h$

The distance is 10 km. Thus, when $d = 10$,
$$2 \times 10^2 = 25h$$
$$2 \times 100 = 25h$$
$$200 = 25h$$
Divide both sides by 25
$$8 = h$$
The height is 8 metres.

(b) The height is 18 m. Thus, when $h = 18$.
$$2d^2 = 25h$$
becomes $2d^2 = 25 \times 18$
Divide both sides by 2
$$d^2 = 25 \times 9$$
or $d^2 = 225$

Take the square toot of both sides
$$d = \sqrt{25 \times 9} = \sqrt{225}$$
$$= \pm 15$$

In this example the value $d = -15$ would not be sensible; the distance that can be seen is 15 km.

Exercise 3g

1. If $y = 40x^2$, find (a) y when $x = 0, 1, 2, 3, 4, 5$; (b) x when $y = 10, 360, 1000, 4000$.

2. If $y = 16 - x^2$, find (a) y when $x = -4, -2, 0, 2, 4$, (b) x when $y = 0, 7, 12, 15$.

3. If $y = 2x^2 - 5x - 3$, find the value of y when (a) $x = -1$, (b) $x = 0$, (c) $x = 1$, (d) $x = 2$, (e) $x = 3$.

4. The formula $A = P\left(1 + \dfrac{RT}{100}\right)$ gives the total money, A, that a principal, P, amounts to in T years at $R\%$ simple interest per annum. Find the amount that a principal of $750 becomes if invested for 5 years at $6\frac{1}{2}\%$ simple interest per annum.

5. If $m = \dfrac{100}{n^2}$, find (a) m when $n = 1, 5, 10, 20$; (b) n when $m = 1, 4, 9, 25$.

6. The area, A square units, of a circle of radius r units, is given by the formula $A = \pi r^2$, where $\pi = \frac{22}{7}$.
   (a) What is the area of a circle of radius 7 m?
   (b) What is the radius of a circle of area 616 cm²?
   (c) What is the radius of a circle of area 38.5 m²?

7. The volume, V cm³, of a cylinder of base radius r cm and height h cm is given by the formula $V = \pi r^2 h$. Use the value $\frac{22}{7}$ for π to find
   (a) the volume of a cylinder of base radius $3\frac{1}{2}$ cm and height 8 cm;
   (b) the height of a cylinder of volume 1100 cm³ and radius 5 cm;
   (c) the base radius of a cylinder of volume 770 cm³ and height 45 cm.

⑧ If a stone is dropped, the distance, d m, which it falls in t seconds is given by the formula $d = 4.9t^2$.
  (a) How far does it fall in 3 seconds?
  (b) How far does it fall in $1\frac{1}{7}$ seconds?
  (c) How long does it take to fall 490 m?
  (d) How long does it take to fall $122\frac{1}{2}$ m?
  (e) How far does it fall in the fifth second?

⑨ The time, t minutes, taken over a committee meeting, is given by the formula
$$t = 5n^2 + 15 \text{ when } n \text{ people are present.}$$
  (a) How long does the meeting take if there are 4 people?
  (b) How many people are present if the meeting takes 2 h 20 min?

⑩ The visible distance, D km, of the horizon from a height of h m is given by the formula $h = \frac{2}{25}D^2$.
  (a) How high must a cliff be if a ship $12\frac{1}{2}$ km away is visible from it?
  (b) How high is an observation tower on the top of the cliff in (a) if a ship 15 km away is visible from it?
  (c) What is the distance of the visible horizon from the top of a wireless mast 200 m high?

⑪ In the right-angled triangle in Fig. 3.1, the length of the hypotenuse, x cm, is given by the formula
$$x = \sqrt{y^2 + z^2}.$$
  (a) Find the value of x when $y = 2\frac{1}{2}$ cm and $z = 6$ cm.
  (b) Find the value of y when $x = 16$ and $z^2 = 60$.

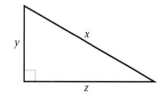

Fig. 3.1

⑫ If there are n numbers in a uniformly increasing series (like 2, 5, 8, 11, ...), starting with a and ending with l, the sum of the numbers is S, where $S = \frac{1}{2}n(a + l)$.
  (a) What is the sum of all whole numbers from 1 to 50 inclusive?
  (b) How many numbers are there in the series 2, ..., 20 if the sum is 143?
  (c) What is the last term of a series of 10 numbers beginning with 7 if their sum is 85?

Summary

In solving a **linear equation**, always carry out, on both sides of the equation, the same arithmetic operations; that is, add or subtract the same number, multiply or divide by the same constant.

A **formula** is an algebraic statement of a relation between quantities denoted by letters. This statement is true for all **corresponding values** of the quantities **in the correct units**.

A **formula** may state a useful result in Mathematics or Science, for example, the area of a circle, A, is given by the formula
$$A = \pi r^2.$$

A is called the **subject** of the formula.

Practice exercise P3.1

Solve the following equations.

① $3a - a + 2a + a = 1$

② $d - 2d + 1 = d + 2d - 1$

③ $2(2w + 3) = \dfrac{3w}{2}$

④ $-3(v - 2)(v + 2) + 3v^2 = 2v$

⑤ $\dfrac{x + 2}{2} + \dfrac{1 - 3x}{3} = \dfrac{1}{3}$

⑥ $\dfrac{x - 2}{x + 2} = \dfrac{3}{5}$

⑦ $(b - 4)(b - 3) = b(b - 5)$

⑧ $c(c - 5) + 6 = (c - 1)(c - 2)$

Practice exercise P3.2

1. A team played 20 matches. The team won 5 more matches than it lost. One match was drawn. Find the number of matches the team lost.

2. One-third of a number n is added to twice the number and then the total is subtracted from 20. If the final number is 6, find n.

Practice exercise P3.3

In each of the following,
(a) write an equation to represent the data;
(b) solve the equation to find the value of the unknown;
(c) check your solution.

1. A rectangle is w cm wide. Its length is 3 cm more than its width. Its perimeter is 70 cm.

2. $500 was shared among three students. George received $x. Ann received twice as much as George and Ben got $40 less than George.

3. A fruit vendor bought m oranges at 75¢ each. There were 12 bad oranges. He sold the remainder at $1.10 each and made a profit of $28.80.

4. In a basket, there is a total of 34 citrus fruits. There are n grapefruits, $3n$ oranges, $2n$ tangerines, $(n + 2)$ lemons and 4 limes.

5. A boy jogs for $\frac{1}{2}$ hour at 8 km/h and then walks for t hours at 6 km/h. He travelled a total distance of 13 km.

Practice exercise P3.4

In each of the following questions,
(a) identify the unknown and define it with a letter;
(b) write an equation to represent the data;
(c) solve the equation;
(d) check your solution.

1. Notebooks with ruled pages cost $3.60 each. A notebook with plain pages costs 80¢ less. Brenda spent $32.80 and bought a total of 10 notebooks. Find how many of each kind she bought.

2. The price of a pack of nuts increases by 15¢. The old price was 90% of the new price. Calculate the old and the new prices.

3. A woman spent $\frac{1}{5}$ of her salary on food and gave $\frac{1}{10}$ to charity. She had $4060 remaining. Calculate her salary.

4. In a particular isosceles triangle, the third side is half the length of one of the equal sides. The perimeter of the triangle is 15 cm. Calculate the length of the sides.

5. Find two consecutive odd numbers such that seven times the smaller is 17 more than six times the larger number.

6. Two cars, A and B, moved towards each other from two stations 240 km apart. The average speed of car A was twice the average speed of car B. The cars met after $2\frac{1}{2}$ hrs. Calculate the speeds of the cars.

7. A column is 30 cm taller than another column. The taller column is one-and-a-quarter times as tall as the shorter column. Find the heights of the two columns.

8. A rectangular room is 1.5 m longer than it is wide. There are two doorways, each 1.3 m wide. The length of skirting board required for the base of the walls is 27.2 m. Find the length of each wall.

Practice exercise P3.5

Write each of the following as a formula using the letters shown in **bold,** as in the given example.

Example   The **d**istance around the sides of a square, being 4 times the length of a **s**ide.

Answer $d = 4s$

1. The **w**ages Bill earns if he is paid $7 for each **h**our he works.

2. The number of **s**tamps Ann has if 45 are added to her **c**ollection.

3. The length **l**eft on a carpet when 5 m is **c**ut off.

4. The **m**oney raised in a sponsored swim at $10 for each **l**ength.

⑤ The number of 5 m pieces that can be cut from a length of carpet.

⑥ The total time on some cassettes of 90 minutes each.

⑦ The length remaining on a 20 m hose when a piece has been cut off.

⑧ The number of cans if there is the same number on each of 5 shelves.

⑨ The sale price of a radio with $35 off the usual price.

⑩ The total length of a car and a truck.

Practice exercise P3.6

Write each of the following as a formula. Select your own letters and make it clear what the letters mean, as shown in the example.

Example   The total cost of a number of books if they are $8 each.

Answer c is the total cost of the books, b is the number of books: $c = 8b$

① The total weight of books weighing 250 g each.

② The amount each person gets when a prize is shared equally between 10 people.

③ The weight of a man if he is 8 kg above his old weight.

④ The total length of a train with 8 carriages.

⑤ The sale price of a pair of shoes with $20 off the usual price.

Practice exercise P3.7

① For each of the following, use the given values to find the value of the letter within the [ ] brackets.

(a) $a = b + cd$ [a]
when $b = -3$, $c = 5$ and $d = -2$

(b) $a = b(1 + d)$ [d]
when $a = 3$ and $b = 5$

(c) $A = 2\pi rh$ (take $\pi = \frac{22}{7}$)   [h]
when $A = 35$ and $r = 14$

(d) $\sqrt{(x + 1)} = A$ [x]
when $A = 2\frac{1}{2}$

(e) $y = \dfrac{3x - 1}{2x}$ [x]
when $y = 1\frac{1}{3}$

② You calculate the surface area, A, of a cylinder as follows.
- Multiply the radius, r, by π, then double the answer.
- Add the radius, r, to the perpendicular height, h.
- Multiply the last two answers together.

(a) Write down the formula for the surface area.

(b) Calculate the surface area of the cylinder in Fig. 3.2. Give your answer correct to 3 s.f. (Take $\pi = 3.142$.)

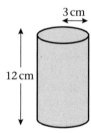

Fig. 3.2

③ A bus sets off with a full tank of petrol. The amount of petrol, v litres, left in the tank is given by the formula $v = 64 - 10t$, where t hours is the time the bus has been travelling.

(a) Find the amount of petrol left after 3 hours.

(b) If there are 19 litres left in the tank, how long has the bus been travelling?

(c) How much longer can the bus travel before it runs out of petrol?

④ The volume of a square-based pyramid is $V = \frac{1}{3}l^2h$, where h is the height of the pyramid and l is the length of one edge of the base. Find l when $V = 196$ and $h = 12$.

Pre-requisites
■ parts of a circle; geometrical constructions

Arcs and chords

A **chord** of a circle is a straight line joining any two points on its circumference. A **diameter** is a chord which passes through the centre of the circle. A chord which is not a diameter divides the circumference into two **arcs** of different sizes, a **major arc** and a **minor arc** (Fig. 4.1).

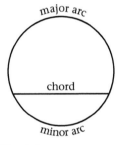

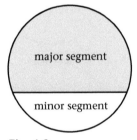

Fig. 4.1

Fig. 4.2

The chord also divides the circle into two **segments** of different sizes, a **major segment** and a **minor segment** (Fig. 4.2).

Angles in a segment

In Fig. 4.3, P, Q and R are points on the circumference of a circle. $A\widehat{P}B$, $A\widehat{Q}B$, $A\widehat{R}B$ are angles **subtended** at the circumference by the chord AB or by the minor arc AB. $A\widehat{P}B$, $A\widehat{Q}B$, $A\widehat{R}B$ are all **angles in the same segment** (here the major segment) APQRB.

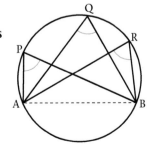

Fig. 4.3

Similarly, in Fig. 4.4, $A\widehat{X}B$ and $A\widehat{Y}B$ are angles subtended by the chord AB or by the major arc AB in the minor segment AXYB.

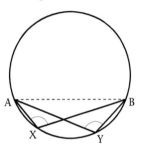

Fig. 4.4

Measure each of the angles in Fig. 4.3 and Fig. 4.4. What do you notice?

Group work

❶ Draw and cut out circles of different radii.

❷ Carefully mark the centre O, of the circle and draw any quadrilateral as shown in Fig. 4.5.

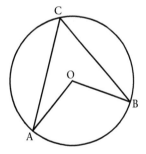

Fig. 4.5

$A\widehat{O}B$ is the angle at the centre, and $A\widehat{C}B$ is the angle at the circumference. Both angles $A\widehat{O}B$ and $A\widehat{C}B$ are subtended by the same arc AB.

❸ Carefully cut out the obtuse angle $A\widehat{O}B$ and fold it in two equal parts.

You should now be able to fit the folded section exactly over the angle at the circumference.

What can you now say? A basic fact about circles is as follows.

The angle which an arc of a circle subtends at the centre is twice that which it subtends at any point on the remaining part of the circumference.

This is sometimes very simply stated as the angle at the centre is twice the angle at the circumference subtended by the same arc (chord).

Example 1

In Fig. 4.6, O is the centre of the circle. If $A\widehat{C}B = 61°$, find $A\widehat{O}B$, $A\widehat{D}B$.

$A\widehat{O}B = 2 \times 61°$ (angle at centre =
    $= 122°$ $2 \times$ angle at circumference)

$ADB = \frac{1}{2} \times 122°$ (angle at centre =
    $= 61°$ $2 \times$ angle at circumference)

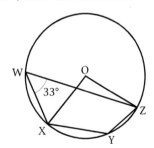

Fig. 4.6

Example 2

In Fig. 4.7, O is the centre of circle WXYZ. If $X\widehat{W}Z = 33°$, find $X\widehat{O}Z$ and $X\widehat{Y}Z$.

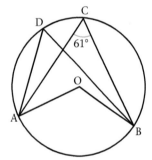

Fig. 4.7

$X\widehat{O}Z = 2 \times 33°$ (angle at centre = 2 ×
    $= 66°$ angle at circumference)
reflex $X\widehat{O}Z = 360° - 66°$ (angles at a point)
    $= 294$
$X\widehat{Y}Z = \frac{1}{2}$ of $294°$ (angle at centre = 2 ×
    $= 147°$ angle at circumference)

① Find the lettered angles in Fig. 4.8. (O is the centre of each circle.)

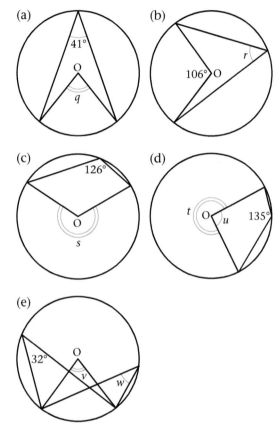

Fig. 4.8

② In Fig. 4.9, P, R and S are points on a circle, centre O. If $P\widehat{O}R = 150°$, calculate $P\widehat{S}R$.

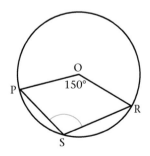

Fig. 4.9

③ In Fig. 4.10, P, Q and R are points on a circle, centre O. If OP̂Q = 36°, what is the size of PR̂Q?

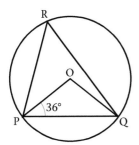

Fig. 4.10

④ Find the lettered angles in Fig. 4.11. (O is the centre of each circle.)

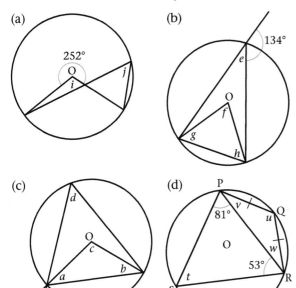

Fig. 4.11

⑤ In Fig. 4.12, L, M, N are points on a circle, centre O. If NM̂O = *a* and NL̂O = *b*. Find the obtuse MÔL in terms of *a* and *b*.

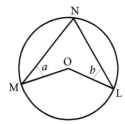

Fig. 4.12

Group work

① Using Example 1, what can you say about angles AD̂B, AĈB?

② Write three statements about the angels p_1, p_2 and p_3 in Fig. 4.13.

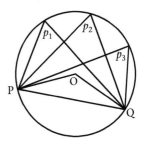

Fig. 4.13

Angles p_1, p_2 and p_3 are in the same segment of the circle. They are subtended by the same arc PQ; they are all half of the same angle PÔQ, the angle at the centre.

You should now be able to say that

angles in the same segment of a circle are equal.

Would this be true for angles in the minor segment?

Example 3

Calculate the size of the angle AX̂B in Fig. 4.14. O is the centre of the circle.

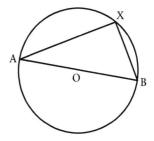

Fig. 4.14

Since AB is the diameter of the circle
AÔB = 180°   (*straight angle*)
AX̂B = ½ × 180°   (*angle at centre = 2 ×*
        = 90°   *angle at circumference*)

This is true for all circles. It is true to say that the angle in a semicircle is a right angle.

Example 4

In Fig. 4.15, PQ is a diameter of circle PMQN, centre O. If $\widehat{PQM} = 63°$, find $\widehat{QNM}$.

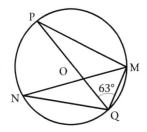

Fig. 4.15

In $\triangle QPM$, $\widehat{PMQ} = 90°$ *(angle in semicircle)*
  ∴ $\widehat{QPM} = 180° - 90° - 63°$ *(angle sum of $\triangle$)*
           $= 27°$
  ∴ $\widehat{QNM} = 27°$ *(in same segment as $\widehat{QPM}$)*

Exercise 4b

Find the lettered angles in each of the parts of Fig. 4.16. Where a point O is given it is the centre of the circle.

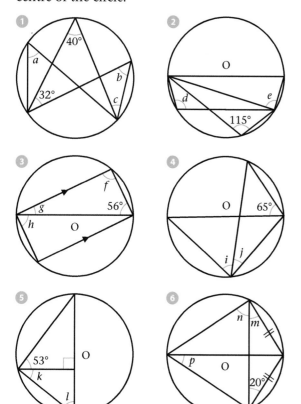

7 and **8**

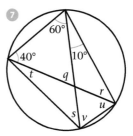

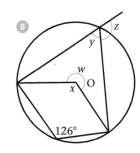

Fig. 4.16

9   AC and BD are diagonals of a quadrilateral in the circle in Fig. 4.17.

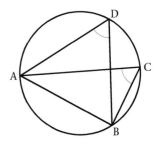

Fig. 4.17

If $\widehat{ACB} = 62°$, $\widehat{ABD} = 54°$ and $\widehat{AED} = 88°$, find $\widehat{ADC}$.

Summary

A chord which is not a diameter divides the circumference of the circle into a major arc and a minor arc.

The chord also divides the region enclosed by the circle into a major segment and a minor segment.

Angle properties of the circle

1   Angles in the same segment, or subtended by the same chord, are equal: $\widehat{ADB} = \widehat{ACB}$

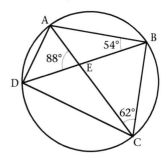

2 The angle in a semicircle is a right angle: AĈB = 90°

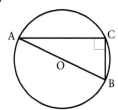

3 The angle at the centre of the circle is twice the angle at the circumference subtended by the same chord (arc): AÔB = 2 × AĈB

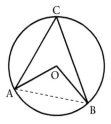

Practice exercise P4.1

1 In Figs. 4.18–4.20, find the angles marked with letters.

(a)

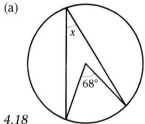

Fig. 4.18

(b)

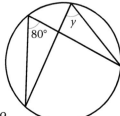

Fig. 4.19

(c)

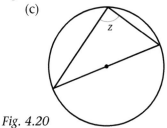

Fig. 4.20

2 In Figs. 4.21–4.23, find the angles marked with letters, in terms of x.

(a)

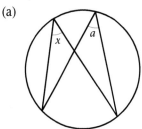

Fig. 4.21

(b)

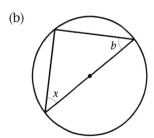

Fig. 4.22

(c)

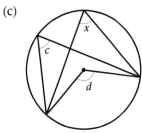

Fig. 4.23

3 In Fig. 4.24
  (a) What is angle x? Why?
  (b) What is true about angles y and z?
  (c) Find angle a in terms of y and z.

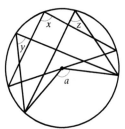

Fig. 4.24

Chapter 5

Consumer arithmetic (3)
Utilities, bills, investments and depreciation

Utilities and bills

Services which contribute to the health and comfort of people (the consumers) are called **utilities**. The most widely used utility services in the Caribbean are the electricity supply, the water supply and the telephone service. These services may be provided by a Government department, or by a private company.

The user of the service is billed for the amount used at regular intervals which may be monthly, quarterly or annually. The rates charged are usually regulated by a public utilities commission appointed by the Government. Rates may vary for different types of consumers, for example, individual householders (domestic users) or industrial companies and commercial enterprises (non-domestic users).

Electricity bills

The electric power used is measured by a meter, which is calibrated in units called **kilowatt hours**.

Fig. 5.1 shows the scales on a sample electric meter. The reading on the meter in Fig. 5.1 is 68 531 kWh.

The number of units used is calculated by taking the difference between the present reading on the meter and the previous reading. If the previous meter reading was 68 124 kWh, then the consumer has used (68 531 − 68 124) kWh, that is, 407 kWh. Note that the numbers, 68 124 and 68 531, are readings on the meter and do not indicate units of electric power.

Generally, the total amount charged depends on the unit cost or rate, and the number of units used. The rate may vary as the number of units used increases – generally this is a lower rate, but there may be exceptions (see Fig. 5.5). There may also be extra charges such as a rental/customer charge, a fuel surcharge and Government tax. In many Caribbean countries, an amount is also included to adjust for changes in the foreign exchange rate. This adjustment amount may be charged as a rate of the units used, or as a percentage of the total amount from energy, fuel and customer charges.

Fig. 5.2 shows a sample of one type of invoice/bill giving details of the amounts charged. Note that for the rate type (Rate A),
- the rate for the energy charge is the same for all the units used;
- the foreign exchange adjustment amount is billed as a rate;
- a Government tax is also included.

The charges for the different rate types and other conditions of the service (listed on the back of the bill) are shown in Fig. 5.3.

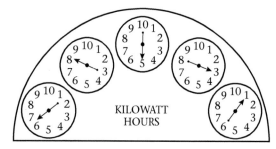

Fig. 5.1

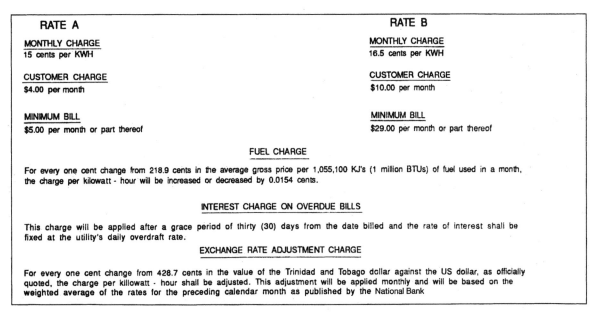

Fig. 5.2

METER NO.	METER READINGS		METER CONSTANT	UNITS USED	NO. OF PERIODS	AMOUNT	
	PREVIOUS	PRESENT				$	¢
F192304	012656	013360	1.0	704	1.00	105	60

Fig. 5.3

RATE A

MONTHLY CHARGE
15 cents per KWH

CUSTOMER CHARGE
$4.00 per month

MINIMUM BILL
$5.00 per month or part thereof

RATE B

MONTHLY CHARGE
16.5 cents per KWH

CUSTOMER CHARGE
$10.00 per month

MINIMUM BILL
$29.00 per month or part thereof

FUEL CHARGE

For every one cent change from 218.9 cents in the average gross price per 1,055,100 KJ's (1 million BTUs) of fuel used in a month, the charge per kilowatt - hour will be increased or decreased by 0.0154 cents.

INTEREST CHARGE ON OVERDUE BILLS

This charge will be applied after a grace period of thirty (30) days from the date billed and the rate of interest shall be fixed at the utility's daily overdraft rate.

EXCHANGE RATE ADJUSTMENT CHARGE

For every one cent change from 428.7 cents in the value of the Trinidad and Tobago dollar against the US dollar, as officially quoted, the charge per killowatt - hour shall be adjusted. This adjustment will be applied monthly and will be based on the weighted average of the rates for the preceding calendar month as published by the National Bank

Fig. 5.4 is another type of bill. Note that

- the units of energy charge are billed at different rates; the rate for the first hundred units is lower than for the remaining units used for residential customers;

- the foreign exchange adjustment amount is calculated as a percentage of the amount due from the other charges;

- there is no Government tax.

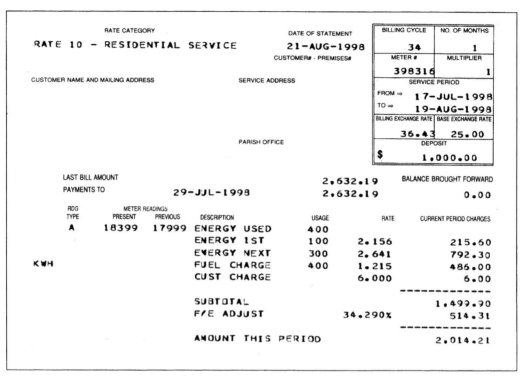

Fig. 5.4

Information about the charges, and the conditions for the service, are given in Fig. 5.5.

BILL STATEMENT INFORMATION

RATE SCHEDULE

THE APPROVED SCHEDULE WHICH SETS OUT THE RATES USED IN CALCULATING THE BILL. RATE AND BILLING INFORMATION IS AVAILABLE UPON REQUEST AT GECO'S BUSINESS OFFICES.

CUSTOMER CHARGE

A FIXED MONTHLY CONTRIBUTION TOWARDS THE COST OF PROVIDING YOUR SERVICE. THIS CHARGE IS APPLIED REGARDLESS OF THE LEVEL OF CONSUMPTION.

ENERGY RATE

THE CHARGE FOR EACH KILOWATT-HOUR (kWh), EXCLUSIVE OF FUEL, FOR PRODUCING AND DELIVERING YOUR ELECTRICITY. A SPECIAL LOWER RATE IS OFFERED FOR THE 1st 100 kWh PER MONTH FOR ALL RESIDENTIAL CUSTOMERS. THIS IS REFERRED TO AS THE LIFE-LINE RATE.

BALANCE BROUGHT FORWARD

LAST BILL AMOUNT LESS RECORDED PAYMENT(S).

READING TYPE

THE LETTER APPEARING BESIDE THE PRESENT READING(S) ON THE BILL REPRESNTS ONE OF THE FOLLOWING.
A - ACTUAL READING   E - ESTIMATED READING

FUEL RATE

THE TOTAL COST OF FUEL REQUIRED TO PRODUCE AND DELIVER EACH KILOWATT-HOUR.

DEMAND CHARGE

THE DEMAND CHARGE IS DESIGNED TO RECOVER COSTS WHICH RELATE TO THE CAPACITY AMOUNT WHICH THE COMPANY RESERVES TO SERVE THE MAXIMUM (PEAK) ELECTRIC REQUIREMENTS FOR EACH CUSTOMER. IT IS LISTED AS A SEPARATE ITEM FOR RATES 40 AND 50 CUSTOMERS ONLY, BUT IT IS INCLUDED IN THE ENERGY RATE FOR OTHER RATE CLASSES.

BASE EXCHANGE RATE

THIS REFERS TO THE RATE OF EXCHANGE THAT WAS USED IN THE DEVELOPMENT OF THE RATES PRINTED IN THE RATE SCHEDULE.

BILLING EXCHANGE RATE

THIS IS THE RATE USED FOR BILLING. WHERE THE BILLING PERIOD EXCEEDS 1 MONTH THIS WILL REPRESENT THE AVERAGE OF THE MONTHLY RATES.

FOREIGN EXCHANGE ADJUSTMENT (F / E ADJUST.)

THE CURRENT PERIOD'S SUB-TOTAL WILL BE INCREASED OR DECREASED BY A PERCENTAGE COMPUTED AS FOLLOWS:

$$\left(\frac{\text{BILLING EXCHANGE RATE - BASE EXCHANGE RATE}}{\text{BASE EXCHANGE RATE}} \right) \times 75 = \% \text{ F/E ADJUST.}$$

NOTE:
i)   75 REPRESENTS THE PERCENTAGE OF GECO'S COSTS WHICH IS DIRECTLY FOREIGN RELATED.
ii)   BASE EXCHANGE RATE IS THE EXCHANGE RATE USED IN THE PUBLISHED TARIFF.

Fig. 5.5

Example 1

A householder is billed for 1260 units of electric power. Use the following rates and charges:

> unit cost = 12 cents per kWh
> meter rental = $5.00
> fuel surcharge = 2.9512 cents per kWh

A Government tax of $7\frac{1}{2}$% is added to the total charges. Calculate the total amount due.

Charge on 1 260 units = $(1260 × 0.12)
                     = $151.20

Meter rental	= $5.00
Fuel charge	= $(1260 × 0.029 512)
	= $37.19
Total charges	= $(151.20 + 5.00 + 37.19)
	= $193.39
Government tax	= $(193.39 × 0.075)
	= $14.50
Total amount due	= $(193.39 + 14.50)
	= $207.89

Example 2

The rates charged by an electric company are as follows:

> the first 300 units cost 20 cents per unit
> the next 800 units cost 15 cents per unit
> the next 1200 units cost 10 cents per unit
> the remainder costs 5 cents per unit.

A Government tax of 12% is added to the charges. Calculate the total bill when 1410 units are used.

Charge on first 300 units = $(300 × 0.20)
                          = $60.00
Charge on next 800 units = $(800 × 0.15)
                         = $120.00
Charge on remaining 310 units = $(310 × 0.10)
                              = $31.00

Total charges	= $(60.00 + 120.00 + 31.00)
	= $211.00
Total amount due	= $(211.00 × 1.12)
	= $236.32

Exercise 5a

1 An electricity meter reading changes from 023 456 to 024 664. The first 100 units cost 15 cents per unit and the remaining units cost 10 cents per unit. Calculate the total amount to be paid for energy charges.

2 An electricity bill includes energy charges and a rental charge. A Government tax is also added to the total charges. The rental charge and the Government tax amount to $45.60.
   (a) The total amount to be paid by the consumer is $248.10. Calculate the amount due for energy charges.
   (b) The number of units used was 1620. Calculate the rate for the energy charge.

3 A customer who is billed at Rate A as shown in Fig. 5.2 used 760 units of electric energy. Use the rates and other charges shown in Fig. 5.2 and Fig. 5.3 to calculate the total amount to be paid.

4 If the customer in question 3 is a Rate B user, calculate the total amount to be paid.

5 A householder is charged for electricity at Rate 10 as shown in Fig. 5.4. The meter reading changed from 15 260 to 15 830. Calculate the total amount to be paid, including all charges and the foreign exchange adjustment.

6 The meter readings for the kWh used for October and November are given in Table 5.1.

Table 5.1

Previous reading 30 September	Present reading 30 November
019802	020137

   (a) Calculate the number of kWh units used. The charges for electricity are given in Table 5.2.

Table 5.2

Item	Rate
Rental	$4.00 per month
Energy charge	15c per kWh
Fuel charge	3.27c per kWh

A Government tax of 15% of the total charges is added to the bill.
   (b) Calculate
        (i)   the fuel charge
        (ii) the total amount to be paid.

Water rates

The amount charged for providing water to a consumer may be calculated by using a unit cost or rate, and the number of units used, that is, the number of cubic metres as measured by meter readings (similar to that in the electric power supply).

Note: 1 cubic metre ≈ 220 gallons

Another method of calculating the amount due for the water supply is based on the assessed or rateable value of the premises to which the water is supplied. This rateable value is called the Annual Taxable Value (ATV). A percentage of this rateable value is used to calculate the annual water rates. The charges calculated may be paid in equal monthly or quarterly amounts.

Fig. 5.6 shows a sample bill. Note that the charges for sewerage services are included on the water supply bill. Fig. 5.7 is a copy of the rate schedule used to calculate the amount to be paid.

PERIOD : 97/09/01 - 97/11/30

WATER AND SEWERAGE

ACCOUNT NO.	L & B	A.T.V.
CLASS A3	METER NO.	
	METER READINGS	
PRESENT READING		
PREVIOUS READING		
USAGE THIS PERIOD		

MAILING ADDRESS

ARREARS: PREVIOUS YEAR(S)	
CURRENT YEAR	794.50
TOTAL (PREVIOUS BILL)	794.50
LESS PAYMENTS	397.25CR
Payments after 97/08/13 not included	
SUB TOTAL	397.25
CURRENT CHARGES	
WATER	304.00
SEWERAGE	93.25
OTHER	
TOTAL AMOUNT DUE NOW ⇨	794.50

RATES FROM	WATER		IF YOU WISH TO PAY FOR THE ENTIRE YEAR
	SEWERAGE		
TO END OF YEAR ⇨	TOTAL		PAY THIS AMOUNT ⇨

Fig. 5.6

SCHEDULE OF RATES

WATER

DOMESTIC - *UNMETERED* **A1**
Land zoned for domestic use.
Domestic premises without a connection.
All land not zoned within 400 meters radius of a public standpipe - $33.75 per quarter

DOMESTIC - *UNMETERED* **A2**
Premises served by a single yard tap - $67.50 per quarter

DOMESTIC - *UNMETERED* **A3**
Premises with internal plumbing

ATV (Annual Taxable Value)	%	Minimum
up to $500	95	$108.00
from $501 to $1,000	81	$118.50
from $1,001 to $2,000	54	$203.00
over $2,000	47	$270.00
Maximum bill per quarter -		$304.00

DOMESTIC - *METERED* **A4/A6**
Up to 150 cu. meters $1.75 per cu. meter/quarter. All over 150 cu. meters $3.50 per cu. meter/quarter.
Minimum bill per quarter - $30.00

DOMESTIC - *UNMETERED* **A5**
Churches
Registered Charitable Organisations $108.00 per quarter

NON DOMESTIC - *UNMETERED* **B3/C3**
Industrial, Commercial, Public Buildings $474 per month

NON DOMESTIC - *METERED* **B4/C4**
Flat rate of $3.50 per cu. meter / per month.
Minimum bill per month - $35.00

COTTAGE - *UNMETERED* **D3**
VAT registered premises used partly for business and partly for domestic. (Domestic part must be at least 50%) $300.00 per month

COTTAGE - *METERED* *W.E.F.- 1 / 8 / 93* **D4**
Up to 150 cu. meters - $2.50 per cu. meter per month. All over 150 cu. meters - $3.50 per cu. meter per month.
Minimum bill per month - $25.00

AGRICULTURAL - *UNMETERED* **E3**
15% of ATV (Annual Taxable Value) per month
Minimum bill per month - $105.00

AGRICULTURA - *METERED* **E4**
Flat rate of $2.25 per cu. metre per month
Minimum bill per month - $20.00

NOTE: cu. metre = 219.969 gallons

SEWERAGE

DOMESTIC - *UNMETERED* **A3/A5**
Applicable only to internally serviced premises within 45 metres of a street or collecting sewer.
Water bill of less than $202.50 per quarter - $75.50 per quarter
Water bill of more than $202.50 per quarter - $93.25 per quarter

DOMESTIC - *METERED* **A4/A6**
Applicable only to internally serviced premises within 45 metres of a street or collecting sewer.
50% of Water Bill.

NON DOMESTIC - *METERED* **B4/C4/D4/E4**
Applicable to All premises within 45 metres of a street or a collecting sewer.
50% of Water Bill.

NON DOMESTIC - *UNMETERED* **B3/C3/D3/E3**
Applicable to All premises within 45 metres of a street or a collecting sewer.
50% of Water Bill.

Fig. 5.7

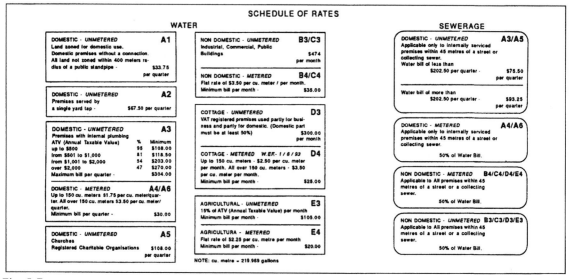

Example 3

Fig. 5.8 shows a bill for water rates. Calculate the unit cost.

METER READINGS ... '000			METER NUMBER
PRESENT	PREVIOUS	CONS	235479
413	401	12	READING DATES
			FROM TO
			15/07/97 22/08/97
01 DOMESTIC			$
Balance BF			1960.51
Payments			3004.51 CR
			1044.51 CR
Water Charges			857.91
Service Charge			35.34
Tax 31.99%			285.76
PLEASE PAY THIS AMOUNT			135.01

Fig. 5.8

Water charges = $857.91
Number of units used = 12 000

$\therefore$ unit cost = $\dfrac{85\,791}{12\,000}$ cents = 7.149 cents

Example 4

A household used 70 cubic metres of water for the first half of 1997.

(a) Use the annual water rates for domestic users shown in Fig. 5.7 to calculate the amount due.

There is a discount of 5% if the bill is paid within two weeks of billing.

(b) Calculate the amount saved if the bill is paid within two weeks.

There is a penalty of 10% if the bill is not paid before the end of the period.

(c) Calculate the total amount to be paid if the bill was not paid on time.

(a) Amount due   = $(70 × 1.75 × 2)
                 = $245
(b) Amount saved = $(245 × 0.05)
                 = $12.25
(c) Total amount = $(245 × 1.1)
                 = $269.50

Example 5

The ATV of a house is assessed to be $2950. The householder was billed the maximum charge of $304 per quarter for water supplied to un-metered domestic premises.

The total amount of water used for the quarter was 22 000 gallons. Use the rate in Fig. 5.7 under the heading for Domestic – Metered 'A4/A6' to calculate the difference in the amount due if the water supply was metered.

[Take 1 cu metre ≈ 220 gallons]

220 gallons ≈ 1 cu metre
$\therefore$ 22 000 gallons ≈ 100 cu metres
Charge for 100 cu metres = $(100 × 1.75)
                         = $175
Difference in charges   = $(304 − 175)
                         = $129

① The rate charged for the water supply to commercial premises is $3.25 per cubic metre per month. Calculate the rate per gallon per month.

[Take 1 cu metre ≈ 220 gallons]

② If the rate for the water supplied to a household is 0.80 cents per gallon per quarter, calculate the rate per cubic metre per quarter.

③ A meter used to measure the water supply is calibrated in cubic metres. The readings of the meter change from 08 240 on November 7th to 08 268 on December 6th. Calculate

(a) the number of cubic metres of water used;
(b) the number of gallons used for the thirty days;
(c) the average number of gallons used per day to the nearest gallon.

④ A household used an average of 6 600 gallons of water per month. Calculate, using 1 cu metre = 220 gals,

(a) the average number of cubic metres used per month;
(b) the average number of cubic metres used per quarter;

(c) the average water charges per quarter, if the rate is $1.80 per cubic metre per quarter.

5 A commercial building used an average of 36 000 gallons of water per quarter.

(a) If the rate is 1.75 cents per gallon per month, calculate the water charges per quarter.

There is a monthly customer service charge of $40.00. A Government tax of 30% of the total charges is included in the total amount due.

(b) Calculate the amount to be paid by the customer.

6 The meter readings for the water supply to a manufacturing business showed that 20 000 cu metres were used for one month.

(a) If the water rates are $3.40 per cubic metre per month, calculate the water charges for the month.

There is also a service charge of 4% of the water charges.

(b) Calculate the service charge.

The Government tax due is 15% of the total water and service charges.

(c) Calculate the total amount due to be paid for the month.

Telephone bills

The total bill for the telephone service usually includes a basic rental fee and service charges. The service charges are calculated at different rates for local calls, for long-distance inland calls and for overseas calls. There may also be special call/service charges and a Government tax.

Fig. 5.9 shows the details given in calculating the total amount due for telephone service. Note that the monthly rental is charged for the current month, October, while the call charges apply to the previous month(s), September. The rates also vary according to the Class Code. These rates depend on the time when the call is made, that is, the rates may be *Full* or *Reduced*; and also on the service used, that is, *Person to Person*, *Station to Station* or *Collect Call*.

RENTAL AND NON CALL CHARGES

From	To	Description	$
1 Oct 97	31 Oct 97	Rental Charges	336.00
	Total		336.00

FREE USAGE

Local calls	60 Free Minutes		−9.00
	Total		−9.00

INTERNATIONAL CALLS

Date/Time		Mins	Cls	$
19 Sep 97	03:56	4	DF	95.12
23 Sep 97	20:52	17	DF	404.26
27 Sep 97	07:30	5	DF	241.00
27 Sep 97	07:35	13	DF	626.60
28 Sep 97	19:31	2	DF	47.56
	Total			1414.54

LOCAL CALLS

From	To	Description	$
15 Sep 97	15 Oct 97	297 Calls	
	1103 mins @ $0.15 per min		165.45

LONG DISTANCE CALLS

Date/Time		Mins	Cls	$
15 Sep 97	19:01	5	DR	1.90
16 Sep 97	15:39	16	DF	12.16
16 Sep 97	19:22	14	DR	5.32
17 Sep 97	19:36	1	DR	0.38
17 Sep 97	19:37	4	DR	1.52
	Total			21.28

CLASS CODES (Cls)

DF – Direct Full	PF – Person to Person Full
DR – Direct Reduced	PR – Person to Person Reduced
CF – Collect Call	SF – Station to Station Full
	SR – Station to Station Reduced

ACCOUNT STATEMENT

1 Nov 1997	$
Amount Brought Forward	2 475.93
Payment 1997/09/1	−2 475.93
OUTSTANDING	0.00
THIS PERIOD	
Rental and Non Call Charges	336.00
Total Call Charges	1 601.27
Free Usage	−9.00
Total Charges This period	1 928.27
Tax @ 15%	289.24
TOTAL NOW DUE	2 217.51

Fig.5.9

Example 6

The charges on a telephone bill for August 1997 include a monthly rental fee of $30.00 which covers the charges for 30 local calls. A charge of 25 cents per call is made for each local call in excess of 30 calls. The number of local calls made in August 1997 was 210, and the charges for overseas calls were $284.60. A tax of 50% is payable on overseas calls.

Calculate

(a) the amount due for local calls;
(b) the tax on overseas calls;
(c) the total amount of the telephone bill.

Number of local calls charged = (210 − 30)
$$= 180$$

(a) Charges for local calls = $(180 × 0.25)
$$= \$45.00$$

(b) Tax on overseas calls = $(284.60 × 0.5)
$$= \$142.30$$

(c) Total amount of bill
= $(30.00 + 45.00 + 284.60 + 142.30)
= $501.90

Exercise 5c

① Use the charges given in Fig. 5.10 on page 58 to calculate the rates paid per minute for an international call to

(a) Jamaica, DD (b) Barbados, DN
(c) Florida, DN (d) St Lucia, DD

② Using the Account Statement given in Fig. 5.11, calculate

(a) the tax on the total charges;
(b) the 'TOTAL NOW DUE' on the bill.

ACCOUNT STATEMENT

	$
Total month rentals	36.00
Total International/Special Call Charges	106.20
Total Local Call Charges	107.18
Concessions	−4.60
Total Charges excluding Tax	244.78
Tax @ 15%	_____
TOTAL NOW DUE	_____

Fig. 5.11

③ Use the amounts listed under the charges for Long Distance Calls in Fig. 5.9 to calculate

(a) the full rate, DF;
(b) the reduced rate, DR;
(c) the difference as a percentage of the full rate.

④ In January, a total of 430 local telephone calls was made. The amount due for overseas calls was $935.20. Use the rental fee and rates given in Example 6 to calculate

(a) the amount due for local calls;
(b) the tax on overseas calls;
(c) the total amount due to be paid.

⑤ The telephone bill for Mary James for April 1992 is given in Fig. 5.12. Telephone subscribers are charged a monthly service fee of $27.50 which covers up to a maximum of 30 local calls per month. A charge of 20 cents per call is made for each local call in excess of 30 calls. A tax of 75% is payable on all overseas calls.

(a) Calculate

Telephone Bill

Name: Mary James   Account No. J0052

Previous Reading March 31, 1992	Present Reading April 30, 1992	Number of local calls
4325	4402	

CHARGES		
	$	¢
Arrears		
Service fee	27	50
Local calls		
Overseas calls	80	00
Tax on overseas calls		
TOTAL	209	40

Fig. 5.12

(i) the number of local calls made in April;
(ii) the amount due for local calls in April;
(iii) the tax on overseas calls in April;
(iv) the arrears from March.

(8 marks)

(b) Mary was charged $13.00 for local calls in May 1992. Calculate the total number of local calls she made in May 1992.

[CXC June 92]   (2 marks)

BILLING DETAILS

Telephone Number: _____

RENTAL CHARGES

From	To	Description	$
1-Nov-1997	30-Nov-1997	Residence Line	29.00
1-Nov-1997	30-Nov-1997	2500 Handset – Touchtone	0.00
1-Nov-1997	30-Nov-1997	Hold Number Delivery – Unlisted	0.00
1-Nov-1997	30-Nov-1997	Tone Facility	2.00
1-Nov-1997	30-Nov-1997	Ex-Directory Listing	5.00
		Total	36.00

CONCESSIONS

		Local Call Credit	−1.60

CALL CHARGES (See below for class codes)

International

Date/Time		Country/Destination	Called No.	Min	Cls	$
29-Sep-1997	16:53	Jamaica	18769621234	2	DD	4.10
1-Oct-1997	06:56	Barbados	12464381234	1	DN	1.70
4-Oct-1997	20:21	Jamaica	18769441234	17	DD	34.85
4-Oct-1997	20:44	Jamaica	18759351234	10	DD	20.50
4-Oct-1997	20:59	Florida	19545701234	3	DN	12.00
10-Oct-1997	08:08	Jamaica	18769621234	4	DD	8.20
15-Oct-1997	20:59	St. Lucia	17584501234	10	DD	10.50
27-Oct-1997	07:06	Jamaica	18769621234	3	DD	6.15
27-Oct-1997	22:00	Jamaica	18769271234	1	DD	2.05
27-Oct-1997	22:01	Jamaica	18099271234	1	DD	2.05
27-Oct-1997	22:06	Jamaica	18099271234	1	DD	2.05
27-Oct-1997	22:06	Jamaica	18099271234	1	DD	2.05
		Total				106.20

Local Calls

From	To		No. of Calls	Min.Sec	$
30-Sep-1997	31-Oct-1997	Local Calls	266	1317.32	107.18
		Total			107.18

Class Codes

DD	Direct Day		PD	Operator Assisted Person to Person Day
DN	Direct Night		PN	Operator Assisted Person to Person Night
SD	Operator Assisted Station to Station Day		CD	Collect Day
SN	Operator Assisted Station to Station Night		CN	Collect Night

Fig. 5.10

Investments

Investments including ordinary savings in a bank, credit union or other financial institution, are a possible source of increasing one's money or financial assets. You may invest a portion of the income earned from your salary or use a part of your savings, usually for some special purpose such as building a retirement fund.

Investments are not a guaranteed form of increasing your assets but do involve a certain degree of risk. You may gain or lose depending on the state of the public economy and the particular investment opportunity. Examples of investments include the purchase of

(a) land or houses;
(b) Government savings bonds;
(c) fixed savings deposits;
(d) stocks and shares;
(e) annuities, insurance or pension plans through a financial institution, for example, a bank or a unit trust company.

The agreement about the amount to be paid by the financial institution at the end of a certain period is stated in a document or contract called the **policy**. The amount that the investor pays is called the **premium**.

Generally, the financial returns depend partly on the amount invested and may be calculated as a percentage of the amount invested, or may be based on the value of unit costs, that is, the cost of the smallest unit of the investment, whether one house or one share in a company. For example, in the case of Government bonds, and stocks or shares, the interest may be based on the value of units of $10, units of $100, units of $10 000, and so on.

In the case of savings bonds, fixed deposits and certain pension and insurance plans, the period for which the money is being invested also affects the money received. As a rule, the longer the period of the policy before payment must be made by the financial institution, the smaller the premium that the investor has to pay. Age is also an important factor in the returns obtained from annuities, insurance and pension plans.

Example 7

Calculate the interest on $640 000 invested for 3 months at $7\frac{1}{2}$% per annum, when interest is added to the principal monthly.

$$\text{Interest for 1st month} = \$\frac{640\,000 \times 7\frac{1}{2} \times \frac{1}{12}}{100}$$
$$= \$\frac{15 \times 1 \times 640\,000}{2 \times 12 \times 100}$$
$$= \$4000$$

Principal for 2nd month = $644 000

$$\text{Interest for 2nd month} = \$\frac{644\,000 \times 7\frac{1}{2} \times \frac{1}{12}}{100}$$
$$= \$\frac{15 \times 1 \times 644\,000}{2 \times 12 \times 100}$$
$$= \$4025$$

Principal for 3rd month = $648 025

$$\text{Interest for 3rd month} = \$\frac{648\,025 \times 7\frac{1}{2} \times \frac{1}{12}}{100}$$
$$= \$\frac{15 \times 1 \times 648\,025}{2 \times 12 \times 100}$$
$$= \$4050.15$$

Total interest on investment = $12 075.15

Example 8

Richard invested $200 000 for 6 months at a rate of 8% per annum. At the end of the six months, he then re-invested the total amount including the interest at 9% per annum for another 6 months.

Alana invested her $200 000 for 12 months at $8\frac{1}{2}$% per annum.

Calculate the difference in the amounts paid to Richard and to Alana at the end of 12 months.

After 6 months, Richard receives

$$\text{Interest} = \$\frac{200\,000 \times 8 \times \frac{6}{12}}{100}$$
$$= \$\frac{200\,000 \times 8 \times 6}{100 \times 12}$$
$$= \$8000$$

At the end of 6 months, Richard receives
$$\text{Amount} = \$(200\,000 + 8000)$$
$$= \$208\,000$$

After the next 6 months, Richard receives

Interest $= \$\dfrac{208\,000 \times 9 \times \frac{6}{12}}{100}$

$= \$\dfrac{208\,000 \times 9 \times 6}{100 \times 12}$

$= \$9360$

After 12 months, Richard receives

Total amount $= \$(208\,000 + 9360)$

$= \$217\,360$

At the end of 12 months, Alana receives

Interest $= \$\dfrac{200\,000 \times 8\frac{1}{2} \times 1}{100}$

$= \$\dfrac{200\,000 \times 17 \times 1}{2 \times 100}$

$= \$17\,000$

After 12 months, Alana receives

Total amount $= \$(200\,000 + 17\,000)$

$= \$217\,000$

∴ Richard receives $360 more than Alana.

Example 9

An investor bought 10 000 shares at $1.50 per share. She later sold 5000 shares at $1.85 per share and then the remaining 5000 shares at $1.10 per share. Calculate the final amount earned on her investment.

Cost price of 10 000 shares at $1.50 per share
$= \$15\,000$

Selling price of 5000 shares at $1.85 per share
$= \$9250$

Selling price of 5000 shares at $1.10 per share
$= \$5500$

Total money received from sale of shares
$= \$(9250 + 5500)$
$= \$14\,750$

∴ investor lost $250

Depreciation

Assets in the form of buildings, vehicles, machinery and other equipment lose value over time due to wear and tear, or to the development of more efficient and up-to-date models. The current value of the asset is called its **book value**, and the loss in value is referred

to as the **depreciation** of the item. Depreciation is usually stated as a percentage of the current value and is taken into account for resale of items. The depreciation varies from year to year depending on the age and condition of the items.

Businesses are allowed to claim for tax concessions on their assets for depreciation of items related to the management of the business. This can result in a significant difference in the amount of tax due to the Government.

Example 10

The original value of a new car is $90 000. The depreciation at the end of the 1st year is set at 2% of the value when new, and at the end of the 2nd year is set at 3% of its value at the start of the 2nd year. Calculate the book value of the car at the start of

(a) Year 2, (b) Year 3.

At the end of Year 1,
Depreciation $= \$\frac{2}{100} \times 90\,000$
$= \$1800$

At the start of Year 2,
Value $= \$(90\,000 - 1800)$
$= \$88\,200$

At the end of Year 2,
Depreciation $= \$\frac{3}{100} \times 88\,200$
$= \$2646$

At the start of Year 3,
Value $= \$(88\,200 - 2646)$
$= \$85\,554$

Exercise 5d

1 (a) (i) Calculate the simple interest earned if $4000 is placed in a bank at 8% per annum for 6 months. (3 marks)
(ii) How long should the $4000 be kept in the bank to earn an interest of $240? (2 marks)
(b) Calculate the compound interest on $1800 for 2 years at 10% interest per annum. (5 marks)
[CXC June 91]

2 Ann bought 1000 shares at $2 per share. She later sold 800 shares at $2.05 per share. Calculate the lowest price that she must get for each of the remaining shares so that she does not lose money on her investment.

3 The cost of a new machine is $12 000. The depreciation at the end of the 1st year is set at 1.5% of the value when new.

(a) Calculate the book value of the machine at the start of Year 2.

At the end of the 2nd year the depreciation is 2% of its value at the start of the 2nd year.

(b) Calculate the book value of the machine at the start of Year 3.

4 Robert put $50 000 in a fixed deposit for 3 months at a rate of 12% per annum. At the end of three months, he re-invested $50 000 at 15% per annum for 1 year. Andrew put $50 000 in a fixed deposit for 15 months at the rate of 15% per annum. Calculate the difference in the total amounts each received at the end of 15 months.

5 After 12 years of paying annual premiums of $15 000, an annuity guarantees 10% interest on the investment. At that time there is a lump sum payment of 40% of the total amount paid-in plus interest and monthly payments of $1100.

Calculate the number of monthly payments before the investor gets back the money invested.

6 Machinery costing $25 000 depreciates to a value of $23 125 at the end of the first year. Thereafter, depreciation is calculated at the rate of 10% per annum.

Calculate

(a) the percentage depreciation for the first year;

(b) the value of the machinery after two years;

(c) the average rate of depreciation over the first three years as a percentage of $25 000. Give your answer to 1 decimal place. (10 marks) [CXC June 90]

Summary

Using percentages and rates (or unit costs), we calculate the charges for **utility services**, that is, for electric power, for telephone services and for a water supply.

We use the simple interest formula (usually compounded), and the profit (or loss) made in buying and selling items to calculate the financial returns on **investments**, for example, saving deposits, pension plans and annuities, and stocks and shares. Using the same formulae, we may calculate the **depreciation** and the **book value** of assets due to wear and tear or being out-of-date. However, in these cases, we always have to deduct the change in value. Such assets include equipment, vehicles, machinery and, at times, buildings.

Practice exercise P5.1

1 Complete the table and calculate the amount owed to the electricity board in each of these cases.

	Previous	Present	kWh used	Rate ($ per kWh)	Amount
(a)	00356	00478		0.73	
(b)	01295	01566		0.55	

2 The table shows how *Sparky Electricity* calculate their bills.

Fixed rate	$11.75
First 100 kWh	$0.44 per kWh
Next 250 kWh	$0.39 per kWh
Remaining units 350 kWh	$0.30 per kWh
VAT	17.5%

Calculate the amount owed in the following readings.

(a) Previous: 248135   Present: 248201

(b) Previous: 017862   Present: 018281

③ Mr Weekes receives the following phone bill:

Line rental:	$15.50
Tariff charge:	$40.00
Local calls:	513 mins
International calls:	36 mins
Arrears:	$7.43
VAT:	17.5%

Local calls cost 22¢ per minute, international calls $2.50 per minute. He pays a tariff charge of $40 per month. This gives him the first hour of local calls and the first 15 minutes of international calls free. How much is his bill, including the arrears from his previous bill?

Exercise P5.2

① Honour has $500 that she wants to invest in a savings account for three years. She has to choose an account.

A1 Savings pays 8% compound interest per annum.
Easycash pays 8.5% simple interest per annum.

(a) Which account pays the larger sum of money?
(b) Calculate the difference in the amounts paid.

② A car is worth $25 456. It depreciates in value by 15% every year. Calculate the value of the car after three years (to the nearest cent).

③ Carl bought a television for $850. It lost 10% of its value in the first year.

(a) Calculate its value after one year.

The television depreciates 12% every year after the first year.

(b) Calculate its value three years after buying it.
(c) What percentage of its original value has been lost?

④ Calculate how much each of these items is worth after three years.

(a) A vintage car currently worth $24 000 that appreciates at 15% per annum.
(b) A computer worth $700 that depreciates at 30% per annum.

⑤ Theresa has taken a loan of $2000 from her bank. The bank charges 25% interest at the end of each year. She repays $100 per month.

(a) How much has she repaid by the end of the first year?
(b) How much interest does the bank charge at the end of the year?
(c) How many months into the second year does she pay off the loan?

Pre-requisites

■ universal sets; subsets; Venn diagrams

Sets (revision)

Example 1

U = {animals}, H = {animals with horns},
W = {wild animals}. Show on a Venn diagram
the set of wild animals that do not have horns.

H = {animals with horns}

H′ = {animals which do *not* have horns}

W ∩ H′ = {wild animals which do not have horns}

In Fig. 6.1, the region with horizontal shading
represents W. The region with vertical shading
represents H′. The region which is cross-shaded
represents W ∩ H′, the set of wild animals
which do not have horns.

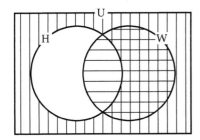

Fig. 6.1

Example 2

Given that x is an integer, list the members of
$\{x: x > -4\} \cap \{x: x \leq 1\}$.

If x is an integer, then

$\{x: x > -4\} = \{-3, -2, -1, 0, 1, 2, 3, ...\}$

and

$\{x: x \leq 1\} = \{... -4, -3, -2, -1, 0, 1\}$

Hence

$\{x: x > -4\} \cap \{x: x \leq 1\} = \{-3, -2, -1, 0, 1\}$

In Example 2, $\{x: x > -4\}$ and $\{x: x \leq 1\}$ are
examples of sets written in **set builder notation**.
The expression $\{x: x > -4\} \cap \{x: x \leq 1\}$ can be
shortened to $\{x: -4 < x \leq 1, x$ is an integer$\}$,
i.e. 'the set of values x, such that x is greater
than -4 and less than or equal to 1, where x
is an integer'. The expression $-4 < x \leq 1$ is a
range of values of x.

Example 3

A survey of 100 families showed that 32 had
a TV set and 51 had a gas cooker. If 40 had
neither, how many had both?

Let U = {all families in the survey}, T = {families
with TV sets}, G = {families with gas cookers}. it
is required to find n(T ∩ G).

In Fig. 6.2, x = n(T ∩ G).

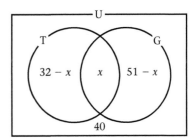

Fig. 6.2

Since n(T) = 32, n(T ∩ G′) = 32 − x

Similarly n(G ∩ T′) = 51 − x

and n(T′ ∩ G′) = 40

The totals for the regions must add up to the
number of families in the universal set, n(U).

$$x + (32 - x) + (51 - x) + 40 = 100$$
$$123 - x = 100$$
$$x = 23$$

23 families had both a TV set and a gas cooker.

Table 6.1 contains the symbols and language of
sets that have been explained in Books 1 and 2.

Table 6.1

Symbols	Meaning
$P = \{a, b, c\}$	P is the set a, b, c
$Q = \{1, 2, 3, \ldots\}$	... means 'and so on'
$A = \{x: x$ is a natural number$\}$	Sets A, B, C are various examples of set-builder notation: sets A and C
$B = \{(x, y): y = mx + c\}$	give values of x; set B is a set of points (x, y)
$C = \{x: a < x < b\}$	
$\varnothing$ or $\{\ \}$	the empty set
U	the universal set
$\in$	is a member of
$\notin$	is *not* a member of
$A \subset B$	A is a subset of B
$A \supset B$	A contains B
$\not\subset, \not\supset$	the negations of $\subset, \supset$
$A \cup B$	union of A and B
$A \cap B$	intersection of A and B
$n(A)$	number of elements in set A
A'	complement of set A

Exercise 6a

1 If $U = \{f, a, c, t, o, r, i, s, e\}$, $P = \{r, a, t, i, o\}$ and $Q = \{s, e, t\}$ write down the members of the following sets.
(a) P' (b) Q'
(c) $(P \cap Q)'$ (d) $(P \cup Q)'$

2 Use the sets of question 1 to demonstrate that $P' \cup Q' = (P \cap Q)'$ and $P' \cap Q' = (P \cup Q)'$. Illustrate these results by suitable shading on a Venn diagram.

3 If $U = \{c, h, i, c, k, e, n\}$, $P = \{n, i, c, e\}$ and $Q = \{h, e, n\}$ list the elements of the following:
(a) $P \cap Q$ (b) $P \cup Q$
(c) $(P \cup Q)'$ (d) $(P \cap Q)'$
(d) $P' \cap Q$ (e) $P \cup Q'$

4 If $U = \{$integers$\}$, list the members of the following sets using ... where appropriate.
(a) $\{x: x \leqslant 9\}$ (b) $\{x: x \geqslant -3\}$
(c) $\{x: x > 5\}$ (d) $\{x: x < 0\}$
(e) $\{x: x - 4 = 0\}$ (f) $\{x: 2x + 3 = 15\}$
(g) $\{x: -6 \leqslant x < 2\}$ (h) $\{x: -8\frac{1}{2} < x < -1\frac{1}{4}\}$

5 If $U = \{x: 1 \leqslant x \leqslant 10, x$ is an integer$\}$, $A = \{x: x$ is a perfect square$\}$ and $B = \{x: x$ is a factor of 20$\}$,
(a) find $n(A')$,
(b) find $n(A \cup B)'$,
(c) list the members of the set B',
(d) list the members of the set $A \cap B'$.

6 If $U = \{1, 2, 3, 4, 5, \ldots, 20\}$ list the members of the following sets.
(a) $\{x: x$ is a square number, $x \in U\}$
(b) $\{x: x + 2 > 15, x \in U\}$
(c) $\{x: x$ is a factor of 40, $x \in U\}$
(d) $\{(x, y): y = 3x + 1, x \in U, y \in U\}$
(e) $\{(x, y): y = 2x^2 - 1, x \in U, y \in U\}$
(f) $\{(x, y): y > 17 + x, x \in U, y \in U\}$

7 $U = \{$countries of Africa$\}$, $N = \{$countries which lie wholly or partly north of the equator$\}$, $S = \{$countries which lie wholly or partly south of the equator$\}$. If $n(U) = 47$, $n(N) = 32$ and $n(S) = 21$, through how many countries of Africa does the equator pass? [What is $n(N \cap S)$?]

8 In a choir of 38 students, 22 are girls. 17 of the students are at least 160 cm tall. If 14 of the girls are less than 160 cm, how many of the boys are also less than 160 cm?

9 In a class, $2x$ students are less than 15 years old, x students are over 13 years old and 17 students are 14 years old. Show this information on a Venn diagram. If there are 37 students in the class, find x.

10 In the Venn diagram of Fig. 6.3, $U = \{$people in a village$\}$, $X = \{$cattle owners$\}$, $Y = \{$car owners$\}$. The letters p, q, r, s represent the numbers of people in the subsets shown.

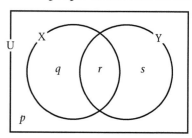

Fig. 6.3

If n(U) = 259, n(X) = 43 and n(Y) = 32,
(a) express *q* in terms of *r*;
(b) express *s* in terms of *r*;
(c) state the greatest possible value of *r*;
(d) find the smallest possible value of *p*;
(e) find *p*, *q* and *r* if *s* = 6.

Equal and equivalent sets

Example 4

A = {letters of the word legal}
B = {letters of the word play}
C = {letters of the word gale}

List the elements of sets A, B and C.

A = {e, g, a, l}
B = {p, l, a, y}
C = {g, a, l, e}

You will notice that
(i) sets A and B have the same number of elements;
(ii) sets A and C have exactly the same elements.

Sets A and B are called **equivalent sets**.
Sets A and C are called **equal sets**.

Group work

Using sets containing the letters of the words *recommend, accommodate, appreciate*, list as many (a) pairs of equal sets, (b) pairs of equivalent sets as possible.

Venn diagrams with more than two subsets

Venn diagrams can be drawn to represent any number of sets, though they become very complicated if the number is large. Fig. 6.4 is a Venn diagram for three subsets of a universal set.

Diagrams such as Fig. 6.4 can be used to simplify problems which at first sight appear to be difficult.

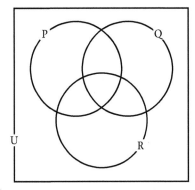

Fig. 6.4

Example 5

In a group of students, 18 play football, 19 play basketball and 16 play volleyball. 6 play football only, 9 play basketball only, 5 play football and basketball only and 2 play basketball and volleyball only. How many play (a) all three games, (b) football and volleyball only, (c) volleyball only? (d) If 8 play no games at all, how many students are there altogether?

In the Venn diagram of Fig. 6.5,
U = {all students}, F = {football players},
B = {basketball players}, V = {volleyball players}.

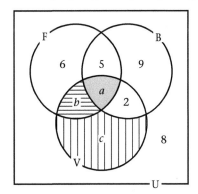

Fig. 6.5

In Fig. 6.5 the 6 who play football only are represented by the region of F which does not lie within B or V. Similarly, the given values 9, 5 and 2 are written in the appropriate regions in Fig. 6.5.

(a) Those who play all three games are represented by the blue region. This region lies in F, B and V, i.e. the set F ∩ B ∩ V.

From the first sentence of the question
$$n(B) = 19$$
Also, from Fig. 6.5,
$$n(B) = 9 + 5 + 2 + a$$
Hence, $19 = 9 + 5 + 2 + a$
$$a = 3$$

3 students play all three games.

(b) Those who play football and volleyball only are represented by the region with horizontal shading. This region lies within F and V but not within B, i.e. the set $F \cap V \cap B'$.

From the first sentence of the question
$$n(F) = 18$$
Also, from Fig. 6.5,
$$n(F) = 6 + 5 + a + b$$
$$= 6 + 5 + 3 + b \; (\textit{since } a = 3)$$
Hence, $18 = 6 + 5 + 3 + b$
$$b = 4$$

4 students play football and volleyball only.

(c) The region with vertical shading represents those playing volleyball only, i.e. the set $V \cap (F \cup B')$.

Hence, $16 = 2 + a + b + c$
$$= 2 + 3 + 4 + c \; (\textit{since } a = 3, b = 4)$$
$$c = 7$$

7 students play volleyball only.

(d) The total number of students
$$= 6 + 5 + 9 + 2 + a + b + c + 8$$
$$= 6 + 5 + 9 + 2 + 3 + 4 + 7 + 8$$
$$= 44$$

Example 6

In a group of 10 university students 6 are taking mathematics, 5 philosophy and 7 economics. 3 take maths and philosophy, 2 philosophy and economics, and 4 economics and maths. Each student is taking at least one of these subjects. How many of the students are taking all three subjects?

Let M = {students taking mathematics},
P = {students taking philosophy},

E = {students taking economics}.

Note that since each students is taking at least one of the subjects, $M \cup P \cup E = U$. Fig. 6.6 is the appropriate Venn diagram. The centre section represents $M \cap P \cap E$, showing x students taking all three subjects.

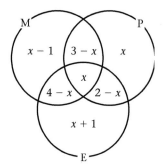

Fig. 6.6

Altogether, 3 students take maths and philosophy, i.e. $n(M \cap P) = 3$. Hence $(3 - x)$ students take maths and philosophy only. $3 - x$ is written in the appropriate region of Fig. 6.6.

Similarly, since $n(P \cap E) = 2$ and $n(E \cap M) = 4$, $2 - x$ and $4 - x$ are written in the appropriate regions of Fig. 6.6.

Since $n(M) = 6$, the number taking maths only is $6 - (3 - x) - x - (4 - x)$
$$= 6 - 3 + x - x - 4 + x$$
$$= x - 1$$

Similarly, the number taking philosophy only is x and the number taking economics only is $x + 1$. The values $x - 1$, x and $x + 1$ are written in the appropriate regions of Fig. 6.6.

As there are 10 students, $n(M \cup P \cup E) = 10$. Hence
$$(x - 1) + (3 - x) + x + (4 - x) + x$$
$$+ (2 - x) + (x + 1) = 10$$
$$x + 9 = 10$$
$$x = 1$$

1 student takes all three subjects.

Exercise 6b

1

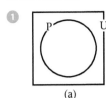
(a)

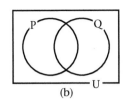
(b)

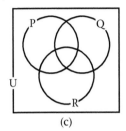
(c)

Fig. 6.7

(a) On a copy of Fig. 6.7(a) shade set P′.
(b) On a copy of Fig. 6.7(b) shade set P′ ∩ Q.
(c) On a copy of Fig. 6.7(c) shade set
(P′ ∩ Q) ∪ R.

2 For each of the following make a freehand copy of Fig. 6.4 and shade the given set.

(a) P ∪ Q ∪ R   (b) P ∩ Q ∩ R
(c) (P ∪ Q) ∩ R   (d) (P ∩ Q) ∪ R
(e) P ∪ (Q ∩ R)   (f) P ∩ Q′ ∩ R
(g) P′ ∪ (Q ∩ R)   (h) Q′ ∩ (P ∪ R)
(i) (P ∪ Q) ∩ (R ∪ Q′)
(j) (P ∩ R) ∪ (Q ∩ R′)

3 Make a freehand copy of the Venn diagram in Fig. 6.8.

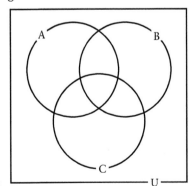

Fig. 6.8

(a) Fill in the members of the sets given that A = {p, q, r, s}, B = {q, r, s, t, u}, C = {p, r, u, y}.
(b) Hence or otherwise list the members of
  (i) (A ∪ B) ∩ C,   (ii) (A ∩ B) ∪ C.

4 The numbers of elements of each region of the Venn diagram of Fig. 6.9 are as shown. If n(U) = 200 find (a) x, (b) n(P ∪ R), (c) n(P ∪ Q ∪ R), (d) n(Q ∩ R).

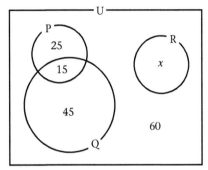

Fig. 6.9

5 The numbers of elements in each region of the Venn diagram of Fig. 6.10 are as shown.

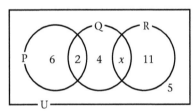

Fig. 6.10

If n(U) = 36, find
(a) x   (b) n(Q)
(c) n(P ∪ R)   (d) n(P ∩ R)
(e) n(R′ ∩ Q′)   (f) n(P′ ∪ Q)

6 In the Venn diagram of Fig. 6.11 the numbers of elements are as shown.

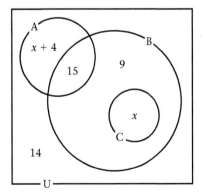

Fig. 6.11

If n(U) = 9x, find
(a) x   (b) n(A)
(c) n(A ∪ B)   (d) n(C′)
(e) n(A ∩ B′)   (f) n(A ∪ C)′

7 All the pupils in a class of 35 play at least one of the games volleyball, netball, hockey. 10 play volleyball only, 5 play netball only and 3 play hockey only.

If 2 play all three games and equal numbers play 2 games, how many altogether play volleyball?

8 20% of the people in a village own a dog, 30% own goats and 40% own cattle, 5% own both a dog and goats, 4% own both a dog and cattle and 3% own both goats and cattle.

If 1% own all three, what percentage of the village own none of the three?

9 150 students were asked what they were doing last night between 8 p.m. and 9 p.m.

50 said they watched TV only; 60 said they listened to the radio only; 5 said they read a book only; 20 watched TV and listened to the radio; 15 watched TV and read a book; 10 listened to the radio and read a book. x students did all three and x students did none of these things.

Illustrate the information on a clearly labelled Venn diagram.

Find x and hence find the total number who read a book at some time between 8 p.m. and 9 p.m.

10 At a Government college, the students can speak at least one of the languages English, Spanish and French.

Form 3A contains 38 students of whom 2 can speak all three languages, 4 speak Spanish and English only, 1 speaks English and French only and no student speaks Spanish and French only. There are x students who speak Spanish only, $5x$ who speak English only and $(x - 4)$ who speak French only.

Illustrate this information on a clearly labelled Venn diagram, showing the number in each separate region.

Find x and hence find the total number of English speakers.

Venn diagrams and statements

Let U = {people}, F = {footballers}, C = {cricketers}.

Each Venn diagram in Fig. 6.12 shows a different relationship between sets C and F.

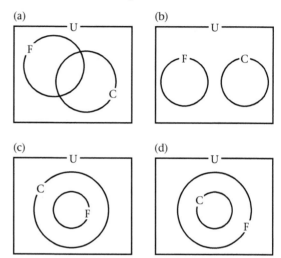

Fig. 6.12

In Fig. 6.12(a), sets F and C intersect. This shows that

some footballers are cricketers;
some cricketers are footballers
not all cricketers are footballers;
not all footballers are cricketers.

Notice the use of 'some' and 'not all' in the above statements. What statements can you make about the information in Fig. 6.12(b)?

No footballer is a cricketer.
No cricketer is a footballer.

In Fig. 6.12(c)

All footballers are cricketers.
Not all cricketers are footballers.

Why are these statements true?
You will notice that set F is a subset of set C:
F ⊂ C.

What can be said from the information in Fig. 6.12(d)?

All cricketers are footballers.
Not all footballers are cricketers.

The information in all these diagrams can be written symbolically.

In Fig. 6.12(a), $F \cap C \neq \varnothing$
In Fig. 6.12(b), $F \cap C = \varnothing$
In Fig. 6.12(c), $F \subset C$
In Fig. 6.12(d), $C \subset F$

Group work

Working in pairs, and using the diagrams in Fig. 6.12, decide where you would put yourself in each of (a), (b), (c), (d) if you were a footballer, a cricketer, a footballer and a cricketer, neither a footballer nor a cricketer.

Example 7

Let U = {students at Farmway High School}
     N = {students who play netball)
     B = {boys at Farmway High School}

Draw a Venn diagram for each of the following statements. State the information symbolically.
(a) All boys play netball.
(b) Some boys play netball.
(c) No boy plays netball.

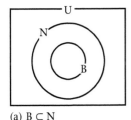
(a) $B \subset N$

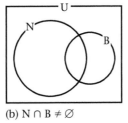
(b) $N \cap B \neq \varnothing$

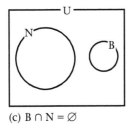
(c) $B \cap N = \varnothing$

Fig. 6.13

Exercise 6c

1. U = {buildings}, B = {buildings with staircases}, H = {houses}. Write down a statement using 'all', 'some', 'not all', 'not' for each of the diagrams in Fig. 6.14.

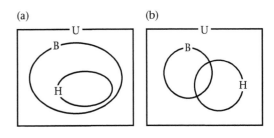

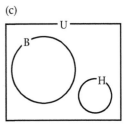

(a)   (b)

(c)

Fig. 6.14

2. U = {quadrilaterals}, P = {parallelograms}, R = {rectangles}, S = {squares}.
   Draw one Venn diagram to show that
   (a) all squares are rectangles;
   (b) all rectangles are parallelograms.

3. If U = {all cars}, B = {modern cars}, C = {cars with seat-belts}.
   Draw Venn diagrams to show
   (a) all modern cars have seat-belts;
   (b) not all cars with seat-belts are modern cars.

4. U = {people}, W = {women}, M = {mothers}, G = {grandmothers}.
   Draw a Venn diagram to show that
   (a) all mothers are women;
   (b) all grandmothers are mothers
   (c) Would it be true to say all mothers are grandmothers?

5. U = {polygons}, T = {triangles}, I = {isosceles triangles}, R = {right-angled triangles}.
   Draw a Venn diagram to show that some right-angled triangles are isosceles triangles. Mark an X to show the set to which a scalene triangle belongs.

6 U = {students at Farmway High School},
S = {students at Farmway who wear glasses},
B = {boys at Farmway who wear glasses}.
Draw a Venn diagram to show that some
boys at Farmway do not wear glasses. Mark
with an X the set to which girls who wear
glasses belong.

Summary

The use of **set building notation** to list the
elements of a set shortens the description
while it provides information about the
range of values of the elements, for example
$\{x: -3 < x \le 7\}$.

Equivalent sets have the same number of
elements. **Equal sets** have identical elements.

Venn diagrams can be used to illustrate
situations involving conclusions made from
statements. These include the representation of
terms such as 'all', 'some', 'not'.

Practice exercise P6.1

For each of the following statements, state
whether it is *true* or *false*, giving your reason if
the answer is *false*.

1 {Friday, Monday} ⊂ {days in a week}

2 −3 ∈ {negative numbers}

3 {units of time} ⊃ {hour, second, gram}

4 {drum, piano} ∈ {musical instruments}

5 {fraction, decimal} ⊃ {integers}

Practice exercise P6.2

1 Describe in words the following sets.
   (a) P = {1, 2, 3, ... 18, 19, 20}
   (b) S = {2, 4, 6, 8, 10, 12, ...}
   (c) T = {4, 9, 16, 25, 36}
   (d) V = {2, 3, 5}

2 For each of the following sets
   (a) use a letter to name the set;
   (b) select, name and describe the members
       of a subset;
   (c) list the members of that subset;
   (d) name the complement of the subset and
       list the members;
   (e) draw a Venn diagram to represent the
       universal set and the subset.
       (i)   {1, 2, 3, 4, ... 7, 8, 9}
       (ii) {plumber, teacher, dentist, barber,
             doctor, beautician, nurse}
       (iii) {5, 10, 15, 20, 25, 30}

Practice exercise P6.3

(a) For each Venn diagram in Figs. 6.15–6.17,
    describe the members of
    (i)   the universal set;
    (ii) the subset;
    (iii) the complement of the subset.

(b) Use symbols to show the relation between
    the universal set and the subset.

1
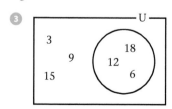

Fig. 6.15

2
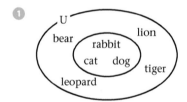

Fig. 6.16

3
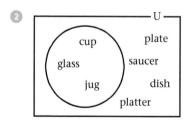

Fig. 6.17

Practice exercise P6.4

1. For each of the following sets, state whether the set is

   (i) infinite(I); (ii) finite(F); (iii) empty(∅).

   (a) {rational numbers}
   (b) {prime numbers between 6 and 12}
   (c) {odd numbers between 10 and 20}
   (d) {multiples of 4 less than 12}
   (e) {fractions that are equal to $\frac{1}{3}$}
   (f) {whole numbers less than 0}

2. For the universal sets given below,
   (i) state the number of subsets;
   (ii) write all the subsets.

   (a) U = {p, q}
   (b) U = {letters of the word nine}

Practice exercise P6.5

For each of the following pairs of sets, state whether the sets are

(i) equivalent; (ii) equal; (iii) not equivalent.

1. (a) {factors of 21}
   (b) {1, 3, 7, 21}

2. (a) {fractions less than 1}
   (b) {fractions between 0 and 1}

3. (a) {odd numbers between 2 and 10}
   (b) {prime numbers less than 10}

4. (a) {multiples of 3 less than 12}
   (b) {rational numbers less than 9}

5. (a) {last five months of the year}
   (b) {March, April, May, June, July}

6. (a) {letters of the word *algebra*}
   (b) {letters of the word *garble*}

7. (a) {1,1, 2, 2, 2, 3, 4, 5, 6, 6}
   (b) {6, 5, 4, 3, 2, 1}

8. (a) {common multiples of 3 and 5}
   (b) {3, 5, 6, 9, 10, 12, 15, 18, 30, ...}

Practice exercise P6.6

Make *six* copies of each of the Venn diagrams in Figs. 6.18–6.20.

In a copy of each Venn diagram, shade in each of the following areas:

(a) M′   (b) H′   (c) M′ ∩ H
(d) M ∩ H′   (e) M′ ∪ H   (f) M ∪ H′

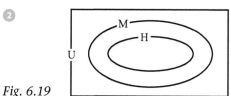

1.

Fig. 6.18

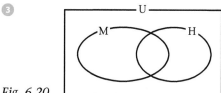

2.

Fig. 6.19

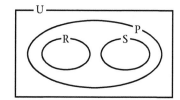

3.

Fig. 6.20

Practice exercise P6.7

Make *seven* copies of each of the Venn diagrams in Figs. 6.21–6.22 and shade the required sets given in (a)–(g).

(a) P ∩ R   (b) R ∪ S   (c) (S ∪ P)′
(d) (P ∪ R) ∩ S   (e) (P ∩ R) ∪ S′
(f) P ∪ (R ∩ S)   (g) P ∩ (R ∪ S)

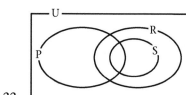

1.

Fig. 6.21

 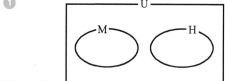

2.

Fig. 6.22

Practice exercise P6.8

In each of the Venn diagrams in Figs. 6.23–6.24, the number of members in each area is shown. For each question

(a) write the equation to show the relation between the members of each area and the number of members in the universal set;

(b) find each value of x;

(c) calculate

(i) $n(A)$ (ii) $n(B)$

(iii) $n(A \cup B)$ (iv) $n(A \cap B)$

1   $n(U) = 20$

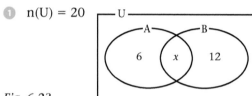

Fig. 6.23

2   $n(U) = 15$

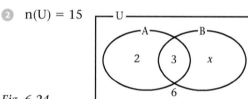

Fig. 6.24

Practice exercise P6.9

Describe the shaded areas in the Venn diagrams in Figs. 6.25–6.26.

1

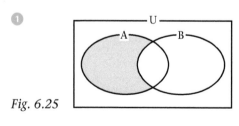

Fig. 6.25

2

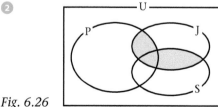

Fig. 6.26

Practice exercise P6.10

In each of the Venn diagrams in Figs. 6.27–6.28

(a) describe the members of the universal set and each subset;

(b) for each question, list the members of

(i) each intersection;

(ii) the complement of the intersection;

(iii) each union;

(iv) the complement of the union.

1

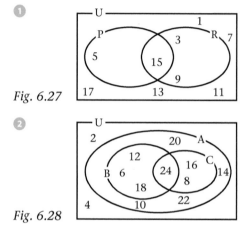

Fig. 6.27

2

Fig. 6.28

Practice exercise P6.11

1   In each of the following problems

(a) use set-builder notation to describe the sets;

(b) list the members of the sets;

(c) solve the problem.

(i) Write the set of prime numbers that are greater than 10 and less than 50.

(ii) Determine the highest common factor of 24 and 30.

2   In a class of 36, some students are

members of the Mathematics Club: 20
members of the Computer Society: 28
members of neither the Mathematics Club nor the Computer Society: 6.

(a) Draw a Venn diagram to represent this information.

(b) Write an expression to find the number of students who are members of both the Mathematics Club and the Computer Society.

(c) Solve the problem.

Linear functions

Pre-requisites
■ straight-line graphs; linear equations in one unknown; rate

Introduction

Example 1

Make a table of the mappings $x \rightarrow 3 - 2x$. Draw the graph of the mapping for value of x from -2 to $+4$.

Table 7.1 $x \rightarrow 3 - 2x$

x	-2	-1	0	$+1$	$+2$	$+3$	$+4$
$3 - 2x$	$+7$	$+5$	$+3$	$+1$	-1	-3	-5

The ordered pairs are $(-2, 7)$, $(-1, 5)$, $(0, 3)$, $(1, 1)$, $(2, -1)$, $(3, -3)$, $(4, -5)$.

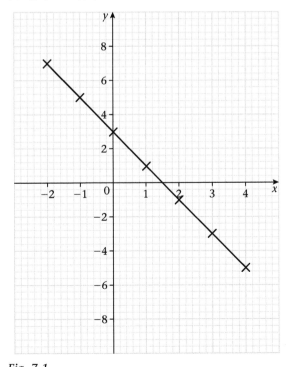

Fig. 7.1

Fig. 7.1 is the graph of the mapping $x \rightarrow 3 - 2x$. Notice that the ordered pairs are the x and y coordinates of each point on the line, with the y-values being the values for $3 - 2x$. The straight line is therefore the graph of $y = 3 - 2x$ which is called the **equation of the line**.

In the equation $y = 3 - 2x$, y is a **function** of x, i.e. the value of y depends on the value of x. $3 - 2x$ is the function of x. The line in Fig. 7.1 is the graph of the equation $y = 3 - 2x$, *or* the **graph of the function** $3 - 2x$. A **linear function** of x is one which contains terms in x of power 1 only. The graph of a linear function is always a straight line.

When plotting the graph of a linear function, two points are sufficient to determine the line. However, in practice it is advisable to plot *three* points. If the three points lie in a straight line it is likely that the working is correct.

Example 2

Table 7.2 gives corresponding values of x and y for the equation $y = 3x - 2$.

Table 7.2 $y = 3x - 2$

x	-1	2	5
y	-5	4	13

(a) Choose suitable scales and draw the graph of $y = 3x - 2$.
(b) Find the value of y when $x = 3.5$.
(c) Find the value of x when $y = 5.5$.
(d) Write down the coordinates of the points where the line cuts the axes.

(a) Fig. 7.2 is the graph of $y = 3x - 2$.
(b) $y = 8.5$
(c) $x = 2.5$
(d) $(0, 22)$, $(0.7, 0)$

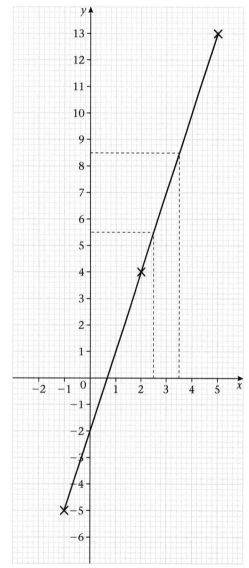

Fig. 7.2

① Using a suitable scale draw the graph of the equation $y = 2x - 3$ for values of x from -2 to $+3$.

Write down the coordinates of the points where the line cuts the axes.

② (a) Given that $y = 5 - 3x$, copy and complete Table 7.3.

Table 7.3 $y = 5 - 3x$

x	-3	0	$+3$
y			

(b) Hence draw the graph of the function $5 - 3x$ for values of x from -3 to $+3$.

③ (a) Draw the graphs of $y = x$, $y = x + 2$, $y = 5 - x$ for values of x from 0 to $+5$ *within the same axes*.

(b) What do you notice about the three lines?

④ Table 7.4 gives corresponding values of x and y for the equation $y = x + 2$.

Table 7.4 $y = x + 2$

x	-4	0	$+4$
y	-2	2	6

(a) Using a scale of 2 cm to 1 unit on both axes, draw the graph of $y = x + 2$.
(b) Find the value of y when $x = 3$.
(c) Find the value of x when $y = 1$.
(d) Write down the coordinates of the points where the line cuts the axes.

⑤ Using a scale of 1 cm to 1 unit on both axes, draw the graphs of the following equations for values of x from -2 to $+2$.
(a) $y = x + 3$
(b) $y = 2x - 1$
(c) $y = 5x$
(d) $y = 3x - 2$

⑥ (a) Draw the graph of $y = 2x - 3$ for values of x from -1 to $+3$. Use a scale of 2 cm to 1 unit on both axes.
(b) On the **same axes**, draw the lines $y = 2x$ and $y = 2x + 1$.
(c) What do you notice about the three lines you have drawn?

Gradient of a straight line

In Fig. 7.3, HG is a horizontal line and HK is a line which makes an angle θ with HG. The $\triangle$s ABC, PQR, UVW are similar.

Hence $\dfrac{BC}{AB} = \dfrac{QR}{PQ} = \dfrac{VW}{UV}$

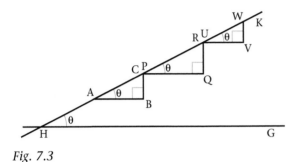

Fig. 7.3

Each of these fractions measures the **gradient** of the line HK. Hence the gradient of a straight line is the same at any point on the straight line.

Also, $\tan \theta = \dfrac{BC}{AB} = \dfrac{QR}{PQ} = \dfrac{VW}{UV}$,

so $\tan \theta$ is also a measure of the gradient.

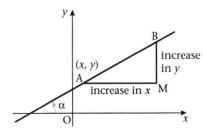

Fig. 7.4

In Fig. 7.4 the point A has coordinates (x, y). In going from A to B the *increase* in x is AM. The *corresponding increase* in y is MB.

$$\text{Gradient of AB} = \frac{\text{increase in } y \text{ from A to B}}{\text{increase in } x \text{ from A to B}}$$

$$= \frac{MB}{AM}$$

Since y *increases* as x increases, the gradient is *positive*. AB makes an acute angle α with the positive direction of the x-axis and $\tan \alpha$ is positive.

In Fig. 7.5, the point C has coordinates (x, y). In going from C to D the *increase* in x is CN. The *corresponding decrease* in y is ND. Consider a decrease to be a negative increase:

$$\text{Gradient of CD} = \frac{\text{increase in } y \text{ from C to D}}{\text{increase in } x \text{ from C to D}}$$

$$= \frac{-ND}{CN}$$

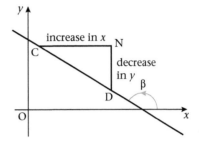

Fig. 7.5

Since y *decreases* as x increases, the gradient is *negative*. CD makes an obtuse angle β with the positive direction of the x-axis and $\tan \beta$ is negative.

In algebraic graphs, the gradient of a straight line is the **rate of change of y compared with x**. For example, if the gradient is 3, then for any increase in x, y increases 3 times as much. Compare this with rates of change in distance–time and velocity–time graphs.

Example 3

Find the gradients of the lines joining (a) A$(-1, 2)$ and B$(3, -2)$, (b) C$(0, -1)$ and D$(4, 1)$.

Fig. 7.6 shows the points A, B, C, D and the lines AB and CD.

(a) Gradient of AB $= \dfrac{\text{increase in } y}{\text{increase in } x} = \dfrac{-PB}{AP}$

$$= \frac{-4}{4} = -1$$

Notice that the increase in y is negative (i.e. a decrease).

(b) Gradient of CD $= \dfrac{\text{increase in } y}{\text{increase in } x} = \dfrac{QD}{CQ}$

$$= \frac{2}{4} = \frac{1}{2}$$

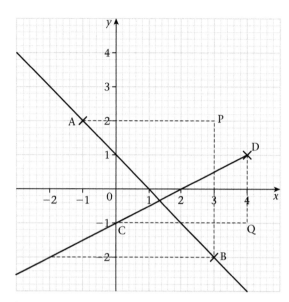

Fig. 7.6

The gradient can also be calculated without drawing the graph. Consider a line which passes through the points (x_1, y_1) and (x_2, y_2). Let the gradient of the line be m.

$$m = \frac{\text{change in } y}{\text{increase in } x}$$

$$= \frac{\text{difference in the } y\text{-coordinates}}{\text{difference in the } x\text{-coordinates}}$$

$$= \frac{y_2 - y_1}{x_2 - x_1}$$

For example, in Example 3,

Gradient of AB $= \dfrac{(-2) - (2)}{(3) - (-1)} = \dfrac{-4}{4} = -1$

Gradient of CD $= \dfrac{(1) - (-1)}{(4) - (0)} = \dfrac{2}{4} = \dfrac{1}{2}$

Exercise 7b

Find the gradients of the lines joining the following pairs of points.

① (9, 7), (2, 5)

② (2, 5), (4, 8)

③ (5, 3), (0, 0)

④ (6, 1), (1, 5)

⑤ (0, 4), (3, 0)

⑥ (−3, 2), (4, 4)

⑦ (2, 3), (6, −5)

⑧ (−4, 3), (8, −6)

⑨ (−4, −4), (−1, 5)

⑩ (7, −2), (−1, 2)

Example 4

Fig. 7.7 shows the graph of the function $2x + 3$. Find (a) the gradient of the line $y = 2x + 3$, (b) the y intercept.

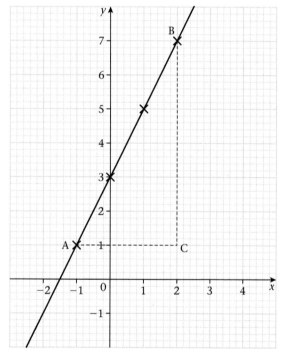

Fig. 7.7

Remember that the graph of the function $2x + 3$ is the straight line whose equation is $y = 2x + 3$. Taking any two points on the line AB

(a) the gradient of AB is given by either

$\dfrac{\text{BC}}{\text{AC}} = \dfrac{6}{3} = 2$ or $\dfrac{7 - 1}{2 - (-1)} = \dfrac{6}{3} = 2$

(b) y intercept $= 3$ from reading the graph.
Or since $x = 0$ on the y-axis
$y = 2x + 3 = 2 \times 0 + 3 = 3$
$\therefore y = 3$

Example 5

Given the linear function $3x - 2$, find (a) the gradient of the straight line representing this function, (b) the y intercept.

$y = 3x - 2$ is the equation of the straight line.

Table 7.5 $y = 3x - 2$

x	0	1	3
y	−2	1	7

(a) From Table 7.5 and using any two points,

$$\text{gradient} = \frac{7 - 1}{3 - 1} = \frac{6}{2} = 3$$

(b) For y intercept,
when $x = 0$, $3x - 2 = 3 \times 0 - 2 = -2$
i.e. y intercept $= -2$

Group work

Working in pairs, find by drawing and calculation (i) the gradient, (ii) the y intercept of each of the following.

❶ $y = 3x + 1$

❷ $y = 2x - 5$

❸ $y = x - 6$

❹ $y - 8 = 2x$

❺ Write down the gradients of the lines (a), (b), (c), (d), (e) in Fig. 7.8.

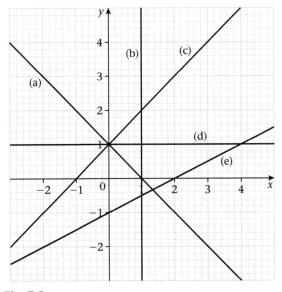

Fig. 7.8

Equation of a straight line

The equation of a straight line gives the relationship between the x-coordinate and the y-coordinate of any point on the line. For example, given the equation of the straight line
$$y = 3x - 2$$
the ordered pairs $(0, -2)$, $(1, 1)$, $(2, 4)$ are the x- and y-coordinates of points on the straight line. Any point on the line must satisfy the condition $y = 3x - 2$.

The equation of a straight line may therefore be found given

(a) the graph of the line;
(b) the gradient and one point on the line;
(c) the coordinates of two points on the line.

(a) Given the graph of the line

Example 6

Write down the equation of the line in the graph of Fig. 7.9.

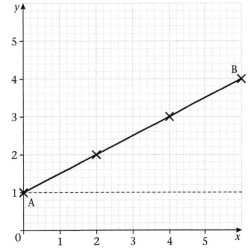

Fig. 7.9

Using points A, B

$$\text{gradient of line} = \frac{3}{6} = \frac{1}{2}$$

y intercept $= 1$

equation of line given by
$y = $ (gradient) $x + $ (y-intercept)

$$y = \frac{1}{2}x + 1 \quad \text{or} \quad 2y = x + 2$$

(b) Given the gradient and a point on the line

Example 7

A straight line of gradient 5 passes through the point B(3, −8). Find the equation of the line.

Fig. 7.10 is a sketch of the line.

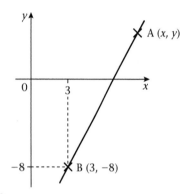

Fig. 7.10

In Fig. 7.10, the point A(x, y) is any general point on the line.

Gradient of AB $= \dfrac{y - (-8)}{x - 3} = \dfrac{y + 8}{x - 3}$

Hence $\dfrac{y + 8}{x - 3} = 5$, since the gradient of AB is 5.

$$y + 8 = 5(x - 3)$$
$$= 5x - 15$$
$$y = 5x - 23$$

The equation of the line is $y = 5x - 23$.

In general, the equation of a straight line of gradient m which passes through the point (x_1, y_1) is given by

$$\frac{y - y_1}{x - x_1} = m$$

(c) Given two points on the line

Example 8

Find the equation of the straight line which passes through the points Q(−1, 7) and R(3, −2).

Fig. 7.11 is a sketch of the line through Q and R.

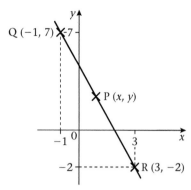

Fig. 7.11

In Fig. 7.11, the point P(x, y) is any general point on the line.

Gradient of QR $= \dfrac{7 - (-2)}{(-1) - 3} = \dfrac{9}{-4} = -2\frac{1}{4}$

Gradient of PR $= \dfrac{y - (-2)}{x - 3} = \dfrac{y + 2}{x - 3}$

But PQR is a straight line, hence gradient PR = gradient of QR

$$\frac{y + 2}{x - 3} = -2\frac{1}{4}$$
$$y + 2 = -2\frac{1}{4}(x - 3)$$
$$y + 2 = -2\frac{1}{4}x + 6\frac{3}{4}$$
$$y = -2\frac{1}{4}x + 4\frac{3}{4}$$

The equation of the line is $y = -2\frac{1}{4}x + 4\frac{3}{4}$.

Note: This may be simplified and written as
$$4y = -9x + 19$$

In general, the equation of a straight line which passes through the points (x_1, y_1) and (x_2, y_2) is given by

$$\frac{y - y_1}{x - x_1} = \frac{y_2 - y_1}{x_2 - x_1}$$

Exercise 7c

1. Find the equation of the line which passes through the point
   (a) (4, 9) and has a gradient of 3,
   (b) (0, 0) and has a gradient of 3,
   (c) (−2, 8) and has a gradient of −1,
   (d) (6, 0) and has a gradient of $-\frac{3}{4}$,
   (e) (0, −5) and has a gradient of −4,
   (f) (−1, 2) and has a gradient of $2\frac{1}{2}$.

2 Find the equation of the line which passes through the points
   (a) (0, 0) and (3, 7)
   (b) (0, 0) and (3, −7)
   (c) (−1, 4) and (5, −2)
   (d) (−6, −6) and (4, −3)
   (e) (7, 2) and (−9, 7)
   (f) (2, −11) and (−4, 4).

3 A straight line is drawn through the points (7, 0) and (−2, 3). Find (a) its gradient, (b) its equation.

4 Write down the equations of the lines (a), (b), (c), (d), (e) in Fig. 7.8.

Simultaneous linear equations

Consider the equation $2x + y = 7$.

For any value of x there is a corresponding value of y. If $x = 0$, $y = 7$; if $x = 1$, $y = 5$; and so on. Table 7.6 gives some of these pairs of values.

Table 7.6 $2x + y = 7$

x	0	1	2	3	4	5	...
y	7	5	3	1	−1	−3	...

Similarly, consider the equation $x − y = 2$.

Table 7.7 sets out pairs of values for this equation in the same way.

Table 7.7 $x − y = 2$

x	0	1	2	3	4	5	...
y	−2	−1	0	1	2	3	...

The graphs of these two lines are shown in Fig. 7.12.

The two lines intersect (i.e. cross each other) at the point (3, 1). This is the only point which is on both lines. The coordinates of the point of intersection give the solution of the simultaneous equations.

In set language:
$\{(x, y): 2x + y = 7\} \cap \{(x, y): x − y = 2\} = \{(3, 1)\}$

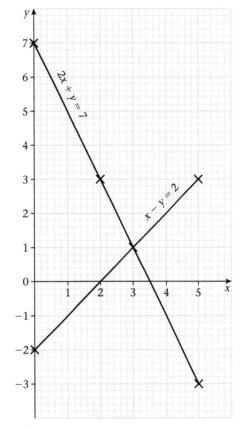

Fig. 7.12

Example 9
Solve graphically the simultaneous equations $2x − y = −1$, $x − 2y = 4$.

1st step: Make tables of values for each equation. Three pairs of values are sufficient for each:

$2x − y = −1$

x	0	1	2
y	1	3	5

$x − 2y = 4$

x	0	1	2
y	−2	−1$\frac{1}{2}$	−1

2nd step: Choose a suitable scale and plot the points. Draw both lines. Extend the lines if necessary so that they intersect. Fig. 7.13 shows the graphs of the two lines.

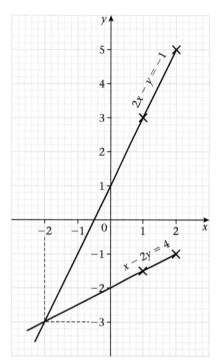

Fig. 7.13

Notice in Fig. 7.13 that it was necessary to extend both lines to find the point of intersection.

3rd step: Find the coordinates of the point of intersection. From the graphs, the lines intersect at $(-2, -3)$.

The solution of the simultaneous equations is $x = -2$ and $y = -3$.

Exercise 7d

Solve graphically the following pairs of simultaneous equations.

① $x + y = 3$
   $3x - y = 1$

② $x - y = 1$
   $x + 2y = 7$

③ $x - 2y = 1$
   $2x + y = 2$

④ $y = 2x + 2$
   $3x + 2y = 4$

⑤ $3y = 2x + 8$
   $x + y = 1$

⑥ $x + 3y = 0$
   $x - 3y = 6$

⑦ $x - y = 0$
   $3x - y + 2 = 0$

⑧ $4x - 2y = 7$
   $x + 3y = 7$

⑨ $x + y = 3$
   $5x - 5y = 1$

⑩ $2x - 2y = 5$
   $2x + 3y + 1 = 0$

Summary

If y is a function of x, then the value of y depends on the value of x.

If $y = a + bx$, where a and b are constants, then y is a **linear function** of x. The graph of a linear function is always a straight line. Hence, to draw the graph of a linear function, the coordinates of two points only are required. However, it is wise to use a third point as a check on the accuracy of the calculations.

Given that P and Q are any two points on a straight line, the gradient of the line may be found from the graph or by calculation. The equation of the straight line may be found.

(a) given the graph;
(b) given the gradient of the line and a point;
(c) given two points on the line.

To solve simultaneous linear equations graphically, the same Cartesian axes are used for both equations. The coordinates of the point of intersection of the lines give the required x and y values.

Practice exercise P7.1

① State, giving reasons, which of the graphs in Figs. 7.14–7.16 does *not* represent a function for the domain {1, 2, 3, 4}.

(a)

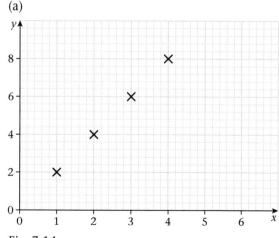

Fig. 7.14

(b)

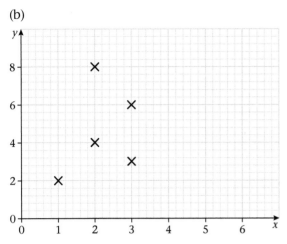

Fig. 7.15

(c)

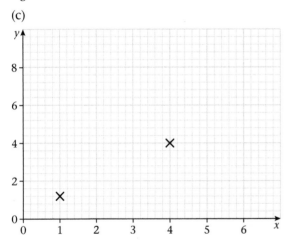

Fig. 7.16

② (a) State clearly the conditions that a relation must satisfy in order to be a function.
(b) Discuss, giving reasons, why the graph of the relation $x \rightarrow y$ shown in Fig. 7.17 would not represent a function.

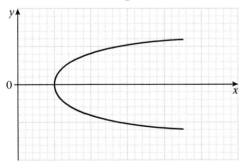

Fig. 7.17

Practice exercise P7.2

In Fig. 7.18, the graph of each linear function is numbered from **1** to **6**. For each line,

(a) write the coordinates of two points on the line;
(b) find the gradient of the line;
(c) find the *y-intercept*.
(d) write the equation of the line.

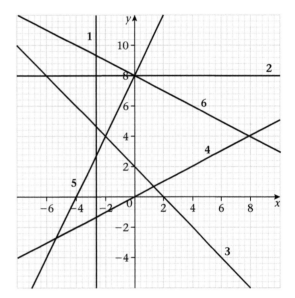

Fig. 7.18

Practice exercise P7.3

For each of the lines joining the following pairs of points,

(a) find the gradient;
(b) write the equation of the line.

① (3, 3) and (−2, 3)

② (4, −3) and (−2, −5)

③ (8, 7) and (2, 5)

④ (2, 5) and (4, −1)

⑤ (5, 3) and (1, −5)

⑥ (6, 2) and (0, 4)

⑦ (4, 0) and (−3, 2)

8 (0, 3) and (−3, 0)

9 (6, 5) and (2, 7)

10 (2, −11) and (−4, −4)

11 (−2, 0) and (4, 0)

12 (−1, 0) and (0, 3)

Practice exercise P7.4

For each of the following, write the equation of the line passing through the given point and having the given gradient, m.

1 (−1, 2), $m = 2$

2 (0, 3), $m = −4$

3 (4, 5), $m = 0$

4 (1, −6), $m = −1$

5 (−7, 5), $m = −3$

Practice exercise P7.5

1 Copy and complete Table 7.8.

Table 7.8

x	−2	0	1	2
3	3	3		
$x + 3$	1			

2 Draw a set of coordinate axes, labelling the x-axis and the y-axis from −5 to +5. Use Table 7.8 to draw the graph of $y = x + 3$.

3 Complete Table 7.9 and use it to draw the graph of $y = 2x + 1$ on the same axes used in question 2.

Table 7.9

x	−2	0	1	2
$2x$				
1				
$2x + 1$				

4 Use your graphs to solve the simultaneous equations $y = x + 3$ and $y = 2x + 1$.

Practice exercise P7.6

For each of the following functions
(a) draw up a table of values;
(b) use suitable scales to represent 1 unit on each axis;
(c) plot the points on graph paper;
(d) draw the graph of the function.

1 $y = 3x − 2$, $−4 < x < 5$

2 $y = x − 2$, $−5 < x \leq 4$

Practice exercise P7.7

1 In exchange for $Y350.00, a customer received $X70.00.
(a) Using a scale of 1 cm to represent $X10 on the x-axis and 1 cm to represent $Y50 on the y-axis, draw a graph to show the relation between $X and $Y.
(b) From your graph, find how many
(i) $Y are received for $X50;
(ii) $X are received for $Y450.

2 2 boxes of nails cost $26 and 5 boxes cost $65.
(a) Using 1 cm : 1 box of nails along the x-axis and 1 cm : $10 on the y-axis, draw a graph to represent this information.
(b) Use the graph to find
(i) the cost of 7 boxes of nails;
(ii) the number of boxes that cost $78.

3 B$10 is equivalent to F$4, and B$15 is equivalent to F$6.
(a) Using 1 cm to represent F$1 along the x-axis and 1 cm to represent B$1 on the y-axis, draw a graph to represent the data.
(b) Find the gradient of the graph.
(c) State the equation of the graph.
(d) Use the graph to find
(i) the number of F$ for B$20;
(ii) the number of B$ for F$3.

4   The monthly charges, D, for a utility are a basic rental of $30 plus 15 cents per unit. Fig. 7.19 shows the charges for 0–400 units.

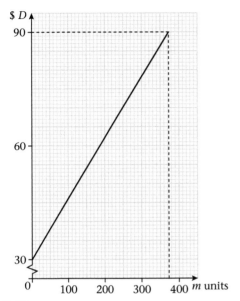

Fig. 7.19

Use the graph to find
(a) the charges for 100 units;
(b) the number of units for a charge of $82.50.

The graph shows the relation between $D and m units.
(c) Write the relation as a linear equation.

5   The information given in Table 7.10 represents a linear function.

Table 7.10

Time [t h]	1	2	3	4	5
Distance from home [d km]	142	197	252	307	362

(a) Show the data on a graph.
   [*Hint: use a broken distance-axis.*]
(b) Write down the relation between the distance, d, and the time, t.
(c) Find the time when the distance moved was 160 km.

Chapter 8

Simultaneous linear equations

Sometimes, instead of the graphical method, it may be quicker and more accurate to use an algebraic method to solve simultaneous equations.

Method of substitution

Example 1

Solve the equations $2x + y = 7$, $x - y = 2$.

Note: When asked to solve two equations with two unknowns, assume that they are simultaneous equations.

Write out the equations, one below the other. Label the equations (1) and (2).

$$2x + y = 7 \qquad (1)$$
$$x - y = 2 \qquad (2)$$

Use one of the equations to write one unknown in terms of the other. Substitute this expression in the other equation.

From equation (2),

$$x = 2 + y \qquad (3)$$

Substitute $(2 + y)$ for x in equation (1)

$$2(2 + y) + y = 7$$

Clear brackets by multiplying and collect terms

$$4 + 2y + y = 7$$
$$3y = 3$$
$$y = 1$$

Substitute the value 1 for y in equation (3)

$$x = 2 + 1 = 3$$

Thus $x = 3$ and $y = 1$.

Check: Substitute 3 for x and 1 for y in LHS of (1) and (2).

(1) $2x + y = 2 \times 3 + 1 = 6 + 1 = 7 = $ RHS
(2) $x - y = 3 - 1 = 2 = $ RHS

As a general rule, it is more useful to apply the method of **substitution** when the coefficient of one of the unknowns in the given equations is 1 (positive or negative).

Example 2

Solve the equations $3a + b = 10$, $2a + 4b = 0$.

$$3a + b = 10 \qquad (1)$$
$$2a + 4b = 0 \qquad (2)$$

from (1),

$$b = 10 - 3a \qquad (3)$$

Substitute $(10 - 3a)$ for b in (2)

$$2a + 4(10 - 3a) = 0$$

hence,

$$2a + 40 - 12a = 0$$
$$-10a = -40$$
$$a = 4$$

Substitute 4 for a in (3)

$$b = 10 - 3 \times 4 = 10 - 12$$
$$b = -2$$

Thus $a = 4$ and $b = -2$.

Check: Substitute 4 for a and -2 for b in LHS of (1) and (2).

(1) $3a + b = 3 \times 4 + (-2) = 12 - 2 = 10 = $ RHS
(2) $2a + 4b = 2 \times 4 + 4(-2) = 8 - 8 = 0 = $ RHS

Always check the accuracy of the answers by substituting the values into the original equations.

Exercise 8a

Use the method of substitution to solve the following pairs of simultaneous equations.

1. $y = x + 1$
    $x + y = 3$

2. $y = 2x - 4$
    $3x + y = 11$

3. $a = 5 - 2b$
    $5a + 2b = 1$

4. $2m + n = 0$
    $m + 2n = 3$

5. $x + y = 4$
    $2x - y = 5$

6. $y - 2x = 1$
    $3x - 4y = 1$

7. $a - 2b = 9$
    $2a + 3b = 4$

8. $3x + 2y = 10$
    $4x - y = 6$

⑨ $x + 2y = 7$
    $3x - 2y = -3$

⑩ $2a + b = 19$
    $3a - 2b = 11$

⑪ $4x - 3y = 1$
    $x - 2y = 4$

⑫ $4x = y + 7$
    $3x + 4y + 9 = 0$

Method of elimination

When none of the coefficients of the unknowns is 1, use the method of **elimination**. The aim of this method is to get rid of one of the unknowns by making its coefficient the same numerical value (negative or positive) in both equations. The equations are then added or subtracted as necessary. Read Example 3 carefully.

Example 3

Solve the equations $3x + 2y = 12$ and $5x - 3y = 1$.

$3x + 2y = 12$ (1)
$5x - 3y = 1$ (2)

The coefficients of y can be made the same if (1) is multiplied by 3 and (2) is multiplied by 2. (Remember that an equation remains the same if every term is multiplied or divided by the same number.) So

(1) ×3: $9x + 6y = 36$
(2) ×2: $10x - 6y = 2$

Adding $19x = 38$
              $x = 2$

Substitute 2 for x in (1)
              $3 \times 2 + 2y = 12$
              $2y = 12 - 6 = 6$
              $y = 3$

$x = 2$ and $y = 3$.

Check: Substitute 2 for x and 3 for y in (1) and (2).
(1) $3x + 2y = 3 \times 2 + 2 \times 3 = 6 + 6 = 12 = $ RHS
(2) $5x - 3y = 5 \times 2 - 3 \times 3 = 10 - 9 = 1 = $ RHS

In Example 3, instead of substituting 2 for x to find y, it may be simpler to start again with the original equations and eliminate x to find y.
For example:

(1) ×5: $15x + 10y = 60$
(2) ×3: $15x - 9y = 3$

Subtracting $19y = 57$
                   $y = 3$

This method can be very useful when the first value found is a fraction, since fractions often give difficult working when substituted.

Example 4

Solve the equations $3f = 4 - 4e$ and $2e = 5f + 15$.

$3f = 4 - 4e$ (1)
$2e = 5f + 15$ (2)

Arrange the equations so that unknowns are in alphabetical order on the LHS and numbers are on the RHS.

$4e + 3f = 4$ (3)
$2e - 5f = 15$ (4)

Multiply (4) by 2
$4e - 10f = 30$ (5)

hence,

$4e + 3f = 4$ (3)
$4e - 10f = 30$ (5)

Subtract (5) from (3) to eliminate terms in e
$13f = -26$
$f = -2$

Substitute -2 for f in (2)
$2e = 5(-2) + 15$
$2e = -10 + 15 = 5$
$e = 2\frac{1}{2}$

(The check is left as an exercise.)

Where necessary, arrange the given equations so that the unknowns are in alphabetical order on the LHS and the numbers are on the RHS.

Notice that to eliminate one of the unknowns:
(a) in Example 3, when the coefficients had opposite signs, the terms ($+6y$ and $-6y$) were **added**;
(b) in Example 4, when the coefficients had the same sign, the terms ($4e$) were **subtracted**.

Exercise 8b

Use the method of elimination to solve the following pairs of simultaneous equations.

① $5a - 2b = 14$
    $2a + 2b = 14$

② $4p + 3q = 9$
    $2p + 3q = 3$

③ $2x + 5y = 4$
　 $2x - 2y = 18$

④ $5x + 2y = 2$
　 $2x + 3y = -8$

⑤ $5x + 3y = 1$
　 $2x + 3y = -5$

⑥ $4x + 3y = 9$
　 $2x + 5y = 15$

⑦ $4a = 5b + 5$
　 $2a = 3b + 2$

⑧ $2x + 5y = 0$
　 $3x - 2y = 19$

⑨ $3x - 2y = 4$
　 $2x + 3y = -6$

⑩ $6h = 2k + 9$
　 $3h + 4k = 12$

⑪ $2p - 5q = 8$
　 $3p - 7q = 11$

⑫ $2r + 3s = 29$
　 $3r + 2s = 16$

⑬ $2x - 5y = -6$
　 $4x - 3y = -12$

⑭ $2x + 5y + 1 = 0$
　 $3x + 7y = 1$

⑮ $3a = 2b + 1$
　 $3b = 5a - 3$

⑯ $5v = 11 + 3u$
　 $2u + 7v = 3$

⑰ $5d = 2e - 14$
　 $5e = d + 12$

⑱ $6x - 5y = -7$
　 $3x + 4y = 16$

⑲ $3f - 4g = 1$
　 $6f - 6g = 5$

⑳ $8y + 4z = 7$
　 $6y - 8z = 41$

Word problems

Example 5

4 pens and 6 pencils cost \$1.36 altogether. 6 pens and 5 pencils cost \$1.64 altogether. Find the cost of one pen and one pencil.

Let one pen cost x cents and one pencil cost y cents.
Then 4 pens cost $4x$ cents
　　　6 pencils cost $6y$ cents
　　and　　\$1.36 = 136 cents
Thus,　　　$4x + 6y = 136$　　　　　(1)
　　　　　　　　(*1st sentence in question*)
Similarly,　$6x + 5y = 164$　　　　(2)
　　　　　　　　(*2nd sentence in question*)

(1) ×3:　$12x + 18y = 408$
(2) ×2:　$12x + 10y = 328$
Subtracting:　　$8y = 80$
　　　　　　　　$y = 10$

Substitute 10 for y in (1)
　　　　$4x + 60 = 136$
　　　　　$4x = 136 - 60 = 76$
　　　　　　$x = 19$

A pen costs 19c and a pencil costs 10c.

Check:

4 pens cost \$0.76	6 pens cost \$1.14
6 pencils cost \$0.60	5 pencils cost \$0.50
\$1.36	\$1.64

Example 6

Karen's age and Rudy's age add up to 24 years. Six years ago, Karen was three times as old as Rudy. What are their ages?

Let Karen's age be a years and Rudy's age be b years.
Then $a + b = 24$　　　　　　　　　(1)
　　　　　　　　(*1st sentence in question*)
Six years ago, Karen was $(a - 6)$ and Rudy was $(b - 6)$. Hence
$(a - 6) = 3(b - 6)$　　(*2nd sentence in question*)
Clear brackets and collect terms
　$a - 6 = 3b - 18$
$a - 3b = -12$　　　　　　　　　　(2)
Subtract (2) from (1)
　　　　　　$a + b = 24$　　　　　(1)
　　　　　　$a - 3b = -12$　　　　(2)
Subtracting:　　$4b = 36$
　　　　　　　　$b = 9$

Substitute 9 for b in (1)
　　　　　$a + 9 = 24$
　　　　　　$a = 15$

Karen is 15 and Rudy is 9.

Check: Sum of ages = $15 + 9 = 24$ years; 6 years ago, Karen was 9 and Rudy was 3; $9 = 3 \times 3$.

Note: In both of these examples, the given facts are checked and not the equations. The units of each term in an equation must be the same so that in Example 5, \$1.36 is changed to 136 cents.

Exercise 8c

① The sum of two numbers is 19. Their difference is 5. Find the numbers.

② A father is 25 years older than his son. The sum of their ages is 53 years. Find their ages.

③ The sum of two numbers is 17. The difference between twice the larger number and three times the smaller is 4. Find the numbers.

4 Victor and Terry have 60c between them. Terry has 24c more than Victor. How much does each have?

5 A newspaper and a magazine cost 55c together. The newspaper cost 35c less than the magazine. Find the cost of each.

6 A pencil and an eraser cost 27c together. 4 pencils cost the same as 5 erasers. Find the cost of each.

7 I have x 5c coins and y 10c coins. There are 8 coins altogether and their total value is 55 cents. How many of each coin do I have?

8 Carol's age and Ron's age add up to 25 years. Eight years ago, Carol was twice as old as Ron. How old are they now?

9 The sides of the rectangle in Fig. 8.1 are given in cm.

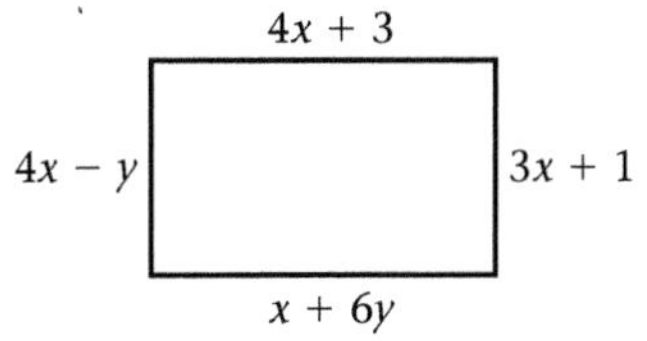

$$4x + 3$$
$$4x - y \qquad\qquad 3x + 1$$
$$x + 6y$$

Fig. 8.1

Find x and y and the area of the rectangle.

10 A boy travels for x hours at 5 km/h and for y hours as 10 km/h. He travels 35 km altogether and his average speed is 7 km/h. Find x and y.

Example 7

A two–digit number is such that the sum of its digits is 11. The number is 27 greater than the number obtained by interchanging the digits. Find the number.

Let the digits be x and y, where x is the tens digit. Then the number is $10x + y$.
From the first sentence of the question,

$$x + y = 11 \qquad\qquad (1)$$

The number obtained by interchanging the digits is $10y + x$.
Hence, from the second sentence of the question,

$$10x + y - (10y + x) = 27$$
$$10y + y - 10y - x = 27$$
$$9x - 9y = 27$$
$$x - y = 3 \qquad\qquad (2)$$
$$x + y = 11 \qquad\qquad (1)$$

Adding (1) and (2),
$$2x = 14$$
$$x = 7$$

Subtract (2) from (1)
$$2y = 8$$
$$y = 4$$

The number is 74.

Check: $7 + 4 = 11$ *(1st sentence)*
         $74 - 47 = 27$ *(2nd sentence)*

Exercise 8d

1 7 mugs and 8 plates cost $63. 8 mugs and 7 plates cost $64.50. Calculate the cost of a mug and of a plate.

2 Four knives and six forks cost $39; six knives and five forks cost $42.50. Find the cost of (a) a knife, (b) a fork.

3 Half of A's money plus one-fifth of B's make $1. Two-thirds of A's money plus two-fifths of B's make $1.50. How much money has each?

4 Divide 75 into two parts so that one part is $\frac{2}{3}$ of the other.

5 A man cycles for x hours at 12 km/h and y hours at 18 km/h. Altogether he cycles 78 km in 5 hours. Find x and y.

6 Fig. 8.2 shows an equilateral triangle with the lengths of its sides given in terms of a and b.

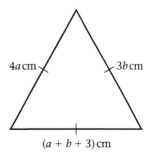

$4a$ cm $3b$ cm

Fig. 8.2 $(a + b + 3)$ cm

Find a and b and hence find the length of the sides of the triangle.

7. The perimeter of the isosceles triangle in Fig. 8.3 is 28 cm.
Find x and y and hence state the lengths of the sides of the triangle.

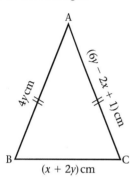

B $(x + 2y)$ cm   C

Fig. 8.3

8. In Fig. 8.3, $\widehat{A} = 2\frac{1}{4}n° + \frac{1}{2}m°$, $\widehat{B} = 2n° - \frac{2}{5}m°$, $\widehat{C} = 2m°$. Find m, n and the sizes of $\widehat{A}$ and $\widehat{B}$.

9. The difference between the digits of a two-digit number is 1. The number itself is 1 more than 5 times the sum of its digits. If the units digit is greater than the tens digit, find the number.

10. In a two-digit number, the sum of the digits is 8. The difference between this number and the number with the digits reversed is 54. What is the number?

11. In a positive number of two digits, the sum of the digits is 15. If the digits are interchanged, the number is increased by 9. Find the number.

12. A man's age and his son's add to 45 years. Five years ago the man was 6 times as old as his son. How old was the man when the son was born?

13. Charles' and Tom's ages add up to 29. Seven years ago Charles was twice as old as Tom. Find their present ages.

14. A girl travels 10 km in 50 min if she runs for 8 km and walks for 2 km. If she runs 4 km and walks 6 km, her time is 1 h 15 min. Find her running and walking speeds.

15. If 1 is added to both the numerator and denominator of a fraction, the fraction becomes equal to $\frac{1}{2}$. If 8 is added to both, the fraction becomes equal to $\frac{2}{3}$. What is the fraction?

Summary

Generally there are two non-graphical methods for solving simultaneous linear equations in two variables.

Substitution: using one of the equations, obtain an expression for one variable in terms of the other. Substitute this expression in the other equation in order to get an equation in only one variable or unknown. This equation may then be solved.

Elimination: one variable is eliminated by addition (or subtraction) of corresponding terms in the two equations leaving an equation in only one variable. This equation may then be solved.

In both methods, substitution of the first value found in either of the original equations gives the value of the second variable.

Practice exercise P8.1

(a) Select the method of solving the equations.
(b) State the reason for your choice.
(c) Solve the equations.

1. $x + y = 6$
   $2x - 3y = 2$

2. $3x - y = 4$
   $x - y = 2$

3. $3x + 2y = 1$
   $2x + 3y = -1$

4. $x + 2y = -4$
   $3x - 2y = 12$

5. $3x - y + 5 = 0$
   $x - 2y = 0$

6. $\frac{2x}{3} + y - 1 = x + \frac{y}{3} = 4$

7. $3x - y = x + y = 3$

8    $\dfrac{x + y}{3} = \dfrac{2y - 1}{2} = 2x - y$

9    $3(x - 2y) = 2(x + 1)$
     $2x - y = 2(2 + y) + 3$

Practice exercise P8.2

Solve the following pairs of simultaneous equations.

1    $x + 3y = -4$
     $x - y = 4$

2    $2x - y = 4$
     $3x + y = 11$

3    $4x + 3y - 2 = 0$
     $x - 2y - 5 = 1$

4    $5x = y + 3$
     $3x - 2 = y$

5    $\frac{1}{2}x - \frac{1}{3}y = \frac{2}{3}$
     $\dfrac{3x}{2} + \dfrac{5y}{3} = \dfrac{2}{3}$

6    $5(x + y) = 2(x + 3y) + 1$
     $3(x + 2y) - 7 = x + 3y + 1$

7    $\frac{3}{4}(4x + y) = 4$
     $2x = \frac{3}{5}(2x - 3y)$

8    $3x - y + 10 = 5x - 2y + 12 = x + y$

9    $\dfrac{x - 3}{2} = \dfrac{3y + 1}{6}$
     $\dfrac{x - 4}{3} = \frac{1}{6} - \dfrac{1 - y}{2}$

Practice exercise P8.3

1   If 2 is added to the numerator and to the denominator of a fraction, the new fraction becomes equal to $\frac{5}{9}$. If 3 is subtracted from the numerator and from the denominator of the fraction, the new fraction becomes equal to $\frac{1}{2}$. Find the fraction.

2   In a test, Dani scored 17 more marks than Fred. If Dani had scored twice as many marks, she would have scored 7 more marks than three times Fred's score. Find the marks each student scored.

3
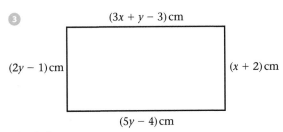

Fig. 8.4

Figure 8.4 shows the dimensions of a rectangle in centimetres.
(a) Write two equations in x and y.
(b) Solve the equations for x and y.
(c) Find the dimensions of the rectangle.

4   The sum of the digits of a two-digit number is 12. When the digits are interchanged, the number increases by 18. Find the original number.

5   The volume of water held in 3 barrels and 5 tanks is 43.6 kl, and the volume held in 2 barrels and 3 tanks is 26.4 kl. Find the volume held in one barrel and in one tank.

6   The weight of 6 televisions and 2 radios is 53 kg; the weight of 3 televisions and 4 radios is 34 kg. Find the average weight of one television and one radio.

7   The total daily wages bill for 3 unskilled and 5 skilled workmen was $780, and for 5 unskilled and 3 skilled workmen was $660. Assuming a 6-hour day, find the hourly rates paid to an unskilled and to a skilled workman.

8   The maximum number of cars that can be parked at a terminus is 60. When a bus parks at the terminus, it uses the same area as 3 cars. The daily rate for parking a bus is $100 and for parking a car is $50. Find the number of buses and cars parked on a day when the terminus was full and the amount of money received was $2600.

9   When a student runs for 1 hour at u km/h and $\frac{1}{2}$ hour at v km/h, he travels a total distance of 12 km. When he runs for $\frac{1}{2}$ hour at u km/h and 1 hour at v km/h, he travels a total distance of 14 km. Find u and v.

⑩ A father is 28 years older than his son. In 4 years he will be twice as old as his son. Find their present ages.

⑪ Find two numbers that have a sum of 98 and a difference of 16.

⑫ Andy travels a total distance of 56 km in 5 hours, moving at 10 km/h for s hours and at 12 km/h for t hours. Find the values of s and t.

⑬ Half of Rob's money plus a third of Dan's is $32.50. Two-fifths of Dan's money plus a quarter of Rob's money is $21.50. Calculate the amount of money that Rob and Dan each has.

⑭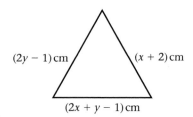

Fig. 8.5

Figure 8.5 shows an equilateral triangle.
(a) Write two equations in x and y.
(b) Solve the equations for x and y.
(c) Find the perimeter of the triangle.

⑮ Ann has $3.20 in 10¢-coins and 25¢-coins. If she has twenty coins altogether, find how many of each type of coin she has.
John also has twenty coins, but has 45¢ more than Ann. Find how many of each type of coin he has.

Revision exercise 1 (Chapters 1, 4)

① Each angle of a regular polygon is 162°. How many sides does the polygon have?

② In △ABC, BÂC = 68° and AB̂C = 30°. BC is produced to X. The bisectors of AB̂C and AĈX meet at P. Calculate BĈP and BP̂C.

③ ABCD is a quadrilateral in which AB = CD and AD = BC. Show that AD is parallel to BC. [Hint: Join AC or BD.]

④ In Fig. R1, AB̂P = 110° and DĈP = 163°. Calculate BP̂C.

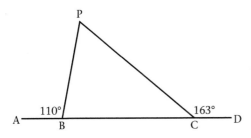

Fig. R1

⑤ The angles of an isosceles triangle are x°, y°, 73°. Find three possible values of x and the corresponding values of y.

⑥ In Fig. R2, O is the centre of the circle. Find the value of x.

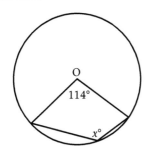

Fig. R2

⑦ In Fig. R3, AB is a diameter of the circle. Calculate the sizes of the following angles.
(a) AB̂P   (b) AB̂Z   (c) AŶZ
(d) BŶZ   (e) AB̂Y   (f) YẐB

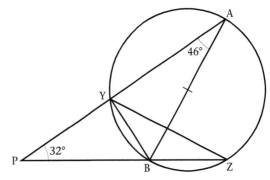

Fig. R3

⑧ In Fig. R4, AB is a diameter and AĈD = 27°. Find (a) AB̂D, (b) BÂD.

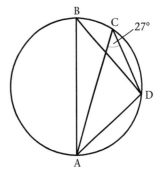

Fig. R4

⑨ L, M, N, P are points on the circumference of a circle, centre O. LN and PM intersect at X. NL̂M = 35° and LX̂P = 98°. Calculate
(a) LN̂P,
(b) LÔP.

10 In Fig. R5, show that $R\widehat{P}Q + P\widehat{Q}S = 180°$. What can be said about the lines RP and SQ?

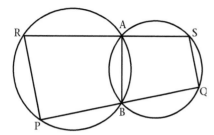

Fig. R5

Revision test 1 (Chapters 1, 4)

1 The angles of a pentagon are $x°$, $2x°$, $(x + 30)°$, $(x - 10)°$ and $(x + 40)°$. $x =$

A 50   B 60   C 80   D 108

2 ABCD is a quadrilateral such that $A\widehat{B}C = A\widehat{D}C = 90°$ and AD = BC. Which one of the following angles is equal to $B\widehat{A}C$?

A $A\widehat{C}B$ B $A\widehat{C}D$ C $A\widehat{D}B$ D $C\widehat{A}D$

3 In Fig. R6, O is the centre of the circle and $A\widehat{B}C = 140°$. Find y.

A 20°   B 70°   C 240°   D 280°

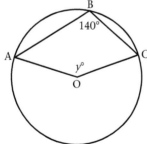

Fig. R6

4 The sum of the angles of an n-sided convex polygon is double the sum of the exterior angles. Calculate n.

A 4   B 5   C 6   D 7

5 In Fig. R7, O is the centre of the circle, $W\widehat{X}Y = 80°$ and $W\widehat{X}Z = 45°$. Calculate $Y\widehat{W}Z$.

A 35°   B 45°   C 55°   D 80°

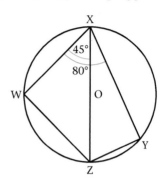

Fig. R7

6 In a quadrilateral WXYZ, XZ bisects $W\widehat{X}Y$ and $W\widehat{Z}Y$. Prove that $X\widehat{W}Z = X\widehat{Y}Z$.

7 In Fig. R8, PQ = PR = PS, and SP∥RQ. If $R\widehat{P}Q = 38°$, calculate a and b.

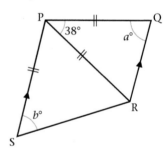

Fig. R8

8 In a quadrilateral ABCD, the lines bisecting $A\widehat{B}C$ and $B\widehat{C}D$ meet at P. Prove that $B\widehat{A}D + A\widehat{D}C = 2B\widehat{P}C$.

9 In Fig. R9, obtain an equation in x. Hence find the value of x.

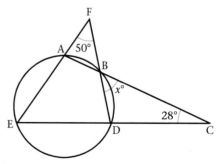

Fig. R9

10 W, X, Y and Z are points on the circumference of a circle such that WZ is a diameter, $X\widehat{W}Y = 36°$ and $Y\widehat{W}Z = 29°$. Calculate
(a) $X\widehat{Y}Z$,   (b) $W\widehat{X}Y$.

Revision exercise 2 (Chapters 2, 3)

1 If $y = 3x - 7$, find y when x is (a) 5, (b) −5.

2 Expand the brackets in
$(x + 5)(x - 3) = x(x - 2)$
and hence solve the equation.

3 Simplify the following.
(a) $\dfrac{2x - 1}{x} - \dfrac{3a \times 5}{a} + 6$

(b) $\dfrac{3t + 1}{10t} + \dfrac{4}{5t} + \dfrac{2t - 1}{3t}$

(c) $3d(1 - 5d) + (2d - 1)(4 - d)$

4 If $c * d = 2c - \dfrac{d}{2}$
calculate the value of
(a) $5 * 4$ (b) $6 * (5 * 4)$

5 Factorise the following, collecting like terms if necessary.
(a) $14a + 21$
(b) $2x + 9z + 7x$
(c) $20 + 7m - n + 5n + m - 12$
(d) $15x^2 + 24x + 9xy$
(e) $35ab^2 - 25a^2b$

6 Solve the following equations.
(a) $3x + 5 = \dfrac{7x + 1}{2}$

(b) $\dfrac{5}{x + 3} - \dfrac{2}{x - 1} = 0$

7 If $A = 2d^2 - 42$, copy and complete Table R1.

Table R1 $A = 2d^2 - 42$

d	3	4	5	6	7	8
d^2						
$2d^2$						
-42						
A						

8 Factorise the following by grouping in pairs.
(a) $(an + am) - (3m + 3n)$
(b) $a^2 - 7a + 3a - 21$
(c) $3xy - 6xz - 5ay + 10az$

9 The formula for converting a temperature in degrees Fahrenheit (F) to degrees Celsius (C) is $C = \frac{5}{9}(F - 32)$.
(a) Find C when $F = 149$.
(b) Find F when C is (i) 35, (ii) −5.

10 Two trains started a journey at the same time from two places 300 km apart. They travelled towards each other, one train travelling twice as fast as the other. They met $1\frac{1}{2}$ hours after they started. Find the speeds of the trains. (*Hint*: let the speed of the slower train be v km/h.)

Revision test 2 (Chapters 2, 3)

1 If $a = 2b - c$, find the value of b when $a = 11$ and $c = -3$.
A 4   B 7   C 8   D 14

2 Given that n is a positive number, write, in a symbolic statement, 'the square of a positive number decreased by 10 equals 6'.
A $10 - 2n = 6$ B $10 - n^2 = 6$
C $n^2 - 10 = 6$ D $2n - 10 = 6$

3 Given that $p * q = \dfrac{\frac{1}{3}p + q}{2(p - q)}$, $3 * 1 =$
A $\frac{1}{3}$ B $\frac{2}{5}$ C $\frac{1}{2}$ D $\frac{4}{5}$

4 One of the factors of $(mn - n^2) + (mq - nq)$ is $(m - n)$. The other factor is
A $(n + q)$ B $(q - n)$
C $(n - q)$ D $(m - q)$

5 Simplify $\dfrac{3}{2xyz} - \dfrac{7}{3xyz}$.
A $\dfrac{-5}{6xyz}$ B $\dfrac{-4}{6xyz}$
C $\dfrac{-3}{xyz}$ D $\dfrac{4}{xyz}$

6 (a) If $y = 3x^2$, find (i) y when $x = 2$, (ii) x when $y = 147$.
(b) If $\dfrac{1}{u} + \dfrac{1}{v} = \dfrac{1}{f}$, find u if $v = 6$ and $f = 2$.

7 Given that $m \square n = m^2 + 3n$, calculate the value of n, if $3 \square n = 42$.

8 A worker spent $\frac{1}{4}$ of his salary on food and gave $\frac{1}{10}$ to charity. He had $1625 remaining. If his salary is s, calculate the value of s.

9 Factorise the following, simplifying brackets where necessary.

(a) $52x - 8x^2$

(b) $5a^2 + a(2b - 3a)$

(c) $(x^2 + 5x) - (9x + 45)$

(d) $6ax - 2by + 4ay - 3bx$

10 The cost, c dollars, of having a car repaired is given by the formula $c = 76 + 16t$ where t is the number of hours the work takes.

(a) Find the cost when the work takes $2\frac{1}{2}$ hours.

(b) If the cost of repairs is $316, how long did the work take?

Revision exercise 3 (Chapters 5, 6)

1 If $U = \{x: 1 \le x \le 20, x \text{ is an integer}\}$, $P = \{x: x \text{ is a factor of } 36\}$ and $Q = \{x: x \text{ is an odd number}\}$,

(a) list the members of set P,

(b) list the members of set Q′,

(c) find n(P′),

(d) find n(P ∪ Q).

2 For each of the following, make a copy of Fig. R10 and shade the given set.

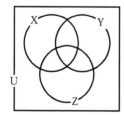

Fig. R10

(a) X ∪ (Y ∩ Z′)   (b) (X ∪ Y) ∩ Z′

(c) (X ∩ Y ∩ Z)′ ∩ (X ∪ Y)

(d) (X ∪ Y′) ∩ (Y ∪ Z′)

3 A woman invests $5000 in government bonds for 5 years at 8% simple interest. Calculate

(a) the total interest she receives for the 5 years

(b) the sum that must be invested in bonds to obtain a total interest of $3600 in 5 years.

[CXC June 1989]   (7 marks)

4 Using the rental charge and the rates for telephone calls given in Fig. 5.9, calculate

(a) the total charges if the amount due for long distance and international calls was $2464.50 and there were 324 local calls totalling 1215 mins;

(b) the total amount due to be paid.

5 Electricity bills are calculated at the following rates

15 cents per kWh for the first 200 units used

20 cents per kWh for the remaining units.

There is also a monthly rental charge of $5.00 and a fuel charge of 3.45 cents per unit used. Calculate

(a) the charge for 1208 units of electricity;

(b) the total amount due including the rental and fuel charges.

6 The numbers of elements in each region of the Venn diagram of Fig. R11 are as shown.

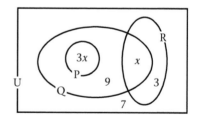

Fig. R1

If n(U) = 43, find

(a) x,   (b) n(Q),

(c) n(R′),   (d) n(P′).

7 In a class of 35 girls, 16 girls wear eyeglasses, and 24 girls ride bicycles to school. René is the only girl who neither wears eyeglasses nor rides a bicycle. Draw a Venn diagram to show this information.

8 A worker had $12 000 to invest.
   (a) Calculate the amount he receives after 3 years at
       (i) 4.5% per annum simple interest
       (ii) 4% per annum when interest is compounded annually.
   (b) State which investment earns more money.
   (c) Calculate the difference in the amounts received.

9 A woman bought a stove for $2800. After using it for 2 years she decided to trade in the stove. The company estimated a depreciation of 15% for the first year of its use and a further 15% on its reduced value, for the second year.
   (a) Calculate the value of the stove after the two years.
   (b) Express the value of the stove after two years as a percentage of the original value.
   [CXC June 88]   (6 marks)

10 100 people in a survey drink at least one of the following every day: tea, coffee, water. Two people drink coffee only, 17 people drink tea and coffee and 23 people drink coffee and water. If 71 people do not drink coffee, how many drink all three?

Revision test 3 (Chapters 5, 6)

1

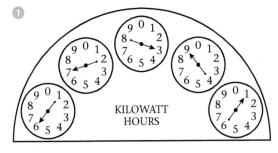

Fig. R12

   The reading on the meter in Fig. R1 is

   A   78 401   B   67 391
   C   10 487   D   09 376

2 A household uses an average of 25 gals of water each day. If 219.9 gals ≈ 1 cu metre, the approximate number of cubic metres used per year is

   A   4.2   B   42
   C   420   D   4200

3 Two members of a co-operative society had shares of $12 000 and £7000. The society paid dividends at the rate of 4.0%. The difference in the amounts they received was

   A $760   B $480
   C $280   D $200

Fig. R13 is a Venn diagram showing the elements p, q, r, s, …, z arranged within sets H, J, K, U. Use Fig. R13 to answer questions 4 and 5.

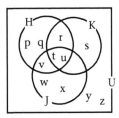

Fig. R13

4 What is n(H ∪ K)'?

   A   2   B   3   C   4   D   7

5 Which one of the following gives the members of the set H' ∩ K ∩ J?

   A   ∅   B   {s}   C   {t, u}   D   {w, x}

6 A lorry costs $60 000 when new.
   (a) The lorry decreases in value by 20% during the first year. (i) Calculate the value of the lorry at the end of the first year. In subsequent years, the value of the lorry decreases by 25% of its value at the beginning of each year. Calculate the value of the lorry at the end of (ii) the second year and (iii) the third year.
   (b) A contractor buys a new lorry on a hire purchase agreement. He pays $22 400 deposit and $1600 every month for 3 years. (i) How much interest does he pay? (ii) Express the interest as a percentage of the original cost of the lorry.

7 At a certain hotel, the cost of an overseas telephone call to Dominica is made up of time charges, a Government tax and a hotel service charge. These are calculated as follows:

Time charges

Minimum 3 minutes $4.50
Each additional minute
or part of a minute beyond
3 minutes $1.50

Government tax   50% of time charges

The hotel service charge is 10% of the sum of the time charges and Government tax.

Calculate the cost of an overseas call to Dominica lasting $18\frac{1}{2}$ minutes.

[CXC Jan 90]   (5 marks)

8 Make a copy of the Venn diagram in Fig. R14.

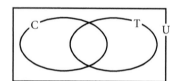

Fig. R14

(a) On your diagram, enter the members of the sets given that
U = {c, o, m, p, u, t, e, r}
T = {t, e, r, m} and C = {c, r, o, p}
(b) List the elements of the following sets.
(i) T′, (ii) C′, (iii) (C ∪ T)′, (iv) C ∪ T′.

9 U = {all teenagers}, B = {all girl teenagers}, C = {all teenagers who wear glasses}.
(a) Draw a Venn diagram to show the above information.
(b) Mark a point X to show a boy teenager who does not wear glasses.

10 A 192-page geography book contains writing, pictures and maps. 40 pages have writing only, 5 pages have pictures only and 12 pages have maps only. n pages have pictures and maps only, $3n$ pages have writing and maps only and $5n$ pages have writing and pictures only.

If 18 pages contain all three, find n and hence find the total number of pages in the book which have maps on them.

Revision exercise 4 (Chapters 7, 8)

1 Table R2 gives corresponding values of x and y for the equation $y = 2x - 5$.

Table R2 $y = 2x - 5$

x	−1	2	4
y	−7	−1	3

(a) Using a scale of 2 cm to 1 unit, draw the graph of $y = 2x - 5$.
(b) Complete the set of ordered pairs (0,), (1,), (3,)
(c) Write down the coordinates of the points where the line crosses the axes.

2 Find the gradients of the lines joining the following pairs of points.
(a) (3, 5), (7, 9)
(b) (0, −2), (3, −8)
(c) (−2, −3), (−3, −5)
(d) (−5, 2), (3, 6)

3 (a) Using a suitable scale, draw the graph of the equation $y = 3x - 4$ for values of x from −1 to 3.
(b) On the same axes draw the straight line joining the points (0, 1) and (2, 7).
(c) State the gradient of both lines.
(d) What can you say about the two lines?

4 (a) Using any convenient scale and on the same axes, draw the graphs of
(i) $y = x + 3$,   (ii) $2y = 5x$,
(iii) $y = 3x - 1$
(b) What do you notice?
(c) Write two pairs of simultaneous equations that could be solved using these graphs.

5 First simplify, then solve, the following simultaneous equations.
$$3(2x - y) = x + y + 5$$
$$5(3x - 2y) = 2(x - y) + 1$$

6 Solve the following pairs of simultaneous equations.
(a) $8y + 4z = 7$, $6y - 8z = 41$
(b) $2y = 3x + 2$, $9x + 8y = 1$

7 Solve the following pairs of simultaneous equations.
   (a) $y = \frac{1}{2}x$, $x + y = -6$
   (b) $\frac{p}{3} + \frac{q}{5} = 4$, $\frac{p}{2} + \frac{3q}{2} + 12 = 0$

8 Find the equation of the line which passes through the points
   (a) P(2, −3) and Q(10, −7)
   (b) A(−7, 3) and B(−3, 11)
   (c) P(−2, −1) and Q(−3, 3)

9 David bought 5 kg of beans and 4 kg of rice for $12.30. At the same market Carol bought 4 kg of beans and 2 kg of rice for $8.40. Calculate the price of 1 kg of rice.

10 The cost of a table and four chairs is $292. The cost of two tables and five chairs is $482. Using x to represent the cost, in dollars, of a table and y to represent the cost of a chair,
   (a) write *two* algebraic equations to represent the information above;
   (b) solve the equations and hence determine the cost of a table *and* the cost of a chair.
   [CXC June 92]   (7 marks)

Revision test 4 (Chapters 7, 8)

1 If $x = 2y - 1$ and $2x = 3y + 2$, then $x =$
   A −9   B −4   C 4   D 7

2 2 pencils and 1 eraser cost $1.65
   2 pencils and 2 erasers cost $2.10
   The cost of 1 pencil is
   A   45 cents
   B   60 cents
   C $1.05
   D $1.20

3 Given that $3x + 7y = 1$ and $x - 7y = 19$, then $x + y =$
   A −2   B −3   C 3   D 7

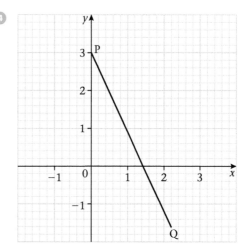

Fig. R15

The gradient of the line PQ in Fig. R15 is
A $\frac{1}{2}$ B $-\frac{1}{2}$ C 2   D −2

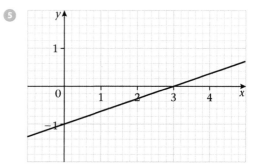

Fig. R16

The y-intercept in Fig. R16 is
A −1   B 0   C 1   D 3

6 Solve the following pairs of simultaneous equations.
   (a) $3x + 2y = 12$, $5x - 3y = 1$
   (b) $2a - 3b = 7$, $4a + 5b = 3$

7 In △ABC, AB = x cm and BC = y cm. AC is 5 cm longer than BC and twice the length of AB. If the perimeter of the triangle is 25 cm, calculate the lengths of the sides of the triangle.

8 Find the calculation (i) the gradient, (ii) the x-intercept and (iii) the y-intercept of each of the following:

(a) $y = 2x - 1$ (b) $y = 3x + 5$
(c) $y + 6 = -2x + 9$ (d) $y + 3x = 0$

9 Solve graphically the pair of simultaneous equations
$2y - x = 8$
$2y = 4x - 1$

10 Given that $2y = 6 - 2x = 6x - 2 = 7 - 3x$, write down any one pair of simultaneous equations and solve graphically or otherwise.

General revision test A
(Chapters 1–8)

1 When $y = -2$, $3y^2 - 5y - 6 =$
A   −8   B   −4   C   8   D   16

2 If $\frac{x + 8}{x} = 5$, $x =$
A   −3   B   −2   C   2   D   3

3 If $x + 3y = 7$ and $x - 3y = 7$, then $y =$
A $-4\frac{2}{3}$ B   0   C $-2\frac{1}{3}$ D $2\frac{1}{3}$

4 If $a \square b = \dfrac{2}{a} - \dfrac{3}{b}$,
then $6 \square 6 =$
A $\frac{5}{6}$ B $-\frac{1}{6}$ C   0   D $\frac{1}{6}$

5 One factor of $(6ax - 10ay) + (3x - 5y)$ is $3x - 5y$. The other factor is
A $2a + 1$ B $2a - 1$
C $2a$ D $6ax - 10ay$

6
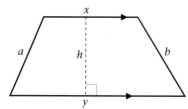

Fig. R17

The area of the trapezium in Fig. R17 is given by the formula
A $\frac{1}{2}(a + b)h$ B $\frac{1}{2}(x + y)h$
C $\frac{1}{2}(a + b)x$ D $\frac{1}{2}(a + b)y$

7 Two successive readings of a water meter are 03 251 and 03 262. If the rate is $12 per unit, the amount due for water charges is
A $781   B $391
C $390   D $132

8 The sum of the interior angles of a polygon is 1080°. How many sides has the polygon?
A   6   B   8   C   10   D   12

9 In Fig. R18, O is the centre of the circle. Find x.
A   75°   B   105°   C   150°   D   190°

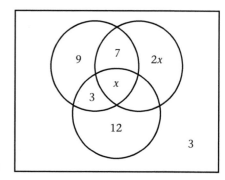

Fig. R18

10

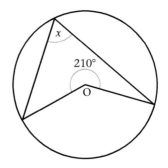

Fig. R19

If $n(U) = 85$ in the Venn diagram, Fig. R19, then $x =$
A   17   B   18   C   51   D   54

11 Factorise the following
(a) $3x^2 - 9y$
(b) $2mn + 4m^2 - 6n^2$
(c) $(3x - xy) + (6 - 2y)$

12 Simplify the following
(a) $(x + 3)(x - 2) - x(x - 1)$
(b) $(a - b)^2 - (a^2 - b^2)$

13 Solve the following equations.

(a) $5x - 6(2 - x) = 32$

(b) $1\frac{1}{3} + 4\frac{1}{3} = \frac{5}{6}y + \frac{2}{3}$

(c) $\frac{x}{3} - \frac{2x - 1}{5} = \frac{1}{3}$

(d) $\frac{3x - 7}{5} - \frac{5x - 1}{3} = 4$

14 Evaluate the following when $x = -3$, $y = 4$ and $z = -2$

(a) $2y + x - 5z$

(b) $2x^2 - 4y$

(b) $x^3 - z^2y$

(d) $\frac{x + y}{y + z}$

15 Simplify the following

(a) $\dfrac{(3 - x)(4 - x) + (1 - x)(6 + x)}{3 - 2x}$

(b) $\dfrac{22ab^3}{3apq} \div \dfrac{33ab^2p^2}{4bp^2}$

(c) $\dfrac{x + 2}{2} + \dfrac{x - 7}{7}$

(d) $\dfrac{3}{x + 2} - \dfrac{4}{2x + 1}$

16 Machinery costing $25 000 depreciates to a value of $23 125 at the end of the first year. Thereafter, depreciation is calculated at the rate of 10% per annum. Calculate

(a) the percentage depreciation for the first year

(b) the value of the machinery after two years

(c) the average rate of depreciation over the first three years as a percentage of $25 000. Give your answer to 1 decimal place.

[CXC June 90] (10 marks)

17 In Fig. R20, AB is a diameter of the circle. Find the value of x, giving a reason for each step in your answer.

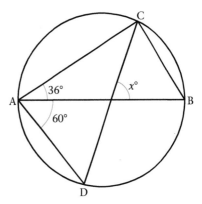

Fig. R20

18 In a test $3n$ students got less than 8 marks, n students got more than 6 marks and 5 students got exactly 7 marks. Show this information on a Venn diagram. Hence find n, given that 39 students took the test.

19 Using ruler and compasses only, construct

(a) triangle ABC with AB = 6.5 cm, ABC = 60°, BC = 8 cm;

(b) the line AD perpendicular to BC to meet BC at D;

(c) a circle on AB as diameter.

Why must the circle pass through D?

20 Use a graphical method to solve the following simultaneous equations.

$y = 3 - x$

$2x + y = 1$

Chapter 9

Non-linear functions

Linear functions (revision)

Example 1

Draw a graph of $y = 3 - 2x$ for values of x from -2 to $+4$. Use the graph (a) to find y when $x = -1.4$, (b) to solve the equation $3 - 2x = -4$.

When $x = -2$, $y = 3 - 2(-2) = 3 + 4 = +7$ and so on. Table 9.1 gives the values of y for values of x from -2 to $+4$.

Table 9.1 $y = 3 - 2x$

x	-2	-1	0	$+1$	$+2$	$+3$	$+4$
y	$+7$	$+5$	$+3$	$+1$	-1	-3	-5

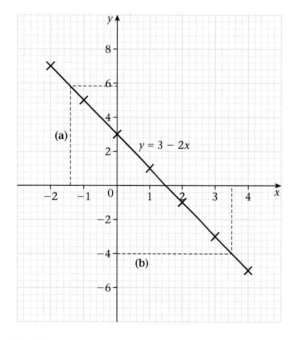

Fig. 9.1

Fig. 9.1 is the graph of the equation $y = 3 - 2x$.
(a) From the graph, $y = 5.8$ when $x = -1.4$.
(b) Since $y = 3 - 2x$, to solve the equation $3 - 2x = -4$ means to find the value of x when $y = -4$. From the graph, $x = 3.5$.

In the equation $y = 3 - 2x$, y is a **function** of x, i.e. the value of y depends on the value of x. $3 - 2x$ is the function of x. The line in Fig. 9.1 is the graph of the equation $y = 3 - 2x$, *or* the **graph of the function** $3 - 2x$. A **linear function** of x is one which contains terms in x of power 1 only. The graph of a linear function is always a straight line.

When plotting the graph of a linear function, two points are sufficient to determine the line. However, in practice it is advisable to plot *three* points. If the three points lie in a straight line it is likely that the working is correct.

Non-linear functions

Example 2

Draw the graph of $y = x^2 + 2x - 3$ for values of x from -4 to $+2$. Use the graph to find (a) the value of y when $x = 1.5$, (b) the values of x when $y = 1$.

Table 9.2 $y = x^2 + 2x - 3$

x	-4	-3	-2	-1	0	1	2
x^2	16	9	4	1	0	1	4
$+2x$	-8	-6	-4	-2	0	2	4
-3	-3	-3	-3	-3	-3	-3	-3
y	5	0	-3	-4	-3	0	5

Fig. 9.2 is the graph of $y - x^2 + 2x - 3$.

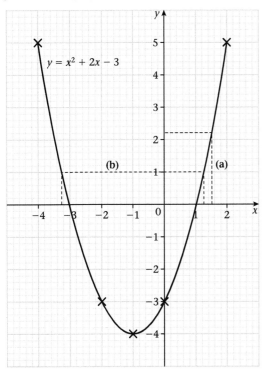

Fig. 9.2

From the graph

(a) $y = 2.25$ when $x = 1.5$,

(b) $x = 3.25$ or 1.25 when $y = 1$.

Note that the graph of this function is a curved line.

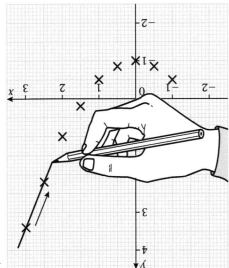

Fig. 9.3N

It takes a lot of practice to be able to draw a smooth curve. It helps to position your hand inside the curve, drawing from left to right as in Fig. 9.3. In many cases this will mean turning the graph paper round.

Example 3

Table 9.3 is a table of values for $y = x^2 + 3x - 1$.

Table 9.3 $y = x^2 + 3x - 1$

x	−4	−3	−2	−1	0	1
y	3	−1	−3	−3	−1	3

(a) Use a scale of 2 cm to 1 unit on both axes to draw the graph of the function
$y = x^2 + 3x - 1$.

Use the graph to find

(b) the values of x when $y = 0$,

(c) the values of x when $y = 3$,

(d) the values of x when $y = -3.5$.

(a) Fig. 9.4 is the graph of $y = x^2 + 3x - 1$.

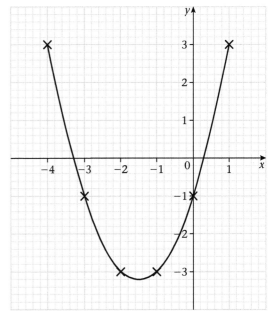

Fig. 9.4

(b) when $y = 0$, $x = -3.3$ and 0.3.

(c) when $y = 3$, $x = -4$ and 1.

(d) when $x = -3.5$, $y = 0.7$.

Example 4

Using the graph in Fig. 9.4, find the range of values for which $x^2 + 3x - 1$ has negative values.

Required range $-3.3 < x < 0.3$.

Example 5

(a) Complete the table of values shown in Table 9.4 for the function $y = x^2 - 3x + 2$.

Table 9.4 $y = x^2 - 3x + 2$

x	-1	0	1	2	3	4
y	6			0		6

(b) Using a scale of 2 cm to 1 unit, draw on graph paper the graph of the function $y = x^2 - 3x + 2$.

(c) Using your graph, determine
   (i)   the solution of $x^2 - 3x + 2 = 0$,
   (ii) the minimum value of the function $y = x^2 - 3x + 2$.

(a)

x	-1	0	1	2	3	4
y	6	2	0	0	2	6

(b) See Fig. 9.5 for the graph of the function.

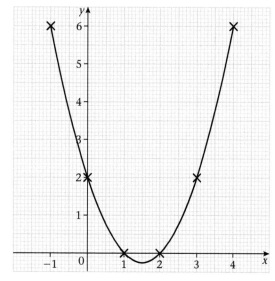

Fig. 9.5

(c) (i) The solutions are determined by the values of x when $y = 0$: $x = 1, 2$.
    (ii) The minimum value of $y = x^2 - 3x + 2$ is $y = -\frac{1}{4}$, at a value of $x = 1\frac{1}{2}$.

Note that the minimum value of the function is also the least value of the function.

Example 6

Fig. 9.6 shows the graph of the function $y = 3 + 2x - x^2$ for $-2 \leqslant x \leqslant 4$.

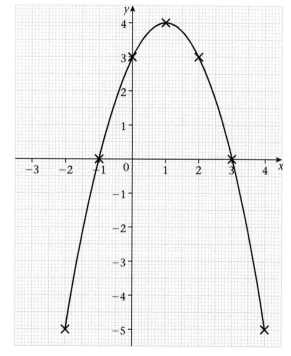

Fig. 9.6

Using the graph determine
(a) the solution of $3 + 2x - x^2 = 0$,
(b) the maximum value of the function $y = 3 + 2x - x^2$.

(a) Solutions are $x = -1, 3$
(b) Maximum value of the function is 4.

Exercise 9a

1 Fig. 9.7 is the graph of $y = x^2 - 5x + 8$.

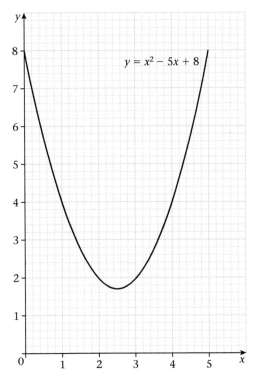

Fig. 9.7

(a) Find the value of y when $x = 4.5$.
(b) Find the values of x when $y = 4.6$.

2 Table 9.5 gives the table of values for $y = x^2 - 3x - 2$.

Table 9.5 $y = x^2 - 3x - 2$

x	−2	−1	0	1	2	3	4
x^2	2	1	0	1	4	9	16
$-3x$	6	3	0	−3	−6	−9	−12
-2	−2	−2	−2	−2	−2	−2	−2
y	8	2	−2	−4	−4	−2	2

(a) Use a scale of 1 cm to 1 unit on both axes to draw the graph of $y = x^2 - 3x - 2$.

Use your graph to find
(b) the value of y when $x = 1.6$,
(c) the values of x when $y = 3$.

3 Table 9.6 gives the table of values for $y = 3 + 2x - x^2$.

Table 9.6 $y = 3 + 2x - x^2$

x	−2	−1	0	1	2	3	4
3	3	3	3	3	3	3	3
$+2x$	−4	−2	0	2	4	6	8
$-x^2$	−4	−1	0	−1	−4	−9	−16
y	−5	0	3	4	3	0	−5

(a) Use a scale of 2 cm to 1 unit to draw the graph of $y = 3 + 2x - x^2$.

Use your graph to
(b) state the range of values of x for which $3 + 2x - x^2$ is positive,
(c) find the value of y when $x = -0.6$,
(d) find the values of x when $y = 2$.

4 Table 9.7 gives the table of values for $y = x^2 - 3x$.

Table 9.7 $y = x^2 - 3x$

x	−1	0	1	2	3	4
x^2	1	0	1	4	9	16
$-3x$	3	0	−3	−6	−9	−12
y	4	0	−2	−2	0	4

(a) Using a scale of 2 cm to 1 unit on both axes, draw the graph of $y = x^2 - 3x$.

Use your graph to
(b) find the values of x when $y = 0$,
(c) find the value of y when $x = 2.5$,
(d) find the values of x when $y = 3$,
(b) state the range of values for which $x^2 - 3x$ is negative.

5 Table 9.8 is a table of values for $y = x - x^2$ from $x = -2$ to $x = 3$.

Table 9.8 $y = x - x^2$

x	−2	−1	0	1	2	3
y	−6	−2	0	0	−2	−6

Choose a suitable scale to draw the graph of $y = x - x^2$. Use the graph to (a) state the range of values of x for which y is positive, (b) find the value of x for which $x - x^2$ is a maximum.

6 (a) Given that $y = x^2 - 2x - 1$, copy and complete Table 9.9.

Table 9.9 $y = x^2 - 2x - 1$

x	-2	-1	0	1	2	3	4
x^2	4	1	0	1	4		
$-2x$	4	2	0	-2	-4		
-1	-1	-1	-1	-1	-1	-1	-1
y	-7	2		-2			

(b) Use a scale of 2 cm to 1 unit on both axes to draw the graph of $y = x^2 - 2x - 1$.
Use your graph to find
(c) the value of y when $x = 1.3$,
(d) the values of x when $y = 6$,
(e) the minimum value of y.

7 (a) Given that $y = 4 + 3x - x^2$, copy and complete Table 9.10.

Table 9.10 $y = 4 + 3x - x^2$

x	-1	0	1	2	3	4
4	4	4	4	4	4	4
$+3x$	-3		3		9	
$-x^2$	-1		-1		-9	
y	0		6		0	

(b) Choose a suitable scale and then draw the graph of $y = 4 + 3x - x^2$.
Use your graph to
(c) state the range of values of x for which $4 + 3x - x^2$ is positive,
(d) find the value of y when $x = -0.6$,
(e) find the values of x when $y = 3$,
(f) find the maximum value of y.

8 (a) Complete the table of values shown in Table 9.11 for the function $y = x^2 - x - 6$.

Table 9.11 $y = x^2 - x - 6$

x	-3	-2	-1	0	1	2	3	4
y	6		-4	-6		-4	0	

(b) Using a scale of 2 cm to 1 unit, draw the graph of the function of $y = x^2 - x - 6$.
Using your graph, find
(c) the solution of $x^2 - x - 6 = 0$,
(d) the minimum value of y.

9 (a) Table 9.12 gives the table of values for $y = 5 - 3x - x^2$.

Table 9.12 $y = 5 - 3x - x^2$

x	-5	-4	-3	-2	-1	0	1	2
y	-5	1	5	7	7	5	1	-5

(b) Use a scale of 2 cm to 1 unit on both axes to draw the graph of $y = 5 - 3x - x^2$.
Use your graph to find
(c) the values of x when $y = 0$,
(d) the maximum value of $5 - 3x - x^2$.

Summary

Sketch graphs of various types of functions are illustrated below.

(a) **Linear functions**
$y = mx + c$

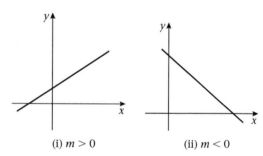

(i) $m > 0$ (ii) $m < 0$

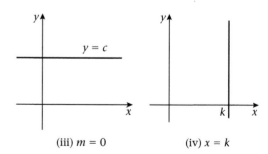

(iii) $m = 0$ (iv) $x = k$

(b) Non-linear functions

$$y = ax^2 + bx + c$$

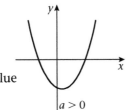

y has a minimum (least) value

$a > 0$

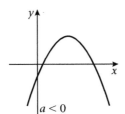

y has a maximum value

$a < 0$

Practice exercise P9.1

1 If $f(x) = x^2 - 8x + 5$, find

(a) $f(4)$ (b) $f(-3)$ (c) $f(\frac{1}{2})$

2 For each of the following functions, find the range of the function.

(a) $g : x \rightarrow 2x(x - 3)$,
for the domain $-2, -1, 0, 1, 2, 3, 4$

(b) $h : x \rightarrow 2x^2 - 3x - 5$,
$x \in \{-2, 0, 2, 4, 6\}$

Practice exercise P9.2

For each of the following functions,

(a) draw up a table of values;
(b) use suitable scales to represent 1 unit on each axis;
(c) plot the points on graph paper;
(d) draw the graph of the function.

1 $y = x^2 + 3,$ $-2 \leqslant x < 4$

2 $y = \frac{1}{2}x^2 - 4,$ $-3 \leqslant x \leqslant 3$

3 $y = 2 - x^2,$ $-2 \leqslant x < 3$

Practice exercise P9.3

1 The information given in Table 9.13 represents a function.

Table 9.13

Length of a side of a square [x cm]	1	2	3	4	5	6
Area of the square [A cm²]	1	4	9	16	25	36

(a) Show this on a graph.
(b) Write down the relation between the length of a side, x, and the area, A.
(c) Find the area of a square of side 2.3 cm.

Chapter 10

Geometrical transformations (2)
Enlargements

Enlargement, scale factor

Fig. 10.1 shows a method of drawing a $\triangle A_1B_1C_1$ which is similar to a given $\triangle ABC$.

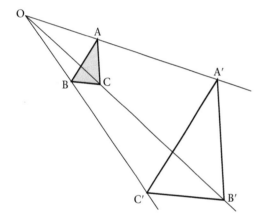

Fig. 10.1

Give $\triangle ABC$, choose any point O. O can be anywhere; in Fig. 10.1, O is outside $\triangle ABC$. From O, draw lines through A, B and C. The point A′ is on OA extended such that OA′ = 3OA. Similarly B′ and C′ are such that OB′ = 3OB and OC′ = 3OC. Join A′B′C′.

We say that $\triangle A'B'C'$ is an **enlargement** of $\triangle ABC$. O is the **centre of enlargement**.

In Fig. 10.1,

$$\frac{OA'}{OA} = \frac{OB'}{OB} = \frac{OC'}{OC} = 3$$

The fraction $\frac{OA'}{OA}\left(= \frac{OB'}{OB} = \frac{OC'}{OC}\right)$ is the **scale factor** of the enlargement. In this case, the scale factor is 3.

The sides of $\triangle A'B'C'$ are 3 times longer than the corresponding sides of $\triangle ABC$.

Thus,

$$\frac{A'B'}{AB} = \frac{B'C'}{BC} = \frac{C'A'}{CA} = 3$$

Thus the ratio of the lengths of corresponding sides is equal to the scale factor.

Example 1

In Fig. 10.2, rectangle OXYZ is enlarged to rectangle OX′Y′Z′ where O is the centre of enlargement. (a) If OX = 2 cm and XX′ = 1 cm, find the scale factor of the enlargement. (b) Hence find OZ′, given that OZ − 5 cm.

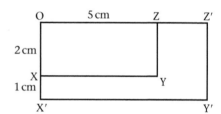

Fig. 10.2

(a) OX′ = 2 cm + 1 cm = 3 cm

$$\text{scale factor} = \frac{OX'}{OX} = \frac{3}{2}$$

(b) OZ′ = OZ × scale factor

$$= 5 \times \frac{3}{2}\,\text{cm}$$

$$= 7\tfrac{1}{2}\,\text{cm}$$

Group work

1. Draw any triangle ABC.
2. Choose any point O inside the triangle.
3. Join OA, OB, OC.
4. From O draw OA′, OB′, OC′ such that OA′ = 2OA, OB′ = 2OB, OC′ = 2OC.
5. Join A′, B′, C′.
6. Write down the ratios of the lengths
   $$\frac{OA'}{OA},\ \frac{OB'}{OB},\ \frac{OC'}{OC}.$$
7. What is the scale factor of the enlargement?

Exercise 10a

① In Fig. 10.3, APQR is an enlargement of ABCD.

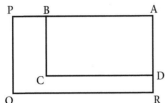

Fig. 10.3

(a) Which point is the centre of enlargement?
(b) Use measurement to find the scale factor of enlargement.

② In Fig. 10.4, △OAB is enlarged to △OA′B′.

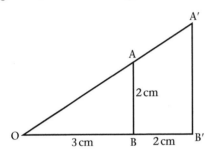

Fig. 10.4

(a) Which point is the centre of enlargement?
(b) If OB = 3 cm and BB′ = 2 cm, find the scale factor.
(c) If AB = 2 cm, calculate the length of A′B′.

③ The coordinates of the vertices of △ABC are A(1, 1), B(3, 1), C(1, 4).

(a) Choose a suitable scale and draw △ABC on graph paper.
(b) With the origin (0, 0) as the centre of enlargement, enlarge △ABC by scale factor 2.
(c) Write down the coordinates of A′, B′, C′, the vertices of the enlargement.

④ The rectangle A(3, 3½), B(2, 3), C(3, 1), D(4, 1½) is enlarged to rectangle A′(2, 5), B′(0, 3), C′(2, −1), D′(4, 0).

(a) Choose a suitable scale and draw these rectangles on graph paper.
(b) Hence find the scale factor and the coordinates of the centre of enlargement.

⑤ The coordinates of the vertices of △PQR are P(0, 2), Q(4, 6), R(8, 1).

(a) Choose a suitable scale and draw △PQR on graph paper.
(b) With the point (4, 4) as the centre of enlargement, enlarge △PQR by scale factor ½.
(c) Write down the coordinates of the vertices of the enlargement, △P′Q′R′.

Similar triangles

Fig. 10.5 shows two similar triangles ABC and DEF.

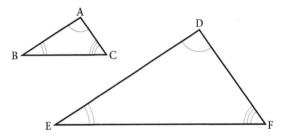

Fig. 10.5

Notice that $\widehat{A}, \widehat{B}, \widehat{C}$ in △ABC are equal respectively to $\widehat{D}, \widehat{E}, \widehat{F}$ in △DEF. The two triangles are **equiangular**. This means that the angles of one are equal to the angles of the other. Equiangular triangles are always similar. This is true even if one of the triangles is turned round as in Fig. 10.6.

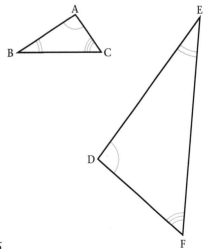

Fig. 10.6

Group work

In Fig. 10.7, △ABC is similar to △DEF.

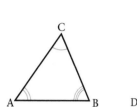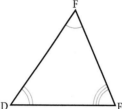

Fig. 10.7

Measure the lengths AB and DE.

Write down the ratio of the lengths $\dfrac{AB}{DE}$.

Measure the lengths AC and DF.

Write down the ratio of the lengths $\dfrac{AC}{DF}$.

Similarly for the ratio $\dfrac{BC}{EF}$.

If you measure correctly you will find that

$$\frac{AB}{DE} = \frac{BC}{EF} = \frac{AC}{DF} = \frac{2}{3}$$

Hence AB and DE, BC and EF, AC and DF are all in the ratio 2 : 3. △ABC can be said to be a scale drawing of △DEF.

Notice that AB and DE are the sides opposite the equal angles. Similarly for AC and DF, BC and EF. Sides AB and DE, AC and DF, BC and EF, are called corresponding sides so that in similar figures: **corresponding sides are in equal ratio.**

In Fig. 10.4, since

△OA'B' is an enlargement of △OAB then

$$\frac{OA'}{OA} = \frac{OB'}{OB} = \frac{A'B'}{AB}$$

∴ △s OAB, OA'B' are similar triangles.

Exercise 10b

1. On graph paper, draw a square ABCD and points X and Y as shown in Fig. 10.8.

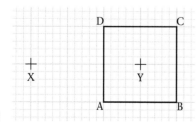

Fig. 10.8

Enlarge the square, (a) by scale factor 3 with centre of enlargement at X,

(b) by scale factor 2 with centre Y,

(c) by scale factor $1\frac{1}{2}$ with centre A.

2. In Fig. 10.9, rectangle B is an enlargement of rectangle A with O as centre of enlargement.

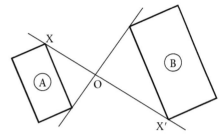

Fig. 10.9

(a) If OX = 8 cm and OX' = 10 cm, find the scale factor of the enlargement.

(b) If the diagonals of rectangle A are 12 cm long, calculate the length of the diagonals of rectangle B.

3. Square A(−2, −1), B(1, −1), C(1, −4), D(−2, −4) is enlarged to square A'(4, 1), B'(3, 1), C'(3, 2), D'(4, 2).

(a) Choose a suitable scale and draw these squares on graph paper, labelling the vertices carefully.

(b) Hence find the scale factor and the coordinates of the centre of enlargement.

4. The coordinates of the vertices △ABC are A(2, 1), B(4, 2), C(0, 5). Enlarge △ABC by scale factor 2 about centre of enlargement A.

5. Rectangles ABCD and WXYZ are such that AB = 5 cm, BC = 15 cm, WX = 8 cm, XY = 18 cm. Is ABCD an enlargement of WXYZ? Give reasons for your answer.

6. In triangle ABC, AB = 3 cm, BC = 4 cm, AC = 6 cm. Triangle ABC is enlarged about A by scale factor 2 to form triangle AB'C'. What is the length of AB', B'C', AC'?

7. The vertices of triangle ABC are A(2, 1), B(9, 5), C(3, 8). Using P(5, 5) as the centre of enlargement and a scale factor 2, enlarge triangle ABC.

③ In Fig. 10.10, △A'B'C' is an enlargement of △ABC.

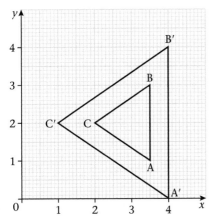

Fig. 10.10

(a) Calculate the scale factor of the enlargement.
(b) State the coordinates of the centre of enlargement.

Summary

Enlargement is a transformation that does not conserve size. In an enlargement each side of the shape is changed by the same factor called the **scale factor**. An enlargement produces a shape that is similar to the original shape. Corresponding sides of similar shapes are in equal ratio.

Practice exercise P10.1

① Given △PQR with P(2, 5), Q(1, 2) and R(3, 1),
(a) on graph paper, using a scale of 1 cm to 1 unit on both axes, draw △PQR;
(b) with centre R and a scale factor 3, draw △P'Q'R', the enlargement of △PQR.

② Draw △ABC with A(−2, −4), B(3, −2) and C(2, −6). Mark E(1, −4). With centre E and a scale factor of 2, draw the enlargement of △ABC. Label the triangle A'B'C'.

③ (a) Draw △ABC on graph paper, with A(3, 1), B(2, −6) and C(1, 2).
(b) With centre O, and a scale factor 2, draw the enlargement of △ABC.
(c) Name the similar triangles and the corresponding equal angles.

④ On graph paper, draw △LMN, with L(0, 3), M(2, 4) and N (−2, 7). With centre, C(−3, 5), and a scale factor 3, draw the enlargement of △LMN. Name the image L'M'N'.

⑤ (a) On graph paper, draw △PQR, with P(0, −2), Q(2, −1) and R(−1, 1).
(b) With centre O and a scale factor 3, draw the enlargement of △PQR to give △P'Q'R'.
(c) Reflect △P'Q'R' in the line x = −3 to form the image P"Q"R".

⑥ The vertices of △ABC are A(2, 1), B(7, 5) and C(3, 6).
(a) Using P(4, 3) as the centre of enlargement and a scale factor 2, enlarge △ABC. Name the image P'Q'R'.
(b) Using P' as the centre of rotation, rotate triangle P'Q'R' through 90° anti-clockwise to form the image P"Q"R".

⑦ The vertices of △ABC are A(3.5, 1), B(3.5, 3) and C(2, 2). The vertices of △A'B'C' are A'(4, 0), B'(4, 4) and C'(1, 2).
(a) Draw triangles ABC and A'B'C' on the same axes and using the same scale.
(b) Calculate the scale factor of the enlargement.
(c) State the coordinates of the centre of enlargement.

⑧ △ABC with vertices A(1, 1), B(3, 1) and C(1, 4) is reflected in the line x = 4.
(a) On graph paper, draw △ABC and the reflection.
(b) With centre of enlargement at A and a scale factor 2, draw the enlargement of triangle ABC.
(c) With centre of rotation O(2, 3), rotate A'B'C' through 90° clockwise to form the image A"B"C".

⑨ (a) On graph paper, using a scale of 1 cm to 1 unit on each axis, draw the △PQR whose vertices are P(3, 2), Q(5, −1) and R(1, −2).
(b) △P'Q'R' is the image of △PQR under an enlargement with centre, C(4, 1), and scale factor 2. Draw △P'Q'R' and state the coordinates of the vertices.

Chapter 11

Measurement (3)
Arcs and sectors of circles, similar shapes and solids

Pre-requisites
- properties of shapes and solids; enlargement; units of measurement

Arcs and sectors of circles

Length of arc

In Fig. 11.1, the arc AB subtends an angle of 90° at O, the centre of the circle. The whole circumference subtends 360° at O. Therefore the length of arc AB is $\frac{90}{360}$ or $\frac{1}{4}$ of the circumference of the circle. Similarly, arc CD is $\frac{45}{360}$ or $\frac{1}{8}$ of the circumference. It can be seen that the length of an arc of a circle is proportional to the angle which the arc subtends at the centre.

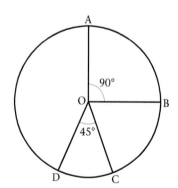

Fig. 11.1

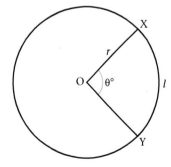

Fig. 11.2

In Fig. 11.2, arc XY subtends an angle of θ° at O. The circumference of the circle is $2\pi r$. Therefore,

in Fig. 11.2, the length, l, of the arc XY is given as

$$l = \frac{\theta}{360} \times 2\pi r$$

Example 1

An arc subtends an angle of 105° at the centre of a circle of radius 6 cm. Find the length of the arc.

It is easier to work out the length of the arc by first making a drawing as in Fig. 11.3.

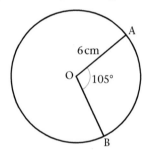

Fig. 11.3

From Fig. 11.3

$$\text{arc AB} = \frac{105}{360} \times 2\pi \times 6 \text{ cm}$$

$$= \frac{105}{360} \times 2 \times \frac{22}{7} \times 6 \text{ cm}$$

$$= 11 \text{ cm}$$

Example 2

Calculate the perimeter of a sector of a circle of radius 7 cm, the angle of the sector being 108°.

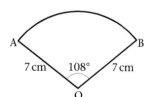

Fig. 11.4

$$\text{arc AB} = \frac{108}{360} \times 2 \times \frac{22}{7} \times 7 \text{ cm}$$

$$= 13.2 \text{ cm}$$

perimeter of sector AOB
$$= (7 + 7 + 13.2) \text{ cm} = 27.2 \text{ cm}$$

Example 3

What angle does an arc 6.6 cm in length subtend at the centre of a circle of radius 14 cm?

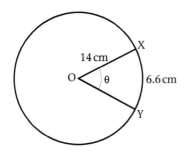

Fig. 11.5

arc XY $= \dfrac{\theta}{360} \times 2\pi \times 14$ cm

$6.6 = \dfrac{\theta}{360} \times 2 \times \dfrac{22}{7} \times 14$

$\theta = \dfrac{6.6 \times 360 \times 7}{2 \times 22 \times 14}$

$= 27$

The arc subtends an angle of 27°.

Example 4

An arc subtends an angle of 72° at the circumference of a circle of radius 5 cm. Calculate the length of the arc in terms of π.

If the arc subtends 72° at the *circumference*, then it subtends 144° at the centre of the circle (angle at centre = 2 × angle at circumference). See Fig. 11.6.

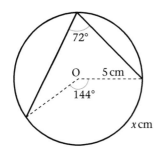

Fig. 11.6

$x = \dfrac{144}{360} \times 2\pi \times 5$ where length of arc $= x$ cm

$= \dfrac{16}{40} \times 10\pi = 4\pi$

The arc is 4π cm long.

Mathematics for Caribbean Schools

Exercise 11a

Where necessary, use the value $3\frac{1}{7}$ for π.

① In Fig. 11.7, each circle centre O is of radius 6 cm. Express the length of the arcs, *l*, *m*, *n*, *p*, *q*, *x*, *y*, *z* in terms of π.

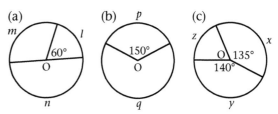

Fig. 11.7

② Complete Table 11.1 for arcs of circles. Make a rough sketch in each case.

Table 11.1

	Radius	Angle at centre	Length of arc
(a)	7 cm	90°	
(b)	35 m	72°	
(c)	4.2 cm	120°	
(d)	5.6 m	135°	
(e)	14 m		11 m
(f)	21 cm		22 cm
(g)		150°	330 cm
(h)		108°	132 cm

③ In terms of π, what is the length of an arc which subtends an angle of 30° at the centre of a circle of radius $3\frac{1}{2}$ cm?

④ What is the length of an arc which subtends an angle of 60° at the centre of a circle of radius $\frac{1}{2}$ m?

⑤ What angle does an arc 5.5 cm in length subtend at the centre of a circle of diameter 7 cm?

6 In Fig. 11.8, O is the centre of the circle.

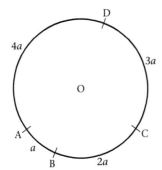

Fig. 11.8

Find the sizes of the following angles subtended by the minor arcs.

(a) AÔB   (b) AÔC   (c) BÔD
(d) AÔD   (e) AD̂C   (f) BD̂C
Note: Angles are less than 180°.

7 In Fig. 11.9, 4 pencils are held together in a 'square' by an elastic band. If the pencils are of diameter 7 mm, what is the length of the band in this position?

Fig.11.9

8 The elastic band in question 7 is used to hold 7 of the same pencils as shown in Fig. 11.10. What is the length of the band in this position?

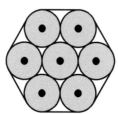

Fig. 11.10

9 A piece of wire 22 cm long is bent into an arc of a circle of radius 4 cm. What angle does the wire subtend at the centre of the circle?

10 In Fig. 11.11, the radius of the circle is 10 cm. Calculate the length of the minor arc XY.

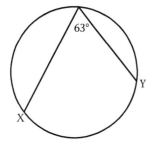

Fig. 11.11

11 The minute-hand of a clock is 6 cm long. How far does the end of the hand travel in 35 minutes?

12 A piece of string is wound tightly round a cylinder for 20 complete turns. The length of the string is found to be 3.96 m. Calculate the diameter of the cylinder in cm.

13 In a circle of radius 6 cm a chord is drawn 3 cm from the centre. (a) Calculate the angle subtended by the chord at the centre of the circle. (b) Hence find the length of the minor arc cut off by the chord.

14 Fig. 11.12 shows a circular wire clip of radius 7 mm with a gap of 7 mm between the ends.

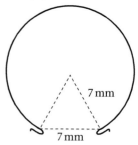

Fig. 11.12

Calculate the total length of wire in the clip to the nearest mm, given that about 4 mm are used altogether in making the turnovers at the ends.

15 Water is taken from a well 11 m deep in a bucket, the rope winding onto a drum 35 cm in diameter (see Fig. 11.13).

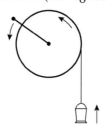

Fig. 11.13

(a) Through what angle does the handle turn in winding up 1 metre of rope?
(b) How many revolutions of the handle does it take to bring the bucket up from the bottom?
(c) If the arm of the handle is 42 cm long, how far does the hand of the winder travel when bringing the bucket up from the bottom?

Area of sector

In Fig. 11.14, the area of sector AOB is $\frac{90}{360}$ of $\frac{1}{4}$ of the area of the whole circle. The area of sector COD is $\frac{45}{360}$ or $\frac{1}{8}$ of the whole circle. The area of a sector of a circle is proportional to the angle of the sector.

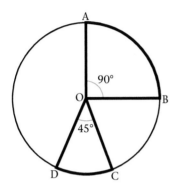

Fig. 11.14

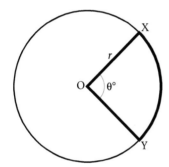

Fig. 11.15

In Fig. 11.15, the angle of the sector is $\theta°$. The area of the whole circle is πr^2. Therefore:

$$\text{Area of sector XOY} = \frac{\theta}{360} \times \pi r^2.$$

Example 5

A sector of 80° is removed from a circle of radius 12 cm. What area of the circle is left?

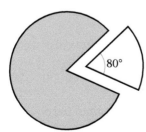

Fig. 11.16

Angle of sector left = 360° − 80° = 280°

Area of sector left = $\frac{280}{360} \times \pi \times 12^2$ cm²
$= \frac{7}{9} \times \frac{22}{7} \times 12 \times 12$ cm²
$= 352$ cm²

Example 6

Calculate the area of the shaded segment of the circle shown in Fig. 11.17. Give your answers correct to 1 decimal place.

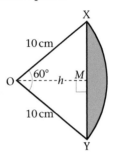

Fig. 11.17

Area of segment
$= $ area of sector XOY $-$ area of $\triangle$XOY
Area of sector XOY $= (\frac{60}{360} \times \pi \times 10^2)$ cm²
$= (\frac{1}{6} \times \frac{22}{7} \times 100)$ cm²
$= 52.38$ cm²

$\triangle$XOY is an equilateral triangle
$\therefore$ XY $= 10$ cm

M is the midpoint of XY
$\therefore$ XM $=$ MY $= 5$ cm
$\frac{h}{10} = \sin 60°,$
where OM $= h$ cm
$h = 10 \times \sin 60° = 8.66$ cm
Area of $\triangle$XOY $= (8.66 \times 5)$ cm²
$\therefore$ Area of segment $= (52.38 − 43.30)$ cm²
$= 9.1$ cm² to 1 d.p.

Exercise 11b

Take π to be $3\frac{1}{7}$ unless told otherwise.

1   In Fig. 11.18, each circle is of radius 6 cm. Calculate the areas of the shaded sectors in terms of π.

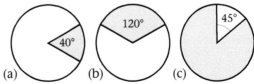

Fig. 11.18

② Complete Table 11.2 for areas of sectors of circles. Make a rough sketch in each case.

Table 11.2

	Radius	Angle of sector	Area of sector
(a)	7 cm	90°	
(b)	6 cm	70°	
(c)	35 cm	144°	
(d)	14 m		462 m²
(e)	2 cm		2.2 cm²
(f)		140°	99 m²

③ Calculate the area of a sector of a circle which subtends an angle of 45° at the centre of the circle, radius 14 cm.

④ The arc of a circle of radius 20 cm subtends an angle of 120° at the centre. Use the value 3.142 for π to calculate the area of the sector correct to the nearest cm².

⑤ The area of circle PQR with centre O is 72 cm². What is the area of sector POQ if PÔQ = 40°?

⑥ A pie chart is divided into four sectors as shown in Fig. 11.19. Each sector represents a percentage of the whole. The two larger sectors are equal and each represents x%. What is the angle subtended by one of those larger sectors?

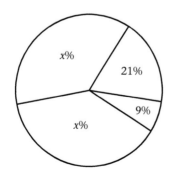

Fig. 11. 19

⑦ Calculate the shaded parts in Fig. 11.20. All dimensions are in cm and all arcs are circular with centres at the corners of the squares.

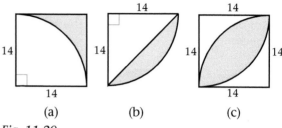

(a)   (b)   (c)

Fig. 11.20

⑧ Calculate the area of the shaded segment of the circle shown in Fig. 11.21.

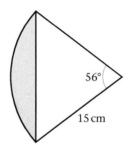

Fig. 11.21

⑨ In Fig. 11.22, ABCD is a rhombus with dimensions as shown. BXD is a circular arc, centre A. Calculate the area of the shaded section to the nearest cm².

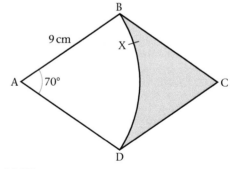

Fig. 11.22

⑩ Fig. 11.23 shows the cross-section of a tunnel. It is in the shape of a major segment of a circle of radius 1 m on a chord of length 1.6 m. Calculate

(a) the angle subtended at the centre of the circle by the major arc correct to the nearest 0.1°,

(b) the area of the cross-section of the tunnel correct to 2 d.p.

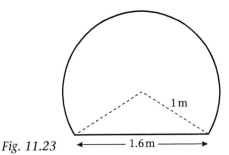

Fig. 11.23

Similar plane shapes and solids

In Chapter 10, we saw that if the corresponding angles of two triangles are equal, then the triangles are similar and the lengths of corresponding sides are in the same ratio.

However, for all other plane shapes and for all solids to be similar, both the corresponding angles are equal and the ratio of the lengths of corresponding sides must be constant.

Areas of similar plane shapes

Fig. 11.24 represents two similar rectangles, one 7 cm × 3 cm, the other 28 cm × 12 cm.

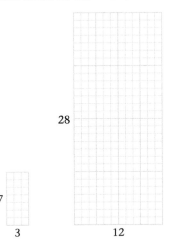

Fig. 11.24

Scale factor of the big rectangle to the small rectangle = ratio of their corresponding sides

$$= \frac{28\,\text{cm}}{7\,\text{cm}} \text{ or } \frac{12\,\text{cm}}{3\,\text{cm}}$$
$$= 4 \text{ (in both cases)}$$

Determine the number of small rectangles needed to completely cover the area of the large rectangle. This number is called the area factor.

Area factor of the big rectangle to the small rectangle = ratio of their areas

$$= \frac{(28 \times 12)\,\text{cm}^2}{(7 \times 3)\,\text{cm}^2}$$
$$= 4 \times 4 = 4^2$$

Hence the area factor is the square of the scale factor of the rectangles: $4^2 = 16$. Notice in Fig. 11.24 that the small rectangle fits 16 times into the big rectangle.

Fig. 11.25 represents two circles with radii 5 cm and 3 cm.

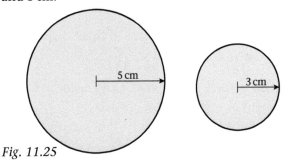

Fig. 11.25

Scale factor of the small circle to the big circle
= ratio of their radii

$$= \frac{3\,\text{cm}}{5\,\text{cm}} = \frac{3}{5}$$

Area factor of the small circle to the big circle
= ratio of their areas

$$= \frac{(\pi \times 3^2)\,\text{cm}^2}{(\pi \times 5^2)\,\text{cm}^2}$$
$$= \frac{3^2}{5^2}$$
$$= \left(\frac{3}{5}\right)^2 \text{ or } \frac{9}{25}$$

Again, the radio of the areas is the square of the scale factor of the given circles.

In general, the ratio of the areas of two similar plane shapes is the square of the scale factor of the two shapes.

Example 7

A map is drawn to a scale of 1 : 5000. On the map, a village has an area of 6 cm². Find the true area of the village in hectares. (1 ha = 10 000 m²)

Scale factor = 5000
Area factor = $(5000)^2$
= 25 000 000
Area of village = 25 000 000 × 6 cm²
$$= \frac{25\,000\,000 \times 6}{100 \times 100} \text{m}^2$$
$$= \frac{25\,000\,000 \times 6}{10\,000 \times 100 \times 100} \text{ha}$$
$$= \frac{25 \times 6}{100} \text{ha} = 1.5 \text{ha}$$

Example 8

A householder uses 5 m² and 3.2 m² of material when making similar drapes for two windows of similar shapes. If the larger window is 165 cm high, what is the height of the smaller window?

Area factor of smaller drape to larger drape
$$= \frac{3.2 \text{ m}^2}{5 \text{ m}^2} = \frac{32}{50} = \frac{16}{25} = \left(\frac{4}{5}\right)^2$$

Scale factor = square root of area factor = $\frac{4}{5}$
Height of larger window = 165 cm
Height of smaller window = $\frac{4}{5}$ of 165 cm
$$= \frac{4 \times 165}{5} \text{cm}$$
$$= 4 \times 33 \text{cm}$$
$$= 132 \text{cm}$$

Exercise 11c

1　Two similar rectangles have corresponding sides in the ratio 10:3. Find the ratio of their areas.

2　Two similar triangles have corresponding sides of length 4 cm and 7 cm. Find the ratio of their areas.

3　Two similar hexagons have corresponding sides of length 2 cm and 5 cm.
　(a) Find the ratio of their areas.
　(b) If the area of the larger hexagon is 150 cm², find the area of the smaller one.

4　The ratio of the areas of two circles is $\frac{4}{9}$.
　(a) Find the ratio of their radii.
　(b) If the smaller circle has a radius of 12 cm, find the radius of the larger one.

5　A map of Plymouth is drawn to a scale 1:50 000. On the map the airport covers an area of 8 cm². Find the true area of the airport in hectares.

6　A sports stadium covers an area of 6 hectares. Find the area in cm² of the sports stadium when drawn on a map of scale 1:5000.

7　A photograph measuring 8 cm by 10 cm costs $4.80. What will be the cost of an enlargement measuring 20 cm by 25 cm?

8　Two square floor tiles are made of the same material. One costs 45c and its edge is 30 cm long. Find the cost of the other if its edge is 50 cm long.

9　Two rectangular flags are similar in shape. Their areas are 5 m² and 0.8 m². If the height of the larger flag is 180 cm, find the height of the smaller flag.

10　The area of the windscreen of a bus is 1.21 m². In a photo of the bus, the windscreen is a rectangle 12 cm by 3 cm. Find the length and breadth of the real windscreen.

Volumes of similar solids

Fig. 11.26 represents two similar cuboids, one 5 cm by 2 cm by 1 cm, the other 15 cm by 6 cm by 3 cm.

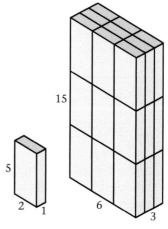

Fig. 11.26

Scale factor of the big cuboid to the small cuboid

= ratio of their corresponding edges

$= \dfrac{15\,\text{cm}}{5\,\text{cm}}$ or $\dfrac{6\,\text{cm}}{2\,\text{cm}}$ or $\dfrac{3\,\text{cm}}{1\,\text{cm}}$

= 3 (in each case)

Volume factor of the big cuboid to the small cuboid

= ratio of their volumes

$= \dfrac{(15 \times 6 \times 3)\,\text{cm}^3}{(5 \times 2 \times 1)\,\text{cm}^3} = 27 = 3^3$

Hence the volume factor is the *cube* of the scale factor of the cuboids. In Fig. 11.26, it should be possible to see that the small cuboid will fit 27 times into the big cuboid.

Fig. 11.27 represents two cylinders, one of height $2h$ and radius $2r$, the other of height h and radius r.

Since the heights and radii are in the same ratio, the cylinders are similar.

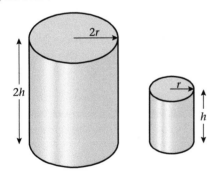

Fig. 11.27

Scale factor of the small cylinder to the big cylinder = ratio of corresponding lengths

$= \dfrac{h}{2h}$ or $\dfrac{r}{2r}$

$= \frac{1}{2}$ (in both cases)

Volume factor of the small cylinder to the big cylinder = ratio of their volumes

$= \dfrac{\pi r^2 h}{\pi (2r)^2 2h} = \dfrac{1}{8} = \left(\dfrac{1}{2}\right)^3$

Again, the ratio of the volumes is the cube of the scale factor of the two shapes.

In general, the ratio of the volumes of similar solids is the cube of the scale factor of the two solids.

Example 9

Two pots, similar in shape, are respectively 21 cm and 14 cm high. If the smaller pot holds 1.2 litres, find the capacity of the larger one.

Scale factor $= \dfrac{21}{14} = \dfrac{3}{2}$

Thus, volume factor $= \left(\dfrac{3}{2}\right)^3 = \dfrac{27}{8}$

The smaller pot holds 1.2 litres
The larger pot holds $1.2 \times \frac{27}{8}$ litres
$= 0.15 \times 27$ litres
$= 4.05$ litres

Example 10

Two bales of rice are of similar shape and contain 128 kg and 250 kg of rice respectively. If the height of the bigger bale is 70 cm, find the height of the smaller one.

Mass is proportional to volume, thus,
ratio of volumes = ratio of masses

Volume factor $= \dfrac{128}{250} = \dfrac{64}{125} = \dfrac{4^3}{5^3} = \left(\dfrac{4}{5}\right)^3$

Scale factor $= \frac{4}{5}$

Height of bigger bale = 70 cm

Height of smaller bale $= \frac{4}{5}$ of 70 cm
$= 4 \times 14\,\text{cm} = 56\,\text{cm}$

Exercise 11d

1. Two similar cups have heights in the ratio $2:3$. Find the ratio of their capacities.

2. Two similar blocks have corresponding edges of length 10 cm and 20 cm. Find the ratio of their masses.

3. A soap bubble 4 cm in diameter is blown out until its diameter is 8 cm. By what ratio has the volume of air in the bubble increased?

4. Two metal bolts are similar in shape and have diameters of 5 mm and 15 mm.
   (a) Find the ratio of their masses.
   (b) If the smaller bolt's mass is 12 g, find the mass of the larger bolt.

⑤ Two similar buckets hold $13\frac{1}{2}$ litres and 4 litres respectively.

    (a) Find the ratio of their heights.

    (b) If the larger bucket is of height 36 cm, find the height of the smaller bucket.

⑥ Two similar pots have heights of 16 cm and 10 cm. If the smaller pot holds 0.75 litres, find the capacity of the larger pot.

⑦ A sports trophy is in the shape of a cup 30 cm high. The winners are each given copies of the cup, $7\frac{1}{2}$ cm high. If one of the copies holds 100 ml, find the capacity of the trophy in litres.

⑧ A tin of beans costs $1.50. How much would a similar tin, 3 times the height and diameter, full of beans cost?

⑨ A pencil manufacturer makes a giant model pencil, 3 m long, as a factory symbol. If a real pencil is 18 cm long and has a volume of 9 cm³, find the volume in m³ of the giant model.

⑩ A builder makes a scale model of a real house. The volumes of air in the scale model and the real house are 27 500 cm³ and 220 m³ respectively. If the height of the door in the real house is 2.4 m, find the height of the door in the scale model.

Example 11

Two similar tins contain 960 g and 405 g of margarine respectively. If the area of the base of the larger tin is 120 cm², find the area of the base of the smaller tin.

Ratio of volumes   = ratio of masses

$$= \frac{960}{405} = \frac{64}{27}$$

$$= \frac{4^3}{3^3} = \left(\frac{4}{3}\right)^2$$

Thus, scale factor $= \frac{4}{3}$

and the area factor $= \left(\frac{4}{3}\right)^2 = \frac{16}{9}$

Area of larger base = 120 cm²

Area of smaller base $= \frac{9}{16}$ of 120 cm²

$$= 67\frac{1}{2} \text{ cm}^2$$

Notice in Example 11 that the scale factor and then the area factor must be found before the areas can be compared.

Exercise 11e

① A metal tray measuring 40 cm by 30 cm costs $12.80. What should be the price of a tray which is similar but 50 cm long?

② Two plastic cups are similar in shape and their heights are 7.5 cm and 12.5 cm. If the plastic needed to make the first cup cost 36c, find the cost for the second cup.

③ In question 2, if the second cup holds $1\frac{7}{8}$ litres, find the capacity of the first cup in ml.

④ A cylindrical oil drum 70 cm long is made of sheet metal which costs $37.50. Find the cost of the metal for a similar oil drum 84 cm long.

⑤ A statue stands on a base of area 1.08 m². A scale model of the statue has a base of area 300 cm². Find the mass of the statue (in tonnes) if the scale model is of mass 12.5 kg.

⑥ A railway engine is of mass 72 tonnes and is 11 m long. An exact scale model is made of it and is 44 cm long. Find the mass of the model.

⑦ In question 6, if the tanks of the model hold 0.8 litres of water, find the capacity of the tanks of the railway engine.

⑧ A garden has an area of 3025 m², and it is represented on a map by an area of 16 cm².

    (a) Find the scale of the map.

    (b) Find the true length of a wall which is represented on the map by a line 2.8 cm long.

⑨ In a scale drawing of a school grounds, a path 120 cm wide is shown to be 15 mm wide.

    (a) Find the scale of the drawing.

    (b) Find the area of the school grounds if the corresponding area on the plan is 2025 cm².

10 A model car is an exact copy of a real one. The windscreen of the model measures 35 cm by 10 cm and the real car has a windscreen of area 0.315 m². If the mass of the model is 25 kg, find the mass of the real car.

Summary

Circles and sectors

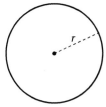 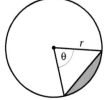

circumference = $2\pi r$

area = πr^2

length of arc = $\frac{\theta}{360} \times 2\pi r$

area of sector = $\frac{\theta}{360} \times \pi r^2$

Corresponding sides of similar plane shapes are in the same ratio. This ratio is called the **scale factor**.

The ratio of the areas of similar plane shapes is the square of the scale factor. This ratio is called the **area factor**.

The ratio of the volumes of similar solids is the cube of the scale factor. This ratio is called the **volume factor**.

Practice exercise P11.1

1 A circular mat has an area of 33.5 m². What is the area of a similar mat of twice the radius?

2 The area representing a circular playground on a map is 154 cm². On a second map, the same playground is represented by an area of 346.5 cm².

(a) Express the ratio of the areas in its simplest form.

(b) What is the ratio of the radii of the playgrounds?

3 The radius of the base of a cone is 4 cm and the height of the cone is 6 cm.

(a) Calculate the volume of the cone.

(b) Calculate the volume of a similar cone of height 12 cm.

4 A spherical gas tank holds 1732.5 litres of gas when it is half-full.

(a) What is, the capacity of the tank?

(b) Calculate the volume of the tank (1000 cm³ = 1 litre).

(c) Calculate the radius of the tank.

(d) What would be the capacity of a tank whose radius is one-third the radius of this tank?

5 A cylindrical drinking-glass of diameter 6 cm and height 14 cm contains milk. If the volume of milk in the glass is 108π cm³, calculate the height of the milk in the glass.

6 (a) The internal dimensions of a cylindrical saucepan are diameter 17.5 cm and height 10 cm. Calculate to the nearest whole number, using $\pi = \frac{22}{7}$,

(i) the area of the bottom of the saucepan;

(ii) the total volume of liquid that the saucepan will hold.

(b) If there is a similar saucepan of height 6 cm, calculate the total volume of liquid that this similar saucepan can hold.

7 (a) A cylindrical container of diameter 1.5 m is filled with water to a height of 2 m. Calculate the volume of water in the container.

(b) If this water is now poured into a similar container, the height of the water there is 8 m. Calculate the diameter of this second container.

8 Water from a full rectangular tank measuring 1 m × 2 m × 0.5 m is poured into a cylindrical tank and fills it to a depth of 1.2 m. Calculate

(a) the volume of water

(b) the diameter of the cylindrical tank.

Chapter 12

Statistics (3)
Statistical graphs, frequency distribution

> **Pre-requisites**
> ■ data collection; averages; margin of error

Bar charts, pie charts, pictograms (revision)

Example 1

Table 12.1 shows how students in class 2A travel to school.

Table 12.1

Method	bicycle	walk	bus	car
Frequency	9	15	7	5

Draw (a) a pictogram, (b) a bar chart, (c) a pie chart to show this information.

Table 12.1 is a **frequency table**. The frequency is the number of students using each method of transport.

Fig. 12.1 shows the required answers.

Revision notes

1 **Pictograms**: each method of transport is represented by a suitable picture. In this case each complete picture represents 5 students.

2 **Bar chart**: each bar represents a method of transport. The height of each bar is proportional to the number of students using that method of transport.

3 **Pie chart**: each sector represents a method of transport. The angle of each sector is proportional to the number of students who use the method of transport shown in that sector.

(a)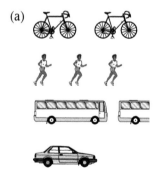

Each symbol represents 5 people

(b)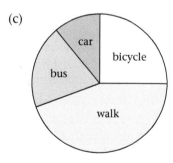

(c)

Fig. 12.1

Fig. 12.2 shows a temperature chart for a hospital patient.

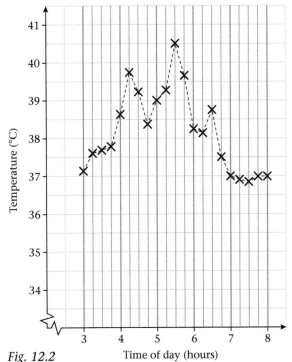

Fig. 12.2 Time of day (hours)

In Fig. 12.2 the temperature is recorded every $\frac{1}{4}$ hour. To make the points easier to see, they are joined by straight lines. In **straight line graphs of this kind**, the lines are given for convenience; they do not tell us what happens between the points.

Frequency polygons

A **frequency polygon** is a line graph in which points represent frequencies. The points are then joined by straight lines to form a polygon.

Example 2

A class of 30 students scored the following marks in a test.

5	4	7	5	6	8	9	5	6	5
9	7	5	6	5	5	7	5	6	7
8	6	5	8	4	7	5	5	4	6

Use tally marks to make a frequency table. Hence draw a frequency polygon.

Table 12.2 shows how the tally marks are used to find the frequencies of the marks.

Table 12.2

Mark	Tally	Frequency
4					3
5	⊮ ⊮		11
6	⊮ ||	7
7						4
8					3
9				2

Fig. 12.3 is the frequency polygon.

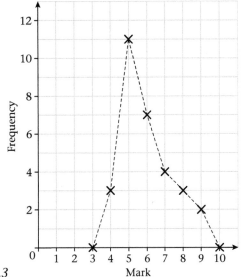

Fig. 12.3 Mark

Exercise 12a

1 Fig. 12.4 is a pictogram of a traffic survey.

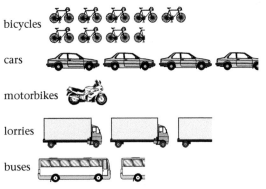

bicycles

cars

motorbikes

lorries

buses

Each symbol represents 10 vehicles

Fig. 12.4

(a) Which kind of vehicle is the most common?
(b) Which kind of vehicle is the least common?
(c) Approximately how many bicycles were counted?
(d) Approximately how many lorries were counted?

② Fig. 12.5 is a bar chart showing the heights and names of the highest mountains of Asia, South America, North America, Africa and Europe respectively.

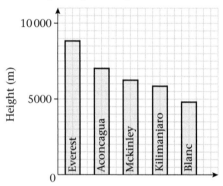

Fig. 12.5

(a) What is the height of Mount Kilimanjaro?
(b) What is the height of Mount McKinley?
(c) Which mountain is 4800 m high?
(d) Find the difference in height between Mount Everest and Mount Aconcagua.

③ Fig. 12.6 is a pie chart showing how a housewife spent the first hour of a day after getting out of bed.

Fig. 12.6

(a) What fraction of the hour was spent eating?
(b) What percentage of the hour was spent reading?
(c) How many minutes were spent cleaning?

④ Use Fig. 12.2 to answer the following.
(a) What is the temperature of the patient at 06:45?
(b) What is the highest temperature shown?
(c) How many of the recorded temperatures are above 39 °C?
(d) At what time is the patient's temperature 39 °C exactly?

⑤ A farmer has 100 cows, 40 sheep and 65 goats. Draw a pictogram to show the animals he has. (Let one symbol represent 10 animals.)

⑥ During a 24-hour period the sun shone for 10 hours, it was cloudy for 6 hours and it was dark for the remainder of the time.
(a) For what fraction of the day was it dark?
(b) For what fraction of the day was it cloudy?
(c) Draw a pie chart to show the periods of sunshine, cloudiness and darkness during the day.
(d) Measure the size of the angle showing the period of sunshine.

⑦ The ages of 36 students are given below.
13	16	16	15	12	14	13	15	16
16	14	15	12	16	13	14	16	12
15	13	15	16	13	13	15	14	16
13	14	16	15	15	12	14	12	13

Make a frequency table and hence represent the age distribution of the students on (a) a bar chart, (b) a frequency polygon.

⑧ Table 12.3 gives a country's production of coffee in tonnes in 2-year intervals from 1980 to 1988.

Table 12.3

Year	Coffee (tonnes)
1980	750
1982	1600
1984	3800
1986	3200
1988	4100

Show the increase in production on (a) a bar chart, (b) a line graph.

Frequency distribution

Very often before a graph can be drawn the information must be organised.

Example 3

The following are the marks of 30 students in a test.

16	13	13	17	14	13
11	13	10	12	15	12
10	12	15	13	14	14
13	17	14	15	13	14
13	14	12	11	16	11

Represent this information on a bar chart.

These data are not organised in any sort of order. The frequency of each mark is needed before the bar chart can be drawn. Table 12.4 shows the organised information.

Table 12.4

Mark	Frequency
10	2
11	3
12	4
13	8
14	6
15	3
16	2
17	2
	30 (total)

Table 12.4 is a *frequency distribution* table in which the number of students obtaining each mark is recorded. Notice that the total of the frequencies equals the number of students who sat the test.

Fig. 12.7 shows a suitable bar chart.

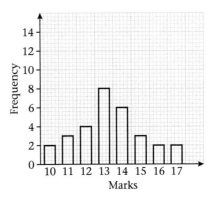

Fig. 12.7

Grouped distribution

Sometimes it is necessary to handle a large amount of numerical data. The frequency distribution table for single observations would be cumbersome. It is useful at such times to group the observations into classes and we can then find the number of observations belonging to each class.

Example 4

The following gives the heights of 50 people in centimetres.

162	165	163	170	157	154	153	150	162	165
168	162	158	159	160	156	154	157	151	168
153	157	150	153	155	160	165	171	172	173
150	162	163	165	170	155	162	158	162	164
154	158	162	157	152	163	165	170	168	167

Draw a tally chart for the classes 150–154, 155–159, 160–164, 165–169, 170–174. Represent the information on a bar chart.

The tally chart is shown in Table 12.5.

Table 12.5

Class (cm)	Tally	Frequency
150–154	JHT JHT I	11
155–159	JHT JHT I	11
160–164	JHT JHT III	13
165–169	JHT IIII	9
170–174	JHT I	6

Fig. 12.8 is a bar chart of the grouped data.

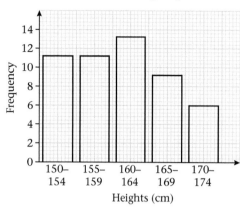

Fig. 12.8

Notes:

1 The class intervals are not always given. Choose convenient class intervals.
2 The class intervals are chosen so that all the data are included.
3 If the class interval is too wide some important features of the information might be lost.
4 If the interval is too narrow, the grouping may be meaningless.

Exercise 12b

① Table 12.6 shows the mark distribution obtained by a class in a test marked out of 24. Represent the data on a bar chart.

Table 12.6

Score	0–4	5–9	10–14	15–19
Frequency	4	5	15	6

② The number of sticks of matches in a box of matches was counted for 50 boxes of matches. The following is the result:

49 41 42 43 48 41 44 42 45 46
45 46 48 44 42 43 44 41 42 49
48 49 47 46 45 41 41 43 48 47
42 44 47 46 49 45 48 47 49 42
41 45 44 43 49 42 46 44 46 43

(a) Make a table showing the frequency distribution for the data grouped in classes: 41–43, 44–46, 47–49.

(b) Represent the information on a bar chart.

③ Table 12.7 is a frequency distribution table giving examination marks out of 100 for a class of 35 students. Construct a suitable graph to illustrate this information.

Table 12.7

Marks	Frequency
30–39	3
40–49	8
50–59	14
60–69	6
70–79	4

④ The following is a list of marks obtained by a group of students in a test marked out of 100.

20 80 88 25   0 15   2 60   3
55 60 59 57 54 51 62 63 70
77 43 55 44 49 81 82 35 36

(a) Make a table of frequency distribution for the data grouped in intervals of 10 marks: 0–9, 10–19, …, 90–99.

(b) Represent the information by an appropriate diagram.

Table 12.8 shows the weekly wages of 50 people.

Table 12.8

Weekly wages of 50 people ($)				
82	132	199	248	300
89	145	200	249	324
94	152	206	255	334
96	156	206	263	348
98	158	214	265	369
108	163	220	270	381
114	176	221	270	401
120	178	232	280	440
125	185	235	288	477
128	189	247	294	485

The data is grouped in equal class intervals of $100 to give the frequency distribution shown in Table 12.9.

Table 12.9

Class interval (weekly wage $)	Frequency (number of people)
0–99	5
100–199	16
200–299	19
300–399	6
400–499	4

It is necessary to define the limits of the class intervals very clearly. Otherwise it may be difficult to decide the class in which to include borderline values, such as $199 and $200.

Table 12.10 shows the number of absentees recorded each day of a school term.

Table 12.10

Number absent	0–9	10–19	20–29	30–39	40–49	50–59
Frequency	5	9	11	8	6	3

These are the results from counting the number of absent students and are exact whole number values. The number of absentees is a **discrete variable**. Discrete variables usually involve counting.

Fig. 12.9 is the bar chart for the data in Table 12.10.

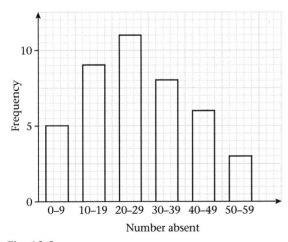

Fig. 12.9

Table 12.11 gives the masses, in kg, of 20 athletes.

Table 12.11

65	56	46	53	48
54	65	63	70	52
63	51	72	69	55
49	53	69	63	70

Any one value may be representative of a range of values. 51 kg for example, represents a range of possible values between 50.5 kg and 51.5 kg. The weight of an athlete is a **continuous variable**. Continuous variables usually arise from situations that involve measuring.

In Table 12.10 the class intervals indicate all the whole numbers in that class. In Table 12.11, since each value represents a range of values, the **true limits**, also called **class boundaries**, of the class includes the range of values which approximate to the limits of the class. For example, the true limits of the class 51–55 kg would be 50.5–55.5 kg (see Fig. 12.10). This includes all the possible values for the class 51–55 kg. For the class 51–55 kg note the following:

> 51 kg and 55 kg are the class limits
>
> 50.5 kg and 55.5 kg are the class boundaries (true limits).

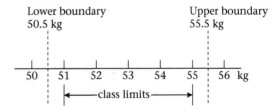

Fig. 12.10

The midpoint of the class is the mean value of the lower and upper class limits.

The midpoint of the class 51–55 kg is

$$\frac{51 + 55}{2} \text{ kg} = 53 \text{ kg}$$

The frequency distribution is shown in Table 12.12.

Table 12.12

Class interval (mass, kg)	Frequency
46–50	3
51–55	6
56–60	1
61–65	5
66–70	4
71–75	1

Exercise 12c

1. State which of the following variables are discrete and which are continuous.
   (a) The time taken to run a 100-metre race
   (b) The number of words on the page of a book
   (c) The number of pages in a book
   (d) The heights of students in the class

2. State the values included in the following data class.
   (a) 10–15 books
   (b) 60–65 kg
   (c) 15–19 kg to the nearest kg
   (d) 20–24 cm

3. Make a frequency distribution of the data in Table 12.8, taking equal class intervals $1–$100, $101–$200, $201–$300, etc. Compare this distribution with that of Table 12.9.

4. Make a frequency distribution of the data in Table 12.8, by taking ten equal class intervals $0–$49, $50–$99, ..., $450–$499.

5. Fifty students were asked to estimate the size of an angle to the nearest degree. Their results, arranged in order of size, are given in Table 12.13.

Table 12.13

Estimates (degrees)									
56	58	58	60	60	61	62	63	64	64
65	65	65	66	66	66	66	67	67	68
68	68	68	69	69	69	70	70	70	70
70	72	72	72	72	73	73	74	74	74
75	75	75	76	78	79	80	80	81	83

Make a frequency distribution table, taking six equal intervals 55–59, 60–64, ..., 80–84.

6. State the class boundaries for the data in Table 12.13.

Histograms

A frequency distribution can be represented by a block graph called a histogram. Fig. 12.11 is the histogram of the frequency distribution in Table 12.12.

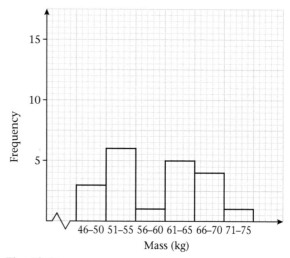

Fig. 12.11

A histogram consists of a number of rectangles. The horizontal width of each rectangle is given by the class interval. The height is such that the area of the rectangle is proportional to the frequency in that interval. Hence the areas of the rectangles show the frequency distribution. The horizontal axis is numbered to give a continuous axis.

Histograms can be used to represent data in which the class intervals are not equal.

Table 12.14 shows a frequency distribution for the information in Table 12.12 with the last two classes combined.

Table 12.14

Class interval (mass kg)	46–50	51–55	56–60	61–65	66–75
Frequency	3	6	1	5	5

The histogram in Fig. 12.12 represents the data in Table 12.14.

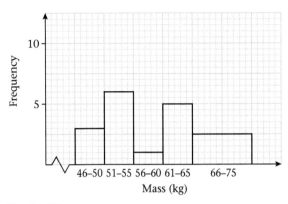

Fig. 12.12

Compare the histograms in Fig. 12.11 and Fig. 12.12. Check carefully to see that the area of the bars for classes 66–70, 71–55 is the same as the area of the bar for class 66–75.

Grouped data can also be shown in a **frequency polygon**. Fig. 12.13 is a frequency polygon of the data shown in the histogram in Fig. 12.11.

In a frequency polygon, the frequencies are plotted at the *midpoints* of each class interval. The points are joined by *straight lines*.

Notice that the frequencies are plotted at the midpoints of the classes.

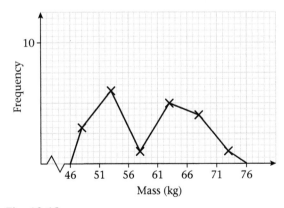

Fig. 12.13

Example 5

Table 12.15 gives the marks of 50 students in a test.

Table 12.15

35	51	83	60	61	73	44	90	70	93
56	34	52	61	43	57	40	58	88	64
52	71	25	86	79	35	73	44	71	95
63	53	48	78	65	98	28	72	67	82
46	54	62	35	70	41	63	73	50	68

(a) Construct the histogram, taking class intervals 21–30, 31–40, …, 91–100. (b) What is the modal class? (c) Find the mean mark.

(a) When the data are not in numerical order, use a tally system to count the frequencies. Take each mark in turn and enter a tally stroke against the proper class interval. The frequency total, 50, gives a check on the accuracy of working. The resulting frequency distribution is given in Table 12.16.

Table 12.16

Class	Frequency	
21–30	||	2
31–40	ЖІ	5
41–50	ЖІ ||	7
51–60	ЖІ ||||	9
61–70	ЖІ ЖІ |	11
71–80	ЖІ |||	8
81–90	ЖІ	5
91–100	|||	3
		total 50

Fig. 12.14 is the histogram of the distribution. Notice that class intervals of $20\frac{1}{2}$–$30\frac{1}{2}$, $30\frac{1}{2}$–$40\frac{1}{2}$, …, have been drawn in Fig. 12.14.

(b) The modal class is 61–70. This can be seen in both the frequency distribution table and in the histogram. The mode of the data is taken to be the central value of the modal class, 61–70, i.e. $65\frac{1}{2}$.

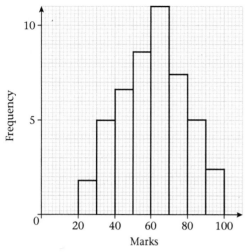

Fig. 12.14

(c) To find the mean mark, choose a working
mean and find the deviations from it.
In this example, $65\frac{1}{2}$ is taken as the working
mean and the marks in each class interval
are represented by the mid-mark of that
class. For example, marks in the class
interval 21–30 are counted as $25\frac{1}{2}$, and so
on. The working is set out in columns as in
Table 12.17.

From Table 12.17
total deviation from working mean $= -190$

mean deviation $= \dfrac{-190}{50}$

$\qquad\qquad\quad = -3.8$

mean mark $\quad = 65.5 - 3.8$

$\qquad\qquad\quad = 61.7$

Table 12.17

Class interval	Class centre	Frequency (f)	Deviation (d)	f × d
21–30	$25\frac{1}{2}$	2	−40	−80
31–40	$35\frac{1}{2}$	5	−30	−150
41–50	$45\frac{1}{2}$	7	−20	−140
51–60	$55\frac{1}{2}$	9	−10	−90
61–70	$65\frac{1}{2}$	11	0	0
71–80	$75\frac{1}{2}$	8	+10	+80
81–90	$85\frac{1}{2}$	5	+20	+100
91–100	$95\frac{1}{2}$	3	+30	+90
			total deviation	−190

The final result in part (c) is likely to be slightly
inaccurate. (In fact, the true mean of the marks
in Table 12.12 is 61.2.) Nevertheless, this
method should be used when a large number of
values is given.

Exercise 12d

1 Draw a histogram and a frequency polygon
for the frequency distribution in Table 12.18.

Table 12.18

Class	1–5	6–10	11–15	16–20	21–25
Frequency	2	4	6	5	3

State the modal class of the distribution.

2 Draw a histogram of the frequency
distribution in Table 12.19.

Table 12.19

Class	1–5	6–10	11–15	16–20	21–25
Frequency	4	6	11	6	3

Calculate the mean of the data.

3 Draw a histogram of the data in Table 12.20.

Table 12.20

Class	1–10	11–20	21–30	31–40	41–50
Frequency	5	12	17	10	6

Estimate the mode of the data.

4 Draw a histogram and frequency polygon of
the frequency distribution in Table 12.21.

Table 12.21

Class	8–14	15–21	22–28	29–35	36–42	43–49
Frequency	3	5	8	18	9	7

Find the mode and the mean of the data.

5 Students taking a teacher-training course are
grouped by age as in Table 12.22.

Table 12.22

Age group	19–20	20–21	21–22	22–23	23–24	24–25
Number in group	4	5	10	16	12	3

Calculate the average age of the students.

6 Table 12.23 shows the numbers of absentees recorded each day of a school term.

Table 12.23

Number absent	0–9	10–19	20–29	30–39	40–49	50–59
Frequency	5	18	23	17	14	3

Calculate the average number of absentees per day.

7 The percentage marks of 100 students in a School Certificate examination are grouped as in Table 12.24.

Table 12.24

Percentage	0–9	10–19	20–29	30–39	40–49
Frequency	1	2	5	17	23

Percentage	50–59	60–69	70–79	80–89	90–99
Frequency	25	18	5	3	1

(a) Estimate the number of students who scored 15% less than the modal mark.
(b) Find the average percentage for the examination.

8 Small nails are sold in packets which have printed on them, 'Average contents 200 nails'. The contents of 100 packets, picked out at random, were counted and the results are given in Table 12.25.

Table 12.25

Nails per packet	185–189	190–194	195–199
Frequency	4	14	32

Nails per packet	200–204	205–209	210–214
Frequency	28	17	5

Is the statement on the packet true or not?

9 Table 12.26 gives the masses, in kg, of 30 students.

Table 12.26

43	45	50	47	51	58	52	47	42	54
61	50	45	55	57	41	46	49	51	50
59	44	53	57	49	40	48	52	51	48

(a) Taking class intervals 40–44, 45–49, ..., construct the frequency distribution of the data.
(b) Draw a histogram of the data.
(c) Calculate the mean mass of the students.

10 Table 12.27 gives the heights, in cm, of the 30 students in question 9.

Table 12.27

145	163	149	152	166	156	159	139	145	141
150	158	150	149	143	159	154	167	146	147
152	162	144	169	162	150	173	160	167	171

(a) Taking class intervals 135–144, 145–154, ..., and construct the table of frequencies.
(b) Calculate the mean height of the students.

Summary

Statistical graphs include pictograms, pie charts, line graphs and bar charts.

A **pictogram** represents data in the form of pictures.

In a **pie chart** the set of data is represented by sectors of the circle. Angles of the sectors in a pie chart are drawn to proportionate sizes. The pie chart serves well for a quick comparison but actual values cannot easily be read.

Data are represented on a **bar chart** by a series of bars, vertical or horizontal. The height (or length) of each bar is proportional to the frequency of the data. The width of the bars is of no significance.

A **frequency distribution** is a table which gives the **frequency** of each observation. The amount of data may sometimes be large and is then grouped into classes. The frequency distribution then gives the frequency for 3each class.

Discrete variables generally have whole number values. **Continuous variables** have rational values.

The **class limits** indicate the lower and upper

values of the class intervals. **Class boundaries** include the range of possible values in the class.

A **histogram** is a column graph in which the area of each column is proportional to the frequency represented. There is no space between the bars of a histogram.

A **polygon** is a closed shape bounded by straight sides.

A **frequency polygon** is a graph in which the frequencies are plotted at the midpoints of the class and the points are joined by straight lines.

When asked to draw a frequency polygon to represent data, you need not draw the histogram.

Practice exercise P12.1

1. Table 12.28 shows the frequency distribution of the marks earned by students in a geography test.

Table 12.28

Class interval	0–4	5–9	10–14	15–19	20–24	25–29
Frequency	3	5	8	17	9	8

(a) Find the mode and the mean of the data.
(b) Draw a histogram and a frequency polygon for the frequency distribution.
(c) Estimate the number of students who scored less than 50% of the marks.

2. Table 12.29 shows the number of absentees recorded each day of a school term. Draw a frequency polygon of the frequency distribution.

Table 12.29

Number of absentees	0–9	10–19	20–29	30–39	40–49	50–59
Frequency	5	16	23	17	14	1

3. Table 12.30 shows the percentage of time that a student spent on studying each subject and on exercise.

Table 12.30

Activities	%
Languages	25
Mathematics	20
Computer Studies	15
Business Subjects	15
Technical Drawing	10
Exercise	15

If the student spends a total of 60 hours on the above activities, draw a pie chart to represent the time spent on each activity (calculate the size of angles to the nearest whole number).

Scale drawing (2)
Bearings and distances

Pre-requisites
■ scales; geometrical constructions

The magnetic compass

Fig. 13.1 is a photograph of a **magnetic compass**.

Fig. 13.1

The magnetic compass is used for finding direction. It has a magnetic needle which always points in the direction north.

Points of the compass

Fig. 13.2 shows the main points of the compass.

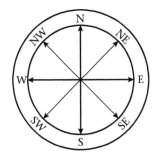

Fig. 13.2

There are 4 main points, or directions: north (N), south (S), east (E) and west (W). There are 4 secondary directions: north-east (NE), south-east (SE), south-west (SW) and north-west (NW). The angle between the directions N and E is 90°. NE is the direction mid-way between N and E. Thus the angle between N and NE is 45°.

Compass bearings

Fig. 13.3 shows two compasses placed at points A and B.

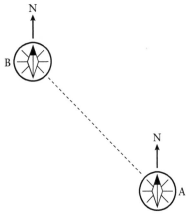

Fig. 13.3

The pointers of both compasses point north-wards (N). The compass at B is in a direction NW of A. We say that the bearing of B from A is NW. Similarly, the compass at A is in a direction SE of B. The bearing of A from B is SE.

Exercise 13a (Oral)

In each of the diagrams (1–6) in Fig. 13.4, state
(a) the bearing of B from A,
(b) the bearing of A from B.

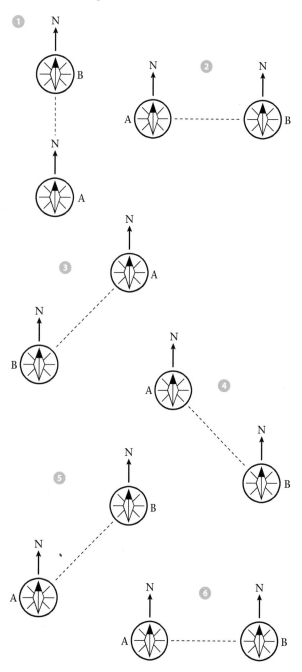

Fig. 13.4

Three-figure bearings

Fig. 13.5 shows the plan of a tree, a church, a well and a flag-pole. Imagine you are standing at A with a compass.

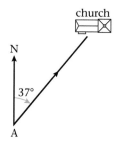

Fig. 13.5

The compass bearing of the tree from A is N. The compass bearing of the well from A is NE. It is not possible to give an exact bearing of the church or the flag-pole in terms of points of the compass.

Fig. 13.6 shows that the direction of the church from A makes an angle of 37° with north.

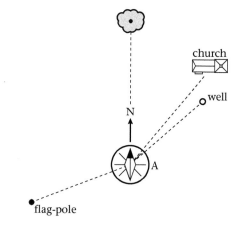

Fig. 13.6

Instead of using the directions N, NE, E, etc., we can give the bearing of the church as a **three-figure bearing**, measured as the number of degrees from north, in a clockwise direction. The three-figure bearing of the church from A is 037°.

Fig. 13.7 shows that the direction of the flag-pole from A makes an angle of 246° with north, measured clockwise.

The three-figure bearing of the flag-pole is 246°.

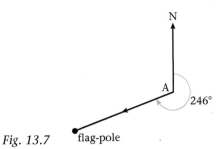

Fig. 13.7   flag-pole

Any direction can be given as a three-figure bearing. Three digits are always given. For angles less than 100°, zeros must be written in front of the digits. For example, the direction east is given as 090°. North is 000° or 360°. Three-figure bearings are often called true bearings.

Note: bearings are sometimes given in terms of acute angles referred to the four main points of the compass. For example 037° can also be given as N37°E. Think of this as 'face north then turn 37° towards east'. Similarly 246° can be given as S66°W; 'face south then turn 66° towards west'. Throughout this course the more common true bearing method will be used.

Exercise 13b (Group work)

Students work in groups of four in two pairs. After the task is completed, each pair compares their results and discusses any differences and the probable reasons.

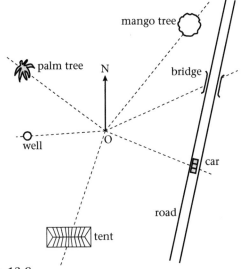

Fig. 13.8

① Use a protractor to find the bearings from O of the objects in the map of Fig. 13.8.

② (a) Mark a point O on a piece of cardboard. With O as centre, draw round a protractor. If necessary, move the protractor so as to make a complete circle with O as centre. Use the protractor to mark off the circle in 30° intervals as shown in Fig. 13.9. Cut out the circle. The circle is a compass face.

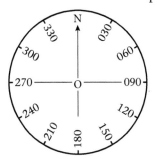

Fig. 13.9

(b) Cut a pointer from a thin strip of cardboard. Push a drawing pin up through the centre of the compass and the pointer as shown in Fig. 13.10. This gives a model compass. The model compass can be used for estimating the sizes of three-figure bearings.

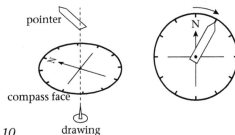

Fig. 13.10   drawing

③ For this exercise, take the front of your classroom to be north. Place your model compass so that N points towards the front of your classroom.

(a) Turn the pointer and estimate the bearings of the four corners of the room to the nearest 10°. See Fig. 13.11.

(b) Estimate the bearings of the following.
   (i)   The centre of the door
   (ii) The centre of each window
   (iii) The teacher's chair

(iv) The friend who is closest to you

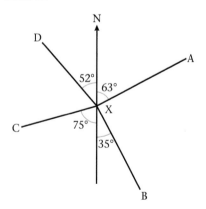

Fig. 13.11

④ Take your desk outside. Place your compass on your desk so that it points northwards. (*to find north without a compass*: face the direction the sun rises (east); make a quarter turn to the left; you are now facing north). Estimate the bearings of some things in your school compound.

Calculating bearings

Example 1

In Fig. 13.12, find the three-figure bearings of A, B, C and D from X.

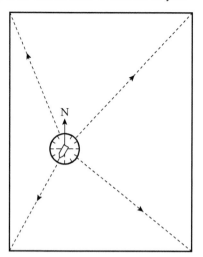

Fig. 13.12

The arrow N shows the direction north.
$N\widehat{X}A = 63°$
The bearing of A from X is 063°.
$N\widehat{X}B = 180° - 35° = 145°$

The bearing of B from X is 145°.
$N\widehat{X}C$ clockwise $= 180° + 75° = 255°$
The bearing of C from X is 255°.
$N\widehat{X}D$ clockwise $= 360° - 52° = 308°$
The bearing of D from X is 308°.

Example 2

If the bearing of X from Y is 247°, find the bearing of Y from X.

The question gives the bearing of X *from* Y. Start by marking a point Y. Draw a line, YN_1, pointing north from Y. (See Fig. 13.13.)

Fig. 13.13

X is on a bearing 247° from Y. Sketch a line YX such that $N_1\widehat{Y}X$ is 247° clockwise from the line YN_1. Mark a point X on this line. From X, draw a line, XN_2, pointing north. See the sketch in Fig. 13.14. The two lines pointing north are parallel. Angle N_2XY is the bearing of Y from X.

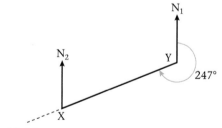

Fig. 13.14

There are many ways of finding the angle. Fig. 13.15 shows two ways.

Since XN_2 and YN_1 are parallel, the angle N_2XY may be found by producing

(a) N_1Y and calculating the value of the alternate angle at Y, that is (247°–180°) in Fig. 13.15(a);

(b) XY and calculating the value of the corresponding angle at Y, that is (247°–180°) in Fig. 13.15(b).

(a)

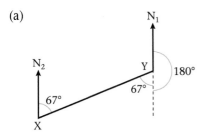

(b)

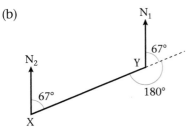

Fig. 13.15

The bearing of Y from X is 067°.

Notice that when making sketches, it is usual to take the top of the page as north.

Exercise 13c

① For each sketch in Fig. 13.16, state the three-figure bearing of B from A.

(a)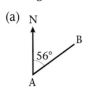

(b)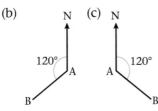

(c)

(d)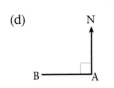

(e)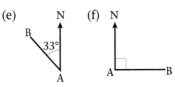

(f)

(g)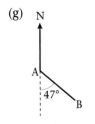

(h)

Fig. 13.16

② In Fig. 13.17, find the bearings of A, B, C and D from X.

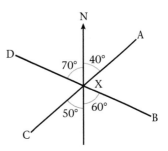

Fig. 13.17

③ In Fig. 13.18, find the bearings of U, V, X, Y, Z from O.

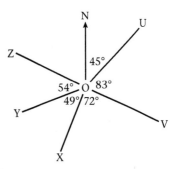

Fig. 13.18

④ Make a sketch of the following bearings. Each sketch should show all the data and must include a line pointing North.
 (a) The bearing of B from A is 040°.
 (b) X is on a bearing 160° from Y.
 (c) P is on a bearing 320° from Q.
 (d) L is on a bearing 200° from K.
 (e) The bearing of G from H is 180°.

⑤ In each diagram in Fig. 13.19, calculate,
 (i) the bearing of A from B,
 (ii) the bearing of B from A.

(a)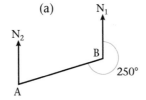

(b)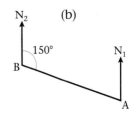

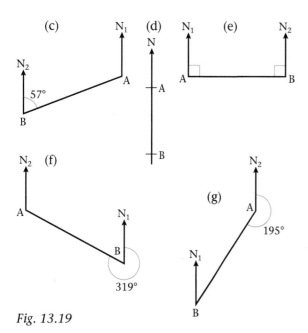

Fig. 13.19

Surveying

To **survey** an area means to take measurements so that a scale drawing of the area can be made. Fig. 13.20 shows a sketch that a student made while surveying a rectangular classroom block and a tree.

The student has measured the bearings and distances of the tree and of two corners of the classroom block from a point P. For the tree, (118°, 15 m) means that the tree is on a bearing of 118° and a distance 15 m from P. The width of the classroom block has also been measured.

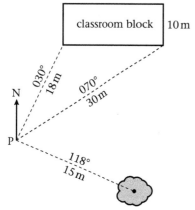

Fig. 13.20

Fig. 13.21 is a scale drawing of the tree and the classroom block.

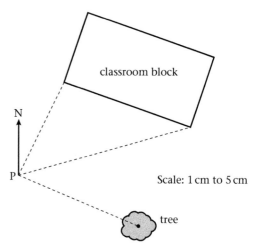

Fig. 13.21

Method: P is any point. A line pointing north is drawn from P. Lines are drawn on bearings 030°, 070° and 118° from P. These are shown dotted in Fig. 13.21. Distances of 18 m, 30 m and 15 m are marked on these lines respectively. These give the positions of the front corners of the classroom block and the centre of the tree. The rest of the classroom block is drawn, using the fact that it is 10 m wide.

Group work

The students work in groups of three or four, and then the groups pair off. Each of the two groups in a pair uses the same classroom block and the same tree to survey an area of the school yard from the same position of P, as described and shown in Fig. 13.21. Students may use the compasses they made in Exercise 13b.

When both groups in a pair have completed the scale drawing, they compare their results and discuss the reasons for any differences.

Exercise 13d

1 Fig. 13.22 is a sketch and notes from a survey taken from a point P of three trees. A, B and C. Choose a suitable scale and draw

an accurate plan of the three trees. Hence find the distance and bearing of A from C.

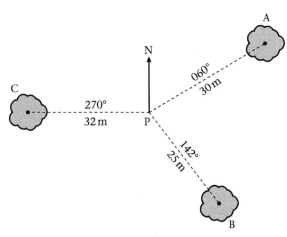

Fig. 13.22

② Fig. 13.23 is a sketch of part of a river. A and B are 100 m apart. The bearing of the tree from A is 000°. The bearing of the tree from B is 290°. The edges of the river are roughly straight and parallel.

Make a scale drawing of points A and B and the tree. Hence find the approximate width of the river.

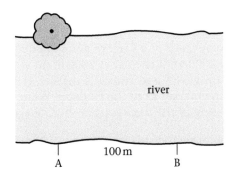

Fig. 13.23

③ Fig. 13.24 is a sketch and notes from a survey of a road. The road has two straight parts, AB and BC. It is 5 m wide.

Draw an accurate plan of the road to a suitable scale. Hence find the bearing of B from A and the bearing of C from B.

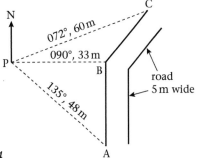

Fig. 13.24

④ Fig. 13.25 is a sketch and notes from a survey of a straight length of railway line.

Draw an accurate plan to show the position of the railway line and the water-tank. Find the bearing of A from B. Find the distance, to the nearest 10 m, of the water-tank from the railway line.

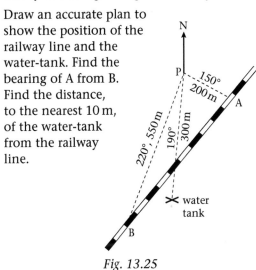

Fig. 13.25

Example 3

A boy starts from A and walks 3 km east to B. He then walks 5 km on a bearing 152° from B. He reaches a point C. Find the distance and bearing of C from A.

First make a sketch of the information. See Fig. 13.26.

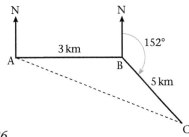

Fig. 13.26

Then make the scale drawing. See Fig. 13.27.

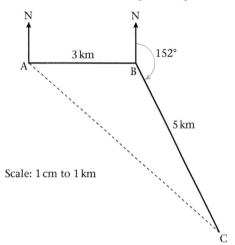

Scale: 1 cm to 1 km

Fig. 13.27

Join AC.
By measurement, AC = 6.9 cm
Distance of C from A = 6.9 × 1 km
              = 6.9 km

By measurement, $N\hat{A}C$ = 130°

Bearing of C from A is 130°.

Example 4

The village, Grange, is 152 km on a bearing 061°
from Balvenie. How far is Grange north of
Balvenie? How far west of Grange is Balvenie?

Make a sketch of the data as in Fig. 13.28.

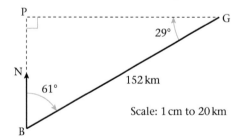

Fig. 13.28

P is the point that is due north of Balvenie and
due west of Grange. BP represents the distance
that Grange is north of Balvenie.
By measurement, BP ≈ 3.7 cm
The true distance BP ≈ 3.7 × 20 km
              ≈ 74 km

GP represents the distance that Balvenie is west
of Grange.
By measurement, GP ≈ 6.65 cm
The true distance GP ≈ 6.65 × 20 km
              ≈ 133 km

Hence Grange is approximately 74 km north of
Balvenie and Balvenie is approximately 133 km
west of Grange.

Exercise 13e

Answer each question by making a scale
drawing. Always make a rough sketch first.

① A boy starts at A and walks 3 km east to B.
He then walks 4 km north to C. Find the
distance and bearing of C from A.

② A girl starts at A and walks 2 km south to B.
She then walks 3 km west to C. Find the
distance and bearing of C from A.

③ An aeroplane flies 400 km west then 100 km
north. Find its distance and bearing from its
starting point.

④ A boy cycles 14 km east and then 10 km
south-east. Find his distance and bearing
from his starting point.

⑤ A road starts at a College and goes due north
for 2000 m. It then goes 2000 m on a
bearing 040° and ends at a market. How far
is the market from the College? What is the
bearing of the market from the College?

⑥ A pilot records his position as 310 km from
Piarco Airport on a bearing 062°. Find (a)
how far north the pilot's position is from
the airport; (b) how far the pilot is east of
the airport.

⑦ Mandeville is approximately 85 km from
Falmouth. The bearing of Mandeville from
Falmouth is 170°.
(a) How far north of Mandeville is
    Falmouth?
(b) How far west of Mandeville is
    Falmouth?

8 A ship is 3 km due east of a harbour. Another ship is also 3 km from the harbour but on a bearing 042° from it.
    (a) Find the distance between the two ships.
    (b) Find the bearing of the second ship from the first.

Summary

The position of one point relative to another is given in terms of the distance between the two points and the **compass bearing**. The compass bearing is the angle between the two points referred to the main points of the compass.

A **three-figure bearing**, or **true bearing** as it is sometimes called, is measured as the number of degrees from north, in a clockwise direction. Bearings are sometimes written in terms of acute angles referred to the four main points of the compass.

Practice exercise P13.1

Answer each question by using a scale drawing showing all the information. Remember to draw a rough sketch first.

1 From the top of a tower 140 m tall, the angle of depression of a house is 16°. What is the distance of the house from the foot of the tower?

2 The bearing of a point B, 27 m from A, is 270°. What is the bearing of A from B?

3 An aircraft is flying at a height of 2400 m. From a point P on the ground, the angle of elevation of the plane is 42°. What is the distance of P from the point on the ground over which the plane is flying?

4 A town P is 125 km on a bearing of 054° from a town Q. How far west of P is Q?

5 An aeroplane flies 500 km west and then 150 km north. Find the distance and bearing of the aeroplane from its starting point.

6 A town P is 95 km from a town R. The bearing of P from R is 054°. How far east of R is P?

7 John works at a pizza shop, making deliveries on his motorcycle. On a certain day, he rides 14 km east to make one delivery and then 10 km south-east to make another delivery. Find his distance and bearing from the pizza shop.

8 A ship is 5 km due east of a harbour. Another ship is 10 km on a bearing 135° from the harbour. Find
    (a) the distance between the two ships;
    (b) the bearing of the second ship from the first ship.

9 The road from Crumstone is 132 km on a bearing of 072° from Blackburn. How far east of Blackburn is Crumstone?

10 The angle of elevation of a helicopter from a point P on the ground is 72°. Given that the helicopter is hovering at a height of 42 m, find the distance between P and the foot of the perpendicular from the helicopter to the ground.

11 From the top of a tree, the angle of depression of a small plant on the ground is 52°. If the plant is 5 m from the foot of the tree, find the height of the tree, giving your answer to one decimal place.

12 A man travels 4 km north, then 7 km south-east, and then 7 km west. Find his distance and bearing from his starting point.

Chapter 14

Probability

Theoretical and experimental probabilities

> **Pre-requisites**
> ■ fractions

Everyday probability

Think of some games you play in which you roll a die. What are the chances of getting a 6 just when you need a 6?

If you take the bus to school you might ask the question 'What are the chances that the bus comes early this morning?'

When the captain of the West Indies cricket team calls 'heads' he asks himself 'What is the chance that I will win the toss?'

The answers to all these questions are based on a branch of mathematics called **probability**.

Probability is a measure of the likelihood of a **required outcome** happening. It is usually given as a fraction:

$$\text{probability} = \frac{\text{number of required outcomes}}{\text{number of possible outcomes}}$$

Experimental probability

Example 1

A student counts the number of cars, trucks and buses he passes on the way to the beach. The results are shown in Table 14.1.

Table 14.1

Vehicles	Number
cars	30
trucks	18
buses	12

(a) What is the total number of vehicles?
(b) Which vehicle is most likely to be the next one? Why?
(c) What is the chance that this will be the next vehicle?

(a) Total number of vehicles is 60
(b) A car, since more cars have been passing than the other vehicles.
(c) Since $\frac{30}{60}$ of the vehicles were cars the chance that the next vehicle is a car is $\frac{1}{2}$.

In Example 1, the required outcomes were cars passing and the possible outcomes were cars, trucks, buses. Thus,

probability of the next vehicle being a car

$$= \frac{\text{number of cars}}{\text{number of cars, buses, trucks}}$$

$$= \frac{30}{60}$$

$$= \frac{1}{2} = 0.5$$

Example 2

A student notes the letters on the registration plates of 210 cars. The results are shown in Table 14.2.

Table 14.2

Registration letters	Number of cars
AA	56
AB	34
AE	31
AI	10
AJ	9
RR	70

(a) Which set of registration letters is most likely to appear on the next car seen by the student?
(b) What is the chance that this will be the next set?

(a) RR.
(b) If the pattern continues the chance of seeing RR next is $\frac{70}{210}$ or $\frac{1}{3}$.

Notice in Examples 1 and 2 that the past is used to predict the future. Events can easily turn out differently. The answers in Examples 1 and 2 are no more than calculated guesses. Both $\frac{1}{2}$ and $\frac{1}{3}$ are based on experimental records and are examples of **experimental probability**.

Exercise 14a (Experiments)

1 A bottle top rests in either a cup or a cap position. See Fig. 14.1.

cup position   cap position

Fig. 14.1

Find a bottle top and drop it 100 times on your desk. Count the number of times it lands in a cup position and the number of times it lands in a cap position. Record your results in a tally table like Table 14.3.

Table 14.3

| Cup position | ЖІ ІІ |
| Cap position | ЖІ ІІІІ |

(a) Copy and complete the following. The bottle top landed in the cup position ... times out of 100 throws. The experimental probability that a bottle top will land in the cup position is ...
(b) Compare your results with other people's results. Give some reasons why your results may be different.

2 A drawing pin either rests with its point up or its point down. See Fig. 14.2.

point up   point down

Fig. 14.2

(a) Repeat the experiment in question 1 using a drawing pin.

(b) Find the experimental probability that the drawing pin will land point up if dropped on the floor.
(c) Compare your results with other people.

3 A coin, when tossed, either lands with its head up or its tail up.
(a) Repeat question 1 using a coin.
(b) Find the experimental probability that if a coin is tossed it will land tails up.
(c) Compare your results with other people.

4 Cut four pieces of string so that three are 15 cm long and one is 10 cm long. Hold all four pieces in your hand so that all four lengths look the same (Fig. 14.3).

Fig. 14.3

Ask a friend to choose one piece. Write down whether a long piece or the shorter piece is chosen. Put the string back, mix them up and ask someone else. Repeat 20 times.

Find the experimental probability that someone will choose the short piece of string.

5 Draw circle of radius 5 cm on a large sheet of paper. Get 10 paper clips. Hold the paper clips about 30 cm above the centre of the circle and drop them onto the paper.

Count the number of paper clips inside the circle and the number outside the circle. (If a paper clip falls on the circumference, count it as being inside the circle.)

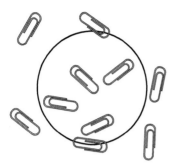

Fig. 14.4

In Fig 14.4, there are 6 paper clips inside the circle and 4 outside. The results can be recorded as shown in Table 14.4.

Table 14.4

Number of paper clips inside	6
Number of paper clips outside	4

Repeat 10 times. Find the experimental probability that a paper clip dropped from a height of 30 cm will stay within 5 cm of the point where it falls.

⑥ Place 3 green marbles and 7 blue marbles in a bag. Take one marble from the bag without looking. Record the colour and replace the marble. Repeat 20 times. Find the experimental probability that a green marble is taken from the bag.

⑦ Make a survey of the first 10 vehicles that pass your school gates. How many were cars? Use your result to predict how many of the next 10 vehicles will be cars. Check your prediction on the next 10 vehicles.

⑧ Open this book at any page. Read the right-hand page number. Write down whether the page number includes a 5 or not.

Repeat 50 times, recording your results as shown in Table 14.5.

Table 14.5

Page number has a 5 in it	
Page number has no 5s in it	

Find the experimental probability that if this book is opened anywhere, the right-hand page number will have a 5 in it.

⑨ Number the faces of a hexagonal pencil from 1 to 6 (Fig. 14.5).

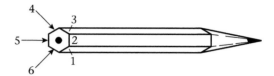

Fig. 14.5

Roll the pencil across your desk. Make a note of the number on the top face when its stops. Repeat 100 times.
How many times did you roll a number 4? What is the experimental probability of rolling a number 4?

Probability of a certain outcome

If you placed 6 Cola bottle tops in a bag, what is the probability of pulling out a Cola bottle top? Since you have only Cola bottle tops you are completely sure that you will pull out a Cola bottle top. The probability of a certain outcome is 1.

Probability of an impossible outcome

What would be the probability of pulling out a Lemonade bottle top in the above exercise? Since there are no Lemonade bottle tops in the bag it is impossible to pull out a Lemonade bottle top. The probability of an impossible event is 0.

From the above it is noticed that the least value of a probability is 0 and the greatest value of a probability is 1. Other values of the probability of any outcome are fractions lying between zero and one.

Exercise 14b

State which of the following has a probability of (a) 0, (b) 1.

① An odd number that is even

② A member of your class being 10 metres tall

③ The sum of the angles of a triangle being 270°

④ A square that is a rectangle

⑤ The sum of the angles of a polygon being $(2n - 4) \times 90°$

⑥ An even number divided by an odd number factor giving an even number

⑦ Drawing a red marble from a bag with 8 red marbles

⑧ A member of the West Indies cricket team being 12 years old

Sum of probabilities of all possible outcomes

Example 3

There are 10 marbles in a bag. Three marbles are green, two are white and five are red.

(a) What is the probability of drawing
    (i) a green marble,
    (ii) a white marble,
    (iii) a red marble?

(b) What is the sum of the probabilities?

(a) (i) Since there are 3 green marbles out of 10 marbles,

probability of a green marble

$$= \frac{\text{number of green marbles}}{\text{number of marbles}}$$

$$= \frac{3}{10}$$

    (ii) probability of a white marble

$$= \frac{\text{number of white marbles}}{\text{number of marbles}}$$

$$= \frac{2}{10}$$

    (iii) probability of a red marble

$$= \frac{\text{number of red marbles}}{\text{number of marbles}}$$

$$= \frac{5}{10}$$

(b) Sum of the probabilities

$$= \frac{3}{10} + \frac{2}{10} + \frac{5}{10}$$

$$= \frac{10}{10}$$

$$= 1$$

Example 4

A card is picked at random* from a pack of playing cards.**

(a) What is the probability of picking
    (i) a club,
    (ii) a spade,
    (iii) a heart
    (iv) a diamond

(b) What is the sum of all the probabilities?

(c) What is the probability of not choosing a spade?

(a) (i) Since there are 13 clubs in the pack of 52 cards,

probability of picking a club

$$= \frac{\text{number of clubs}}{\text{number of cards}} = \frac{13}{52}$$

    (ii) Similarly, probability of picking a spade

$$= \frac{13}{52}$$

    (iii) probability of picking a heart

$$= \frac{13}{52}$$

    (iv) probability of picking a diamond

$$= \frac{13}{52}$$

(b) Sum of the probabilities

$$= \frac{13}{52} + \frac{13}{52} + \frac{13}{52} + \frac{13}{52} = \frac{52}{52} = 1$$

(c) The probability of not choosing a spade is the same as the probability of choosing any of the other 39 cards

$$= \frac{39}{52} = \frac{3}{4}$$

which is the same as the sum of all the probabilities *less* the probability of picking a spade

$$= 1 - \frac{13}{52} = \frac{39}{52} = \frac{3}{4}$$

*'At random' means without carefully choosing.
**A pack of playing cards contains 52 cards in 4 suits: clubs (♣), diamonds (◇), hearts (♡), spades (♠).
There are 13 cards in each suit: A, 2, 3, 4, 5, 6, 7, 8, 9, 10, J, Q, K.
Clubs and spades are black; diamonds and hearts are red.

Example 5

A market trader has 100 oranges for sale. 4 of them are bad. What is the probability that an orange chosen at random is good?

Either:
4 out of 100 oranges are bad,
thus 96 out of 100 oranges are good.
Probability of getting a good orange $= \frac{96}{100} = \frac{24}{25}$

or:
Probability of getting a bad orange $= \frac{4}{100} = \frac{1}{25}$
thus,
probability of getting a good orange $= 1 - \frac{1}{25} = \frac{24}{25}$

Example 6

City School enters candidates for CXC examinations. The results for the years 2002 to 2006 are given in Table 14.6.

Table 14.6

Year	2002	2003	2004	2005	2006
Number of candidates	86	93	102	113	116
Number gaining at least 3 subjects	53	58	59	66	70

(a) What is the probability of a student at City School gaining at least 3 subjects in CXC?
(b) What is the probability of a City School students gaining less than 3 subjects in CXC?

(a) Total number gaining at least 3 subjects
$= 53 + 58 + 59 + 66 + 70$
$= 306$
Total number of candidates
$= 86 + 93 + 102 + 113 + 116$
$= 510$
The probability of a student gaining at least 3 subjects
$= \frac{306}{510} = \frac{3}{5}$

(b) The probability of a student gaining less than 3 subjects
$= 1 - \frac{3}{5} = \frac{2}{5}$

Exercise 14c

1 Statistics show that 4 out of every 100 new radios break down within the first year. What is the probability of buying a radio which does not break down in the first year?

2 It has rained on 5th June 18 times in the last 20 years. What is the probability that it will rain on 5th June next year?

3 The midday temperatures during a week were 26°C, 26°C, 27°C, 27°C, 26°C, 27°C, 27°C. What is the probability that the midday temperature on the next day will be
(a) 2°C, (b) 35°C,
(c) 26°C, (d) 27°C?

4 A matchbox contains 15 used sticks and 25 unused sticks.
(a) How many sticks are in the box altogether?
(b) What is the probability that a stick chosen at random is unused?

5 A statistical survey shows that 4 out of every 10 women wear a size 16 dress. What is the probability that a woman chosen at random does not wear a size 16 dress?

6 An advertisement says, '7 out of every 10 people prefer Green Band margarine.' 50 people were asked which margarine they preferred. If the advertisement is true, approximately how many people will say Green Band?

7 A trader has 100 mangoes for sale. 20 of them are unripe. Another 5 of them are bad. If a mango is picked at random, find the probability that it is (a) unripe, (b) bad, (c) neither unripe nor bad? If 20 of the mangoes were chosen at random, how many would you expect to be (d) unripe, (e) bad?

8 It is known that 1 in 40 of the light bulbs sold by a certain trader is faulty. If one bulb is taken at random from a large number, what is the probability of it being a good one?

9 Ada and Ebenezer play table tennis together. They have already played 10 games and Ada has won 9 of them. What is the probability that she will win the 11th game?

10 Given the data of question 9, Ebenezer wins the 11th and 12th games. What is the probability that he will win the 13th game?

11 A crate contains 15 bottles of Coke and 9 bottles of Sprite. If I choose a bottle at random, what is the probability that it is (a) Coke, (b) Sprite, (c) either a Coke or a Sprite, (d) neither Coke nor Sprite?

12 20 cards are numbered from 1 to 20. A card is chosen at random. What is the probability that it does *not* have the digit 1 in its number?

13 Table 14.7 shows the numbers of pupils getting a place in secondary school from Northside Primary School for the years 2002 to 2006.

Table 14.7

	2002	2003	2004	2005	2006
Number of pupils leaving Northside Primary School	54	58	60	63	65
Number of pupils gaining a secondary school place	25	26	31	28	34

(a) Find Northside Primary School's success rate as a percentage.

(b) Find the approximate probability that a pupil chosen at random from Northside Primary School will gain a secondary school place.

14 Table 14.8 gives the results of a traffic survey on a city road one morning. The table shows the total number of vehicles per hour and the numbers of those that were cars and lorries.

(a) For the whole morning, find the total number of vehicles.

(b) Find the percentage of the vehicles that were cars.

(c) Find the percentage of the vehicles that were lorries.

(d) Find the probability that the next vehicle to come along the road is a car.

(e) Of the next 20 vehicles on the road, how many would you expect to be lorries?

Table 14.8

	Total number of vehicles	Number of cars	Number of lorries
08:00–09:00	46	28	3
09:00–10:00	37	14	10
10:00–11:00	32	13	14
11:00–12:00	35	20	12

Theoretical probability

It is possible to calculate probabilities without doing experiments or keeping records. Consider the following.

Coin tossing

If a coin is tossed, there are only two possible outcomes: a head or a tail. Each is equally likely. The probability of getting a head is $\frac{1}{2}$. The probability of getting a tail is $\frac{1}{2}$. Since these values can be calculated without throwing any coins they are called **theoretical probabilities**.

Die throwing

When a die like that in Fig. 14.6 is thrown, any one of six numbers will come out on top.

Fig. 14.6

Each number is equally likely. Hence the theoretical probability of getting a particular number is $\frac{1}{6}$. For example, the theoretical probability of throwing a 4 is $\frac{1}{6}$.

Theoretical probability is based on the fact that the coin and the die are fair. Each outcome is equally likely.

Example 7

A die is thrown. Find the probability that the outcome is divisible by 3.

There are 6 possible outcomes: 1, 2, 3, 4, 5, 6
Two of these are divisible by 3: 3 and 6

$$\text{probability} = \frac{\text{number of required outcomes}}{\text{number of possible outcomes}}$$

$$= \frac{2}{6} = \frac{1}{3}$$

When a die is thrown there is a $\frac{1}{3}$ probability that the outcome will be divisible by 3.

Probability can also be defined in terms of the language of sets:

$$p(R) = \frac{n(R)}{n(U)}$$

where p(R) is the probability of a required outcome happening, R = {required outcomes} and U is the universal set, {possible outcomes}.

In Example 7:

$$U = \{\text{possible outcomes}\} = \{1, 2, 3, 4, 5, 6\}$$
$$R = \{\text{required outcomes}\} = \{3, 6\}$$
$$p(R) = \frac{n(R)}{n(U)} = \frac{2}{6} = \frac{1}{3}$$

Example 8

Three coins are thrown. What is the probability of getting 2 heads and 1 tail?

$$U = \{\text{HHH, HHT, HTH, HTT, THH, THT,}$$
$$\text{TTH, TTT}\}$$
$$R = \{\text{HHT, HTH, THH}\}$$
$$p(R) = \frac{n(R)}{n(U)} = \frac{3}{8}$$

Example 9

A card is picked at random from a pack of 52 playing cards. What is the probability that it is a 7?

Number of possible outcomes = 52
Number of required outcomes = 4 (7♣, 7◇,
7♡, 7♠)
Probability of picking a 7 = $\frac{4}{52} = \frac{1}{13}$

Exercise 14d

① A fair 6-sided die is thrown. Find the probability of getting
   (a) a 2   (b) a 5   (c) a 6
   (d) a 0   (e) a 1   (f) a 7
   (g) either a 1, 2, 3, 4, 5 or 6
   (h) an odd number
   (i) a number divisible by 7
   (j) either a 1, 2 or 5
   (k) a prime number
   (l) a square number
   (m) a number greater than 2
   (n) a number less than 6

② A card is picked at random from a pack of 52 playing cards. Find the probability of picking
   (a) the 9♠   (b) the 2♣
   (c) the J♡   (d) the 3◇
   (e) a Queen (Q)   (f) an Ace (A)
   (g) a diamond   (h) a black card
   (i) a black 8   (j) a red King
   (k) either a 4 or a 5
   (l) a black diamond

③ Two coins are tossed together. What is the probability of getting (a) a head and a tail, (b) two tails?

④ A fair coin was tossed 3 times. Each time it came up heads. What is the probability that it will come up heads next time?

⑤ A letter is chosen at random from the alphabet. Find the probability that it is
   (a) P   (b) either M or N
   (c) a vowel   (d) either M, Y or Z
   (e) one of the letters of the word MATHEMATICS
   (f) one of the letters of the word PROBABILITY

⑥ A bag contains 2 white balls and 3 red balls. A ball is picked at random. What is the probability that it is (a) white, (b) red?

⑦ A school contains 750 boys and 450 girls. A student is chosen at random. What is the probability that a girl is chosen?

8 The winner of a radio quiz programme is to be chosen from 5000 entries placed in a drum. What is your chance of being the winner if you have 4 correct answers among the 5000?

9 A boy is playing Ludo with a die. He needs a 1 to win. What is the probability that he will win on his next throw?

10 Table 14.9 shows the numbers of pupils in each age group in a class.

Table 14.9

Age (years)	13	14	15
Number of pupils	9	26	5

(a) How many pupils are in the class?
(b) What is the probability that a pupil chosen at random from the class will be (i) 13 years old, (ii) over 13 years old, (iii) 14 years old or less?

11 It is known that 7 out of 10 girls in a school do not wear a necklace. What is the probability that a girl chosen at random from the school is wearing a necklace?

12 Two dice are thrown at the same time and their total is noted.

(a) Copy and complete Table 14.10.

Table 14.10

first die

	+	1	2	3	4	5	6
	1	2	3	4	5	6	7
second	2	3	4	5	6	7	8
die	3	4	5	6			
	4						
	5						
	6						

(b) How many possible outcomes are there?
(c) How many of these outcomes gives a total of 5 for the two dice?
(d) What is the probability of getting a total of 5?
(e) Find the probability that the total for the two dice is
(i) 2   (ii) 4   (iii) 6   (iv) 7
(v) 7 or 11   (vi) a prime number

Summary

Probability is a measure of the likelihood of an outcome occurring and it is usually written as a fraction.

Experimental probability is based on the results of experiments or on records.

Theoretical probability can be calculated without doing experiments or keeping records.

$$\text{Probability} = \frac{\text{number of required outcomes}}{\text{number of possible outcomes}}$$

where the number of possible outcomes is the universal set for a particular event.

For any particular set, the sum of the probabilities of the possible outcomes is 1. The probability of an impossible outcome is 0. The probability of a certain outcome is 1.

Practice exercise P14.1

1 In a class of 40 boys, 30 like mathematics and 15 like history. What is the probability that a boy chosen at random will like both mathematics and history?

2 A bag contains 3 red balls, 5 blue balls and 2 white balls. If a ball is chosen at random,
(a) what is the probability that it is
(i) white,   (ii) red,   (iii) blue?
(b) what is the probability that it is
(i) blue or red,   (ii) red or white?

3 Ten cards are numbered 1 to 10. A card is chosen at random. What is the probability that the number of the card
(a) is prime,
(b) is greater than 4,
(c) is between 3 and 7,
(d) is composite?

4 If two cards are chosen from the set of cards in question 3, what is the probability that the sum of their numbers is
(a) 15,   (b) less than 8,   (c) 19?

5 A survey shows that 1 out of every 8 men wears size 11 shoes. What is the probability that a man chosen at random does not wear size 11 shoes?

6 Two dice are rolled together. What is the probability that the sum of the two numbers shown is
(a) 12,   (b) 6,
(c) 10,   (d) greater than 8?

7 Table 14.11 shows the weight of pupils in a class.

Table 14.11

Weight (kg)	25	28	32	35
No. of pupils	3	12	7	8

What is the probability that a pupil chosen at random will weigh
(a) 25 kg,
(b) more than 25 kg,
(c) less than 35 kg ?

8 One letter is chosen from the letters of the alphabet. What is the probability that the letter is
(a) a vowel,
(b) a consonant?

9 A letter is chosen from the letters of the word ACCOMMODATION. What is the probability that the letter chosen is
(a) A,   (b) C,   (c) O,   (d) M?

10 Suppose your teacher has a case containing a pencil, 3 black pens, 3 red pens and 2 green pens. What is the probability that the teacher will mark your work with
(a) a pencil,
(b) a red pen,
(c) a green pen,
(d) a green or a red pen,
(e) a red or a black pen?

Pre-requisites
■ linear inequalities; directed numbers; Cartesian plane

Inequalities in one variable

Graphical representation

$x > -2$ is an inequality in one variable, x. The inequality can be represented by a simple line graph such as that in Fig. 15.1.

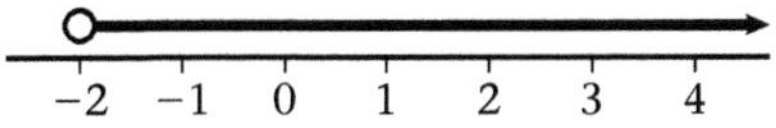

Fig. 15.1

Similarly, Fig. 15.2 is a graph of the set of values given by $x \leqslant 3$.

Fig. 15.2

If $x > -2$ and $x \leqslant 3$ are **simultaneous inequalities**, then the values of x must satisfy both $x > -2$ *and* $x \leqslant 3$. The graphs in Figs. 15.1 and 15.2 can be combined to illustrate the solution set of the simultaneous inequalities as shown in Fig. 15.3.

Fig. 15.3

Note the following:

1 The symbols ● and ○ show whether or not a value is included in the graph. For example, in Fig. 15.3 the 3 is included (●) and the −2 is not included (○).

2 $-2 < x \leqslant 3$ is a short and convenient way of writing the simultaneous inequalities $x > -2$ and $x \leqslant 3$.

Example 1

Illustrate on a single number line the solution set of the simultaneous inequalities $x \geqslant 1$, $-3 < x < 5$.

Fig. 15.4 shows the two given inequalities.

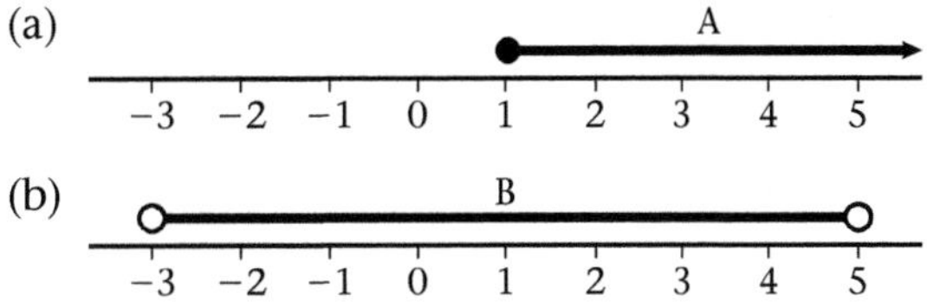

(a)
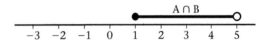

(b)
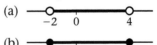

Fig. 15.4

Fig. 15.4(a) shows the set A = {$x: x \geqslant 1$}.
Fig. 15.4(b) shows the set B = {$x: -3 < x < 5$}.
If x belongs to both sets, then $x \in A \cap B$ where A ∩ B = {$x: 1 \leqslant x < 5$}.

Fig. 15.5 shows the required solution set, A ∩ B.

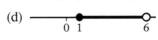

Fig. 15.5

Exercise 15a

1 Fig. 15.6 contains the graphs of six inequalities. Express each inequality in the form $a * x * b$ where a and b are numbers and $*$ may be $<$ or $\leqslant$.

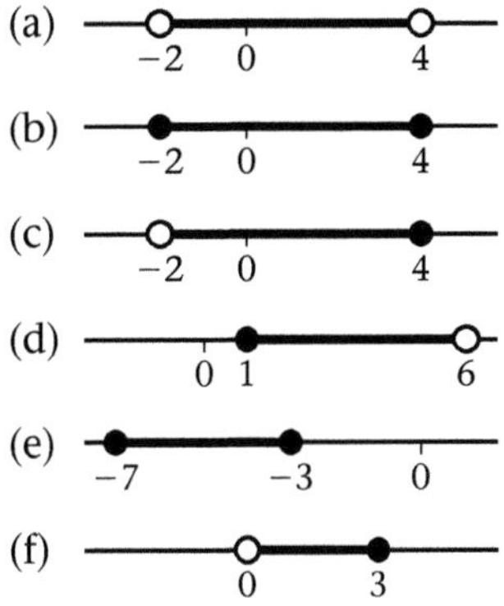

(a) ──○————○──
     −2 0   4

(b) ──●————●──
     −2 0   4

(c) ──○————●──
     −2 0   4

(d) ──────●———○─
          0 1   6

(e) ●————●────
   −7   −3   0

(f) ──────○———●──
            0   3

Fig. 15.6

② Illustrate each of the following inequalities on a number line.

(a) $-4 \leqslant x \leqslant 1$ (b) $-1 < x < 4$
(c) $-5 \leqslant x < -2$ (d) $0 \leqslant x < 3$
(e) $-1 < x < 1$ (f) $7 \leqslant x \leqslant 8$

③ (a) Show the solution sets of $x \leqslant 5$ and $1 \leqslant x \leqslant 7$ on separate number lines.
(b) Hence show the solution set of the simultaneous inequalities $x \leqslant 5$ and $1 \leqslant x \leqslant 7$ on a single number line.

④ Use the number line to illustrate the solution set of the simultaneous inequalities $x \geqslant 1$, and $-3 < x < 5$.

⑤ Illustrate on the number line the set $\{-5 < x < 2\} \cap \{-3 \leqslant x \leqslant 3\}$

Solution by calculation

Example 2

Find the values of x which are multiples of 4 and which satisfy both of the following inequalities:
$2x + 3 < 70, \quad x \geqslant 24$.

$2x + 3 < 70$ and $x \geqslant 24$ are simultaneous inequalities.

If $2x + 3 < 70$
then $2x < 67$ so $x < 33\frac{1}{2}$
Also $x \geqslant 24$

Hence x lies in the range $24 \leqslant x < 33\frac{1}{2}$.

The values of x which lie within that range *and* which are multiples of 4 are 24, 28, 32.

Example 3

List integer values of x which satisfy
$3x - 7 < 24 \leqslant 5x - 8$.

Expressing the inequalities in two separate parts,
     $3x - 7 < 24$ (1)
and $24 \leqslant 5x - 8$ (2)

From (1),
     $3x < 31$ so $x < 10\frac{1}{3}$

From (2),
     $32 \leqslant 5x$ so $x \geqslant 6\frac{2}{5}$

Hence, combining both inequalities,
     $6\frac{2}{5} \leqslant x < 10\frac{1}{3}$

Fig. 15.7 represents the combined inequality.

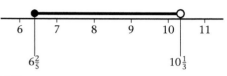

Fig. 15.7

From 15.7, the integer values of x which satisfy both parts of the inequality are 7, 8, 9 and 10.

Exercise 15b

① Express each of the following pairs of simultaneous inequalities in the form $a * x * b$ where a and b are numbers and $*$ may be $<$ or $\leqslant$.

(a) $x \geqslant 3, \; 2x - 3 \leqslant 15$
(b) $25 > 1 - 6x, \; 1 > 3x + 7$
(c) $2x - 7 < 3 < 27 + 4x$
(d) $3x + 8 \leqslant 0 \leqslant 21 + 4x$
(e) $5x - 36 < -1 \leqslant 2x - 1$

② State the integer values of x which are members of the following sets.

(a) $\{x : 2 \leqslant x < 9\}$
(b) $\{x : 1\frac{1}{2} < x < 7\frac{3}{4}\}$
(c) $\{x : -7\frac{3}{4} < x < -1\frac{1}{2}\}$
(d) $\{x : -2\frac{1}{4} \leqslant x \leqslant 3\frac{1}{2}\}$

③ If $6 < x < 7$, which of the following could be values of x?

(a) $+\sqrt{36}$ (b) 2π (c) $\frac{1}{2} \times 14$

(d) $\dfrac{1}{0.15}$ (e) $\dfrac{(-5)^2}{4}$ (f) $\tan 81°$

④ If $6x < 2 - 3x$ and $x - 7 < 3x$, what single range of values of x satisfies both inequalities?

⑤ What is the range of values of x for which $3(1 - x) < 3$ and $3(1 - x) \geqslant 0$ are both satisfied?

⑥ y is such that $4y - 7 \leqslant 3y \leqslant 5y + 8$. Express this inequality in the form $a \leqslant y \leqslant b$ where a and b are both integers.

⑦ Find an integer value of x such that $3x + 5 < 1 < 2x + 6$.

8   List the integer values of x which satisfy
$2(x + 3) \geqslant 20 > 3x - 7$.

9   Express the inequality
$3x - 2 < 10 + x < 2 + 5x$
in the form $a < x < b$ where a and b are
numbers. Hence find the perfect square
which satisfies the given inequality.

10   If x is a multiple of 4, list the integer values
of x which satisfy the inequalities:
$2x + 5 > 37$, $x < 35$.

Word problems involving inequalities

Example 4

A triangle has sides of x cm, $(x + 4)$ cm and
11 cm, where x is a whole number. If the
perimeter of the triangle is less than 32 cm, find
the possible values of x.

Perimeter of triangle $= x + (x + 4) + 11$ cm
Thus $x + (x + 4) + 11 < 32$
$$2x + 15 < 32$$
$$2x < 17$$
$$x < 8\tfrac{1}{2}$$

Also, in any triangle the sum of the lengths of
any two sides must be greater than the length of
the third side.
Thus, $x + (x + 4) > 11$
$$2x + 4 > 11$$
$$2x > 7$$
$$x > 3\tfrac{1}{2}$$

Hence $x < 8\tfrac{1}{2}$ and $x > 3\tfrac{1}{2}$. But x must be a whole
number of cm. Thus the possible values of x are
4, 5, 6, 7, or 8.

Check:
When $x = 4$, perimeter $= 4 + 8 + 11$ cm
$$= 23 \text{ cm}$$
When $x = 8$, perimeter $= 8 + 12 + 11$
$$= 31 \text{ cm}$$

The lowest and highest values of x have been
checked. The perimeters in both cases are less
than 32 cm. There is no need to check the other
values.

Exercise 15c

In each question, first make an inequality, then
solve the inequality.

1   If 9 is added to a number x, the result is
greater than 17. Find the values of x.

2   If 7.3 is subtracted from y, the result is less
than 3.4. Find the values of y.

3   Three times a certain number is not greater
than 54. Find the range of values of the
number.

4   5 times a whole number, x, is subtracted
from 62. The result is less than 40. Find the
three lowest values of x.

5   x is a whole number. If three-quarters of x is
subtracted from 1, the result is always
greater than 0. Find the four highest values
of x.

6   A man gets a monthly pay of x. His
monthly rent is $800. After paying his rent,
he is left with less than $400. Find the range
of values of x.

7   A book contains 192 pages. A boy reads x
complete pages every day. If he has not
finished the book after 10 days, find the
highest possible value of x.

8   A rectangle is x cm long and 10 cm wide.
Find the range of values of x if the area of
the rectangle is not less than 120 cm².

9   A triangle has a base of length 6 cm and an
area of less than 12 cm². What can be said
about its height?

10   Last month a woman had a mass of 53 kg.
She reduced this by x kg so that her mass is
now below 50 kg. Assuming that $x < 6$, find
the range of values of x.

11   A rectangle is 8 cm long and b cm broad.
Find the range of values of b if the perimeter
of the rectangle is not greater than 50 cm
and not less than 18 cm.

⑫ The sides of a triangle are x cm, $x + 3$ cm and 10 cm. If x is a whole number of cm, find the lowest value of x.
Hint: the sum of the lengths of any two sides of a triangle must be greater than the length of the third side.

⑬ An isosceles triangle has sides of length x cm, x cm and 9 cm. Its perimeter is less than 24 cm and x is a whole number.
(a) Find the lowest possible value of x.
(b) Find the highest possible value of x.

⑭ On a journey of 120 km, a motorist averages less than 60 km/h. Will the journey take more or less than 2 hours?

⑮ A man cycles a distance of 63 km in less than 3 hours. Find the range of his average speed.

Summary

An algebraic statement in which quantities are **not equal** is called an **inequality**. The symbols which connect the left-hand side and the right-hand side of an inequality are $\leqslant, <, \geqslant, >, \neq$.

Inequalities in one variable may be represented graphically on a number line.

The solution set of **simultaneous inequalities** is the intersection of the solution sets of the two inequalities, that is, the set of values which satisfy both inequalities. This solution set may be represented by a set of points on a **single** number line. When the set of points is bounded by a shaded circle (●), the end point is included in the solution, that is the inequality symbols is $\leqslant$ or $\geqslant$; when the end point is shown as an empty circle (○) that value is not included in the solution set and the inequality symbol is $<$ or $>$.

When an inequality is multiplied by a **negative number**, the equality sign is **reversed**.

Practice exercise P15.1

Write the following inequalities, using one of the inequality symbols.

① -5 is less than -2.

② h, the height of a cylinder, is greater than or equal to r, the radius of the base.

③ $(x + 2)$ is not equal to $(2x - 3)$.

④ w, the weight of a parcel, must not be greater than 5 kg.

Practice exercise P15.2

Solve the following linear inequalities, where each unknown belongs to the set of integers.

① $x + 3 > 1$ ② $2y < 6$

③ $2y > 6 - y$ ④ $x - 3 < 1$

⑤ $3p \leqslant 15$ ⑥ $2(n + 2) \geqslant 10$

⑦ $\frac{a}{2} > 1$ ⑧ $-1 \leqslant h - 2 < 3$

⑨ $3d - 1 \leqslant 5$ ⑩ $\frac{h}{3} > 2$

⑪ $-2 \leqslant j + 3$ ⑫ $2(m + 4) \geqslant 12$

⑬ $4(x - 2) < 3x - 5$ ⑭ $3 > 4 - h > -3$

⑮ $\frac{x + 7}{2} > 5$ ⑯ $\frac{x}{2} + \frac{3}{4} \geqslant \frac{5x}{6} - \frac{7}{12}$

⑰ $\frac{2x + 3}{3} \leqslant 5$ ⑱ $\frac{3x + 1}{2} > \frac{2x - 1}{3}$

Practice exercise P15.3

For each of the following, write one or more linear inequalities to give the solution set.

① A boy is now n years old. In 5 years' time, he will be over 21 years of age.

② The pass mark in an examination was 45. Andy earned p marks and passed. June got f marks and failed.

③ The passing grade on a test was m%. Ann earned 75% and passed. Ken earned 65% and did not pass.

④ A team had to spend $700 or less to buy 12 medals. n medals cost $70 each, and the remainder cost $50 each.

5 The side of a square is s cm and its perimeter is P cm. The square has an area of less than 100 cm².

6 One truck carries a load of more than 3 tonnes but less than 4 tonnes. The total mass carried by six trucks is t tonnes.

7 A bus must travel a distance of 120 km in not less than $1\frac{1}{2}$ hours and not more than 3 hours. The average speed is v km/h.

Practice exercise P15.4

The diagrams below are the graphs of linear inequalities. Express each inequality in the form $a * x * b$, where a and b are numbers, and $*$ may be $<$ or $\leqslant$.

1
-1 3

2
-3 2

3
-4 4

Practice exercise P15.5

(a) Solve the following linear inequalities for the set of integers.

(b) If there is a finite solution set, list its members.

1 $2 - 3x > 1$ **2** $7 \leqslant 15 + 2y$

3 $4(1 - 2j) < 6$ **4** $5t \geqslant 2t + 6$

5 $-1 < 1 - n < 2$ **6** $-4 < x \leqslant 0$

7 $-2 \leqslant x < 5$ **8** $3 < x \leqslant 11$

9 $0 \leqslant x \leqslant 8$ **10** $-7 < x < 4$

Practice exercise P15.6

1 (a) Find the solution set for *each* of the following inequalities.
   (i) $x + 5 < 9$ (ii) $2x + 1 \geqslant -5$
(b) Use a number line to illustrate each of those solution sets.
(c) Hence, list the members of
   (i) $\{x: x + 5 < 9\} \cap \{x: 2x + 1 \geqslant -5\}$
   (ii) $\{x: x + 5 < 9\} \cup \{x: 2x + 1 \geqslant -5\}$

2 Illustrate on the number line the set $\{-5 < x < 2\} \cap \{-3 \leqslant x \leqslant 3\}$.

3 Use the number line to illustrate the solution set of the simultaneous inequalities $x \leqslant 1$ and $-3 < x < 5$.

Practice exercise P15.7

1 The longest side of a triangle must be 5 cm more than the shortest side and 2 cm more than the third side. The perimeter of the triangle is greater than 21 cm but not greater than 28 cm.
(a) Write algebraic expressions for the sides of the triangle.
(b) Write a linear inequality for the perimeter.
(c) Given that the lengths of the sides are whole numbers, find the corresponding values for the sides of the triangles that satisfy these conditions.

2 A woman's weekly salary is $1500. After paying rent and buying food, she has not more than $600. Her food bill is twice the amount that she pays for rent. Find the maximum amount that she can pay for rent.

3 Three times a whole number is subtracted from 46. The result is less than 20. Work out the four smallest values of the original number.

4 State whether the solution of the following example is correct or incorrect, giving the reason for your answer.
List three members of the solution set of $3x - (x + 1) > 7$, where $x \in \{integer\}$.
Working $2x + 1 > 7$
           $2x > 7 + 1$
           $x > 4$
Answer $x \in \{5, 6, 7, ...\}$

5 The traffic regulations state that the speed of vehicles along a certain road must not be more than 80 km/h but must be less than 40 km/h.
(a) Express this rule using inequality symbols.
(b) Show the range of values using a number line.

Revision exercises and tests

Chapters 9–15

Revision exercise 5 (Chapters 9, 15)

1. Sketch each of the following inequalities on a number line.
   (a) $2 < x < 7$ (b) $-6 \leqslant x \leqslant 0$
   (c) $-4 < x \leqslant -1$ (d) $-9 \leqslant x < 8$

2. Express the inequality $x + 2 < 10 < x + 17$ in the form $a < x < b$ where a and b are integers. Sketch a line graph showing the inequality.

3. (a) If $x - 6 \leqslant 1$ and $2x - 1 > 8$, what is the range of values of x which satisfies both inequalities?
   (b) Sketch the graph of the range of values of x.
   (c) Hence sketch another graph showing values of x which satisfy $x - 6 > 1$ and $2x - 1 \leqslant 8$.

4. (a) Show the solution sets of $3a - 7 < 12$ and $8 - 5a < 20$ on separate number lines.
   (b) Hence show the solution set of the simultaneous inequalities $3a - 7 < 12$ and $8 - 5a < 20$ on a single number line.
   (c) Write down the values of a which are even numbers.

5. A woman had a son when she was x years old. When the son was 12 years old, the woman was less than 40 years old.
   (a) Write this statement as an inequality.
   (b) Determine the range of values of x.

6. Table R3 gives corresponding values of x and y for the relation $y = 2x^2 - 3x - 9$.

   Table R3 $y = 2x^2 - 3x - 9$

x	-2	-1	0	1	2	3	4
y	5	-4	-9	-10	-7	0	11

   (a) Use scales of 2 cm to 1 unit on the x-axis and 1 cm to 1 unit on the y-axis and draw the graph of $y = 2x^2 - 3x - 9$.

Use your graph to find
   (b) the values of x when $y = 0$,
   (c) the values of x when $y = -9$.

7. (a) Copy and complete Table R4, which gives the table of values for $y = x^2 + 3x - 8$.

   Table R4 $y = x^2 + 3x - 8$

x	-5	-4	-3	-2	-1	0	1	2
x^2	25		9		1		1	4
$+3x$	-15		-9		-3		3	6
-8	-8		-8		-8		-8	-8
y	2		-8		-10		-4	2

   (b) Use scales of 2 cm to 1 unit on both axes to draw the graph of the function $y = x^2 + 3x - 8$.
Use your graph to find
   (c) the value of y when $x = 2.5$,
   (d) the values of x when $y = 0$.

8. (a) Calculate the table of values for $y = 2x^2 + 6x + 1$ for $-4 \leqslant x \leqslant 1$.
   (b) Use a scale of 2 cm to 1 unit to draw the graph of $y = 2x^2 + 6x + 1$.
Use your graph to find
   (c) the values of x when $y = 0$,
   (d) the minimum value of $2x^2 + 6x + 1$.

9. (a) Copy and complete Table R5, which gives the table of values for $y = x^2 - 5x + 4$.

   Table R5 $y = x^2 - 5x + 4$

x	-1	0	1	2	3	4	5
y	10	4		-2			4

   (b) Use a scale of 2 cm to 1 unit to draw the graph of $y = x^2 - 5x + 4$.
Use your graph to find
   (c) the solution set for $x^2 - 5x + 4 = 0$,
   (d) the range of values of x for which $x^2 - 5x + 4$ is negative.

10 (a) Copy and complete Table R6, which gives the table of values for
$y = 4 + 5x - x^2$.

Table R6 $y = 4 + 5x - x^2$

x	-1	0	1	2	3	4	5	6
$+4$	$+4$			$+4$	$+4$	$+4$		$+4$
$+5x$	-5			10	15	20		30
$-x^2$	-1			-4	-9	-16		-36
y	-2			10	10	8		-2

(b) Use a scale of 2 cm to 1 unit to draw the graph of $y = 4 + 5x - x^2$.
Use your graph to find
(c) the solution set for $4 + 5x - x^2 = 0$,
(d) the maximum value of $4 + 5x - x^2$.

Revision test 5 (Chapters 9, 15)

1 The range of values of x shown in Fig. R21 is

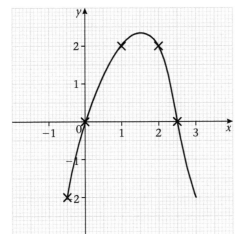

Fig. R21
A $-2 \leqslant x < 5$ B $5 < x < -2$
C $-2 < x \leqslant 5$ D $5 \leqslant x \leqslant -2$

2 If x is an integer, what is the greatest value of x which satisfies $3x + 25 < 2 \leqslant x + 13$?
A -11 B -10 C -8 D -7

Fig. R22 shows the graph of a function y. Use Fig. R22 to answer questions 3 and 4.

Fig. R22

3 The range of values of x for which y is positive is
A $0.5 \leqslant x \leqslant 3$
B $0 \leqslant x \leqslant 2.5$
C $0.5 < x < 3$
D $0 < x < 2.5$

4 Which of the following is *not* true for the graph of the function y in Fig. R22:
A $x = 1, 2$ when $y = 2$
B x has a maximum value
C y has a maximum value
D   when $y = 0$, $x = 0, 2.5$

5 A driver estimated the time required for a 10 km journey at a speed of v km h^{-1}. The actual time taken t min, was longer than the estimate. This statement written in symbols is

A $\dfrac{10}{v} > t$ B $t > \dfrac{10}{v}$

C $\dfrac{10}{v} > \dfrac{t}{60}$ D $\dfrac{t}{60} > \dfrac{10}{v}$

6 If $6(x + 1) > -4$ and $5 > 2(x - 3)$, what range of values of x and which integers satisfy both inequalities?

7 (a) Calculate and complete the table of values for $y = x^2 - 5x + 6$ for $-1 \leqslant x \leqslant 4$.
(b) Use a scale of 2 cm to 1 unit to draw the graph of $y = x^2 - 5x + 6$.
Use your graph to find
(c) the solution set for $x^2 - 5x + 6 = 0$,
(d) the minimum value of $x^2 - 5x + 6$.

8 Table R7 gives corresponding values of x and y for the relation $y = 2x^2 - 5x - 6$.

Table R7 $y = 2x^2 - 5x - 6$

x	-2	-1	0	1	2	3	4
y	12	1	-6	-9	-8	-3	6

(a) Use scales of 2 cm to 1 unit on the x-axis and 1 cm to 1 unit on the y-axis and draw the graph of $y = 2x^2 - 5x - 6$.
(b) Use your graph to find
(i) the least value of y,
(ii) the solutions of $2x^2 - 5x - 6 = 0$.

9 Fig. R23 shows the graph of
(i) $y = 2x^2 - 7x + 6$ and (ii) $y + 4x = 8$.
Using the graph find
(a) the solutions of $x^2 - 7x + 6 = 0$,
(b) the solutions of $x^2 - 7x + 6 = 8 - 4x$.

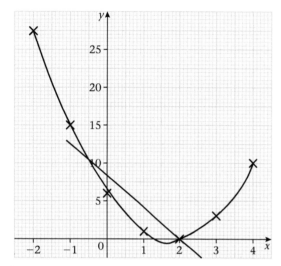

Fig. R23

10 x is any prime number which satisfies both
$x - 1 > 15$ and $4x - 9 < 99$. Find all the
possible values of x.

Revision exercise 6 (Chapters 10, 11)

1 Fig. R24 shows the plan of a corner of a
road.

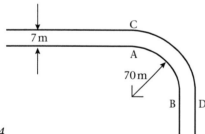

Fig. R24

The road is 7 m wide. The corner is in the
shape of a quarter of a circle. The inside
radius of the corner is 70 m. Carl walks
round the corner on the inside, from A to B.
Tom walks round the corner on the outside,
from C to D. How much further does Tom
walk than Carl? (Use the value $\frac{22}{7}$ for π.)

2 Two friction wheels are of diameter 20 mm
and 200 mm respectively. They touch at P
and rotate without slipping (Fig. R25).

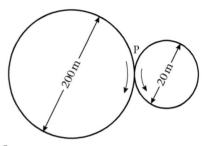

Fig. R25

Calculate the number of turns made by the
small wheel when the large wheel rotates
through 60°.

3 The shortest distance between two towns,
A and B, on a map is 2.5 cm. The map is
drawn to a scale of 1 : 2 500 000. Calculate,
in kilometres, the actual shortest distance
between the two towns A and B.
[CXC June 91]   (3 marks)

4 The ratio of the areas of two similar
rectangles is $\frac{8}{50}$.
(a) Find the ratio of their lengths.
(b) If the width of the smaller rectangle is
    11 cm, find the width of the other
    rectangle.

5 A plan of a house is drawn on a scale of
1 : 80. On the plan, the biggest room has an
area of about 30 cm². Find the true area of
the room to the nearest m².

6 Two circular metal discs are of radius 9.9 cm
and 13.2 cm respectively.
(a) Express the ratio of their areas in its
    simplest terms.
(b) The discs are melted down and recast as
    a single disc of the same thickness as
    before. Find the radius of this disc.

7 A road sign is in the shape of a metal
triangle of height 70 cm and costs $117.6.
How much will a similar road sign of height
1 m cost?

8 (a) State the coordinates of the vertices of triangle OAB in Fig. R26.

(b) Triangle OAB is enlarged by scale factor 2 with O as the centre of enlargement, to give triangle OA′B′. State the coordinates of the vertices of the enlargement, △OA′B′.

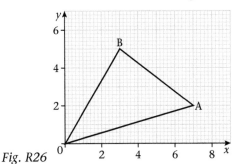

Fig. R26

9 In Fig. R27, △ABC is enlarged to give △PQR.

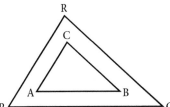

Fig. R27

(a) Copy the diagram and mark the centre of enlargement.

(b) State the scale factor of the enlargement.

10 In Fig. R28, △A₁B₁C₁ is an enlargement of △ABC. Describe the enlargement fully.

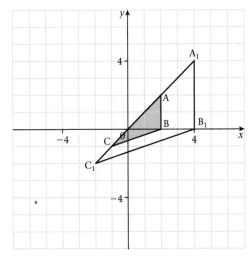

Fig. R28

Revision test 6 (Chapter 10, 11)

Use the following information to answer questions 1 and 2.

In Fig. R29, LM, MN and LN are diameters of the semicircles.

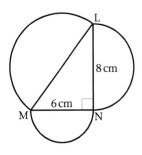

Fig. R29

1 The perimeter, in cm, of the shape in Fig. R29, in terms of π, is

A 24π B 25π

C 48π D 50π

2 The area, in cm², of the shape in Fig. R29, in terms of π, is

A $(24\pi + 24)$ B $(25\pi + 48)$

C $(50\pi + 24)$ D $(50\pi + 48)$

3 The scale of a map is $1:2500$. How many m² does an area of 1 cm² on the map represent?

A 5   B 25

C 625   D 2500

4 Two similar cones have base diameters of 10 cm and 35 cm. The small cone is used to fill the big cone with rice. Approximately how many small cones will it take to fill the big cone?

A 4   B 7

C 16   D 43

5 In Fig. R30, △OAB is transformed to △OCD. Which of the following is *not* true for the transformation:

A △OCD is an enlargement of △OAB

B the centre of enlargement is O(0, 0)

C △OAB is enlarged by a scale factor 2

D AB is parallel to CD

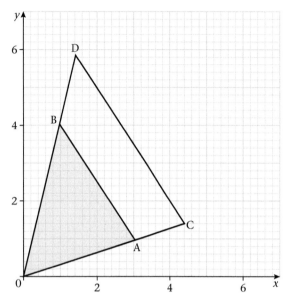

Fig. R30

6 Fig. R31 shows a cross-section of a sheet of corrugated iron.

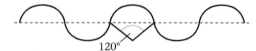

Fig. R31

The sheet is a series of arcs of radius 10 cm, each arc subtending 120° at its centre. If there are 14 such arcs in one sheet, how wide would the sheet be if flattened out?

7 The diagram in Fig. R32, not drawn to scale, represents a flower bed in the shape of a sector BAC of a circle. A is the centre, AB = 10 m and angle BAC = 72°. (Take π as 3.14.)

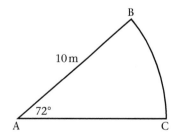

Fig. R32

(a) Calculate
   (i) the length, in metres, of the arc BC,

(ii) the area, in square metres, of the sector BAC.
(b) The flower bed is to be fenced by five strands of wire all around it. The wire is sold in rolls of single strand 20 m long. Calculate the number of rolls needed to fence the flower bed.
(c) The surface of the flower bed is to be covered with topsoil 15 cm deep. Calculate, in cm³, the volume of soil required.
   [CXC June 90]   (10 marks)

8 A gas cylinder is 75 cm long and holds 18 kg of liquid gas when full. How much gas will a similar cylinder, 50 cm long, hold when full?

9 Two similarly shaped cooking pots are made from metal of the same thickness. They have capacities of 20 and 2.5 litres respectively. If the mass of the small pot is 1.5 kg when empty, what is the mass of the big pot when empty? (*Note*: The mass is proportional to the area of metal in the pot since the metal is of the same thickness).

10 (a) On graph paper draw triangle PQR with P(0, 1), Q(5, 3), R(2, 5).
   (b) Enlarge △PQR with centre (2, 0) and scale factor 2, to give △P′Q′R′.
   (c) Write down the coordinates of the vertices of △P′Q′R′.

Revision exercise 7 (Chapters 12, 14)

1 The pie chart in Fig. R33 shows the reading habits of students in a school.

Fig. R33

What percentage of the students read for over 8 hours daily?

2 The bar chart in Fig. R34 gives the estimated costs of providing rural health services for the years 2004 to 2006.

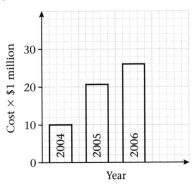

Fig. R34

(a) What was the cost in 2006?
(b) By how much did the 2006 cost exceed the 2004 cost?
(c) What was the average cost per year for the three years?

3 A card is chosen at random from a pack of 52 cards. Find the probability that it it

(a) a six   (b) a red six
(c) a club   (d) a red club

4 Table R8 shows the total number of goals scored by teams during a football competition.

Table R8

Total number of goals	0–2	3–5	6–8	9–11	12–14
Number of teams	5	4	4	3	2

(a) Represent the information in a bar chart.
(b) How many teams participated in the competition?

5 State which of the following variables are discrete and which are continuous.

(a) The number of advertisements during a television show.
(b) The number of cars in a car park.
(c) The weight of students in your class.
(d) The number of books in a library.

6 (a) Draw a frequency polygon for the frequency distribution in Table R9.

Table R9

Class	1–5	6–10	11–15	16–20	21–25
Frequency	5	8	12	7	3

(b) Calculate the mean of the data.

7 There are x black balls and y white balls in a bag. A ball is picked at random.

(a) Write down an expression in x and y which gives the probability of picking a black ball.
(b) If there are 24 balls altogether, find how many are black if the probability of picking a white ball is $\frac{1}{8}$.

8 Table R10 shows the heights of plants to the nearest 5 cm together with the corresponding numbers of plants.

Table R10

Height (cm)	20	25	30	35	40	45
No. of plants	5	3	1	1	3	2

Draw a histogram to illustrate this information.

9 A coin and a die are thrown. What is the probability of getting an even number *and* a tail?

10 Table R11 gives the masses, in kg, of 50 international athletes.

Table R11

67	75	79	56	59	60	64	76	58	80
54	65	78	66	65	65	70	62	70	62
70	61	83	51	74	69	59	73	71	74
73	81	69	82	71	53	67	72	66	70
85	63	58	69	75	61	62	68	52	68

(a) Taking class intervals of 51–55, 56–60, ..., 81–85, construct
(i) the frequency distribution,
(ii) a histogram to show this information.
(b) Choose a suitable working mean and hence find the average mass of the athletes.

Revision test 7 (Chapters 12, 14)

Fig. R35 is a pie chart showing the distribution of education establishments in a country. Use Fig. R35 to answer questions 1 and 2.

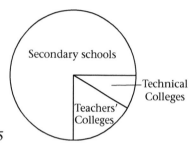

Fig. R35

1. If there are 14 Technical Colleges, how many Teachers' Colleges are there?

   A 7   B 14   C 21   D 28

2. A student doing question 1 in this test does not read the question and just picks one of the answers at random. What is the probability that he picks the correct answer?

   A 0   B $\frac{1}{4}$ C $\frac{3}{4}$ D 1

3. Which of the following statements is *not* true for a histogram?

   A The area of each rectangle is proportional to the frequency represented.
   B The frequency polygon must always be drawn on the histogram.
   C The histogram may be used to represent continuous data.
   D The histogram looks like a bar chart with bars that touch.

4. In a frequency distribution table with a class 30–39 kg,

   A 30 kg and 39 kg are the class boundaries
   B 30.5 kg and 39 kg are the class boundaries
   C 30.5 kg and 39.5 kg are the class limits
   D 29.5 kg and 39.5 kg are the class boundaries

5. Table R12 shows the number of pupils (*f*) scoring a given mark (*x*) in a test.

Table R12

x	2	3	4	5	6	7	8	9	10	11	12
f	3	8	7	10	13	17	15	14	6	2	5

The probability that pupil gets at least 8 marks is

A $\frac{15}{100}$ B $\frac{28}{100}$ C $\frac{43}{100}$ D $\frac{57}{100}$

6. Table R13 gives the number of hours devoted to different exercise activities during a week.

Table R13

Activity	Hours/Week
aerobics	5
swimming	3
dancing	4
running	6
biking	2

Draw a pie chart to show this information.

7. Table R14 gives the time, in hours, that thirty athletes spent running around a field during a one-week period.

   (a) Make a table showing the frequency distribution for the data grouped in intervals: 3–5, 6–8, ….
   (b) Represent the information by an appropriate diagram.

Table R14

8	9	10	11	12	7	6	9	10	11
5	6	7	8	9	7	8	10	10	9
7	6	5	4	4	11	9	6	5	4

8. A bag contains 4 green marbles, 3 red marbles and 5 black marbles. What is the probability that a marble chosen at random will be (a) red, (b) green, (c) black, (d) neither red nor black.

9. Fifty people were asked to estimate the number of bananas on a bunch. The estimates given are listed in Table R15.

Table R15

65	63	72	75	73	71	69	68	63	67
74	74	62	74	70	70	71	66	69	70
78	71	66	69	70	73	61	62	64	68
74	76	63	65	70	69	68	67	62	61
66	72	61	60	67	78	74	75	69	72

(a) Make a frequency distribution table taking equal intervals, 60–62, 63–65, …, 78–80.

(b) Represent this information on a bar chart.

(c) State the modal class of the data.

⑨ Forty persons were asked to state the time, in minutes, spent travelling to work in a day. Table R16 gives the data.

Table R16

50	54	63	51	57	59	82	73	84	92
53	64	72	83	59	53	80	75	76	77
77	78	82	84	86	90	72	76	83	58
92	95	96	50	72	65	66	67	72	84

(a) Taking class intervals of 50–54, 55–59, …, construct the frequency distribution of the data.

(b) Draw a histogram of the frequency distribution.

(c) State the modal class of the data.

(d) Calculate the mean of the data. Give your answer to the nearest minute.

Revision exercise 8 (Chapter 13)

① The scale of the plan of a building is 1 : 50.

(a) What length on the plan represents 12 m?

(b) What length on the building is represented by 9.6 cm?

② In Fig. R36, state the bearing of I, J, K, L, M, from X.

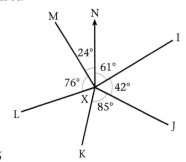

Fig. R36

③ Draw sketches to show the following bearings of X from Y.

(a) 215°   (b) 298°   (c) 018°   (d) 125°

④ (a) Make a scale drawing of the information in Fig. R37.

(b) Hence find the height of the flag-pole to the nearest 0.1 m.

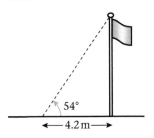

Fig. R37

⑤ From the top of a tower 14 m high, the angle of depression of a man is 32°.

(a) Make a scale drawing, stating the scale used.

(b) Find, to the nearest $\frac{1}{2}$ m, the distance of the man from the foot of the tower.

⑥ A man travels 3 km south, then 4 km south-west and finally 5 km west.

(a) Make a scale drawing, stating the scale used.

(b) Find this final distance and bearing from his starting point.

⑦ A man cycles south for a distance of 4 km. He then cycles 7 km on a bearing 036°. Make a scale drawing of his journey. Hence find

(a) how far east,

(b) how far north

he is from his starting point.

⑧ When the elevation of the sun is 33°, a man has a shadow 2.3 m long. Make a scale drawing and hence find, to the nearest 5 cm, the height of the man.

Revision test 8 (Chapter 13)

① Fig. R38 shows a right square-based pyramid resting so that its base is horizontal.

Which of the points A, B, C, D is vertically below V?

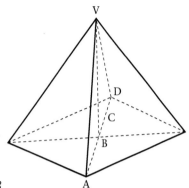

Fig. R38

② The bearing of P from Q is 063°. The bearing of Q from P is

A 027° B 117° C 153° D 243°

③ 3 cm on a map represents a distance of 60 km. If the scale is expressed in the ratio 1 : *n*, then *n* =

A 20   B 2000
C 6000   D 2 000 000

④ The angle of elevation of the top of a tower from a point 23 m from its base on level ground is 50°.
(a) Make a scale drawing, stating the scale used.
(b) Find the height of the tower to the nearest metre.

⑤ A woman walks 2 km N, 3 km NE and finally 2 km E.
(a) Make a scale drawing of her journey.
(b) Find (i) how far north, (ii) how far east she is from her starting point.

⑥ A sphere of radius 13 cm rests on a circular hole of radius 12 cm with a section of the sphere below ground. Calculate the height of the sphere above ground.

General revision test B (Chapters 9–15)

① The length, *d* cm, of the side of a square tile is 5 cm less than the width, *x* cm, of a border. This statement is written in symbols as

A *d* − *x* < 5   B *x* − *d* < 5
C *x* − 5 = *d*   D *d* − 5 = *x*

② What is the least value of *x* for which
$$\frac{5x}{6} \geqslant 10 \geqslant x - 7?$$
A 11   B 12   C 13   D 17

③ The surface area of a cylinder is 31 cm². The surface area of a similar cylinder 3 times as high is

A 62 cm²   B 93 cm²
C 124 cm²   D 279 cm²

④ The scale of a model aeroplane is 1 : 20. If the area of the wings on the model is 600 cm², find, in m², the area of the wings of the aeroplane.

A 12 m²   B 24 m²
C 48 m²   D 72 m²

⑤ Fig. R39 shows two gearwheels arranged so that the small wheel drives the large one.

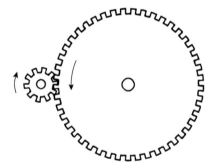

Fig. R39

The diameters of the wheels are 10 cm and 45 cm. Find the angle that the large wheel turns through when the small wheel makes 1 revolution.

A 40°   B 45°   C 60°   D 80°

⑥ The bearing of X from Y is 196°. The bearing of Y from X is

A 016° B 074° C 106° D 164°

Use the data in Table R17 to answer questions 7, 8, 9. Table R17 shows the marks distribution obtained by a class in a test marked out of 30.

Table R17

Score	0–4	5–9	10–14	15–19	20–24	25–30
Frequency	2	4	5	12	4	3

7 The modal class is

   A 0–4   B 5   C 12   D 15–19

8 The frequency of the modal class is

   A 2   B 4   C 12   D 30

9 The probability that a mark chosen at random will be at most 24 is

   A $\frac{27}{30}$ B $\frac{24}{30}$ C $\frac{23}{30}$ D $\frac{4}{30}$

10 The mean of three numbers is 9. The mode is 11. The lowest of the three numbers is

   A 3   B 5   C 8   D 9

11 Calculate the mean of the data in Table R17. Give your answer to 3 significant figures.

12 Express the following inequalities in the form $p * x * q$ where p and q are numbers and $*$ may be $<$ or $\leqslant$. Sketch a line graph of each inequality.

   (a) $7 - 2x \leqslant 5 \leqslant 17 - 3x$

   (b) $4x + 1 > 6x - 8 > 5x - 10$

13 The pie chart in Fig. R40 shows how a company spends its income.

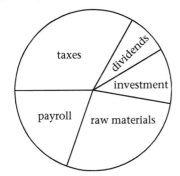

Fig. R40

   (a) Estimate the percentage of the income spent on each item.

   (b) If the company paid dividends of $7500, what was the total revenue?

14 A plane flies 200 km on a bearing 032°. It then flies 350 km on a bearing 275°.

   (a) Find the bearing and distance of the plane from its starting point.

   (b) Find how far north and how far west the plane is from its starting point.

15 Two similar solid cylinders have their heights in the ratio $2:1$. Given that the smaller cylinder has a radius of 3.5 cm and a mass of 3.08 kg, calculate (a) the height and (b) the mass of the larger cylinder. (Density of material is 5 g/cm³.)

16 From the top of a tree, the angle of depression of a stone on horizontal ground is 55°. If the stone is 8 m from the foot of the tree, find, by scale drawing, the height of the tree.

17 Triangle ABC with vertices A(0, −2), B(4, 2) C(−4, 5) is enlarged to give triangle A′B′C′ with vertices A′(0, −1), B′(2, 1), C′(−2, 2½). Using a scale of 2 cm to 1 unit, draw triangles ABC and A′B′C′ on the same axes. State (a) the scale factor (b) the coordinates of the centre of enlargement.

18 A circular tank of diameter 8 m and depth 12 m is built with a conical roof of height 2 m.

   (a) Make a scale drawing of the tank, clearly stating the scale used.

   (b) Calculate

      (i) the slant height of the roof, giving your answer to 2 s.f.

      (ii) the capacity of the tank, giving your answer to the nearest m³.

19 The mean mark in a test taken by twenty students is 65. The pass mark is 46 and four students failed the test. Determine the lowest possible mean mark for the students who passed the test.

20 (a) Calculate and construct the table of values for the function $y = 2x^2 + x - 6$ for $-3 \leqslant x \leqslant 2$.

   (b) Draw the graph of the function $y = 2x^2 + x - 6$.

   (c) Using the graph find the solutions of $2x^2 + x - 6 = 0$.

   (d) State the approximate least value of $2x^2 + x - 6$.

Revision course

The remainder of this book is a revision course for students taking examinations set by the Caribbean Examinations Council at both the Basic and General Proficiency levels in Mathematics.

Content of revision course

Computation and number theory

Factors, multiples, sequences

Example 1

(a) List all the factors of 24, 60.
(b) List the prime factors of 84, 140, 231.

(a) Factors of 24 are 1, 2, 3, 4, 6, 8, 12, 24
     Factors of 60 are 1, 2, 3, 4, 5, 6, 10, 12, 15, 20, 30, 60
(b) Prime factors of 84 are 2, 3, 7
     Prime factors of 140 are 2, 5, 7
     Prime factors of 231 are 3, 7, 11

Example 2

Find the HCF of the numbers in (a) Example 1(a), (b) Example 1(b).

(a) By inspection, HCF of 24 and 60 is 12.
(b) $84 = 2 \times 2 \times 3 \times 7$
     $140 = 2 \times 2 \times 5 \times 7$
     $231 = 3 \times 7 \times 11$
     HCF of 84, 140 and 231 is 7

Example 3

(a) Write down the first five multiples of 2, 3, 6, 8, 12.
(b) State the LCM of these numbers.

Required multiples of 2 are 2, 4, 6, 8, 10
Required multiples of 3 are 3, 6, 9, 12, 15
Required multiples of 6 are 6, 12, 18, 24, 30
Required multiples of 8 are 8, 16, 24, 32, 40
Required multiples of 12 are 12, 24, 36, 48, 60
By inspection, the LCM is 24

Note: since 6 is a multiple of 2 and of 3, multiples of 6 are also multiples of 2 and 3.

Example 4

Evaluate the LCM of 9, 15, 24.

$9 = 3 \times 3$
$15 = 3 \times 5$
$24 = 2 \times 2 \times 2 \times 3$

Since a multiple of a number must have *all* the factors of the number, LCM of 9, 15, 24 is
$3 \times 3 \times 5 \times 2 \times 2 \times 2 = 360$

Example 5

Three signals, A, B and C, are given at the same instant at 10:00. A is repeated in 3 min, B in 4 min and C in 5 min. At what time will the signals next be sounded at the same instant?

Find the LCM of 3, 4 and 5.
LCM is $3 \times 4 \times 5 = 60$
Hence, the signals will next be given at the same instant after 60 mins, that is at 11:00.

Example 6

Write down the next two terms in the sequence
(a) 2, 5, 9, 14, 20, ...
(b) 208, 104, 52, 26, ...

(a) 2, 5, 9, 14, 20, ...
     $= 2, (2 + 3), (5 + 4), (9 + 5), (14 + 6), ...$
     By taking differences it can be seen that the next term is obtained by adding 7 to the previous term, and the next by adding 8 to the previous term, and so on.
     The next two terms are $(20 + 7)$ and $(27 + 8)$, that is 27 and 35.
(b) 208, 104, 52, 26, ...
     $= 208, \frac{208}{2}, \frac{104}{2}, \frac{52}{2}, \frac{26}{2}, \frac{13}{2}, ...$
     The next two terms are 13 and $\frac{13}{2}$.

Exercise 16a

1 List the following.
   (a) Prime numbers between 6 and 30
   (b) Prime factors of 105
   (c) Multiples of 7 less than 50

2 Express each of the following as a product of its prime factors.
   (a) 12, 18, 30
   (b) 28, 63, 98, 147

③ Evaluate the (a) LCM (b) HCF of the numbers in each part of question 2.

④ Find the next four terms of the following sequences.
(a) 6, 10, 14, 18, 22, ...
(b) 5, 10, 15, 20, 25, ...
(c) 1, 4, 9, 16, 25, ...
(d) 12, 6, 3, $\frac{3}{2}$, $\frac{3}{4}$, ...

⑤ Given the sequence 3, 7, 15, 31, 63, 127, 255 list (a) the prime numbers, (b) multiples of 3, (c) the next two terms.

⑥ Bags of different sizes are to be filled with marbles so that each bag must contain either 15, 18, 24 or 42 marbles. The number of different sizes of bags required is not known. Find the smallest number of marbles which must be bought so that there are no marbles left over for any size of bag used.

Number bases

Example 7

State the value, as a decimal number, of the digit 1 in each of the following. (a) 2314_5 (b) $100\,000_2$ (c) 7165_8

(a) In 2314_5, the decimal value of the digit 1 is $1 \times 5^1 = 5$.
(b) In $100\,000_2$, the decimal value of the digit 1 is $1 \times 2^5 = 32$.
(c) In 7165_8, the decimal value of the digit 1 is $1 \times 8^2 = 64$.

Example 8

Calculate the following.
(a) $1101_2 + 11\,001_2$ (b) $11\,010_2 - 1111_2$
(c) $111_2 \times 101_2$

(a)
$$\begin{array}{r} 1\,101_2 \\ +\ 11\,001_2 \\ \hline 100\,110_2 \end{array}$$

(b)
$$\begin{array}{r} 11\,010_2 \\ -\ 1\,111_2 \\ \hline 1\,011_2 \end{array}$$

(c)
$$\begin{array}{r} 111_2 \\ \times\ 101_2 \\ \hline 111 \\ 0\,00 \\ 11\,1 \\ \hline 100\,011_2 \end{array}$$

Example 9

Evaluate the missing digits indicated by ∗ in the following.

(a)
$$\begin{array}{r} 534_8 \\ +\ 1{*}5_8 \\ \hline 721_8 \end{array}$$

(b)
$$\begin{array}{r} 423_5 \\ -\ 1{**}_5 \\ \hline 224_5 \end{array}$$

(a) $534_8 - 1{*}5_8 = 721_8$
implies that
$721_8 - 534_8 = 1{*}5_8$

$$\begin{array}{r} 721_8 \\ -\ 534_8 \\ \hline 165_8 \end{array}$$

Hence, the missing digit is 6.

(b) $423_5 - 1{**}_5 = 224_5$
implies that
$423_5 - 224_5 = 1{**}_5$

$$\begin{array}{r} 423_5 \\ -\ 224_5 \\ \hline 144_5 \end{array}$$

Hence, the missing digits are 4 and 4.

Exercise 16b

① Find the value, in terms of its base, of the digit 3 in each of the following.
(a) 321_5 (b) 430_8 (c) 537_{10} (d) 932_{16}

② Find the value in base 10 of each of the following.
(a) $11\,001_2$ (b) 33_5
(c) 247_8 (d) 198_{16}

③ Evaluate and write down the number 321_{10} in each of the following bases.
(a) base 2 (b) base 5 (c) base 8

④ Evaluate:
(a) $10\,111_2 - 1001_2$
(b) $110_2 \times 100_2$
(c) $11\,111_2 + 10\,001_2$

⑤ Evaluate, in the given base:
(a) $1001_2 \times 11_2$ (b) $234_8 + 51_8$
(c) $613_8 - 474_8$ (d) $789_{16} + 899_{16}$

⑥ Find the value of n in the following:
$25_n = 37_{10}$

Fractions, decimals, percentages

Example 10

Which is the greater, $\frac{5}{6}$ or $\frac{7}{8}$?

24 is the LCM of the denominators, 6 and 8. Express each fraction with a common denominator of 24.

$$\frac{5}{6} = \frac{5 \times 4}{6 \times 4} = \frac{20}{24}$$

$$\frac{7}{8} = \frac{7 \times 3}{8 \times 3} = \frac{21}{24}$$

Since $\frac{21}{24} > \frac{20}{24}$, $\frac{7}{8}$ is greater than $\frac{5}{6}$

Example 11

Evaluate $2\frac{1}{6} + (3\frac{3}{5} \div 1\frac{1}{8})$.

$$\begin{aligned} 2\frac{1}{6} + (3\frac{3}{5} \div 1\frac{1}{8}) &= 2\frac{1}{6} \left(\frac{18}{5} \div \frac{9}{8} \right) \\ &= 2\frac{1}{6} + \left(\frac{18}{5} \times \frac{8}{9} \right) \\ &= 2\frac{1}{6} + \frac{16}{5} \\ &= 2\frac{1}{6} + 3\frac{1}{5} \\ &= 5\frac{5+6}{30} \\ &= 5\frac{11}{30} \end{aligned}$$

Example 12

Evaluate 6.297×0.251 correct to three decimal places.

$$\begin{aligned} & 6.297 \times 0.251 \\ &= \frac{6297}{1000} \times \frac{251}{1000} \\ &= \frac{6297 \times 251}{1\,000\,000} \\ &= \frac{1\,580\,547}{1\,000\,000} \\ &= 1.580\,547 \\ &= 1.581 \text{ to 3 d.p.} \end{aligned}$$

working:
$$\begin{array}{r} 6\,297 \\ 251 \\ \hline 6\,297 \\ 314\,85 \\ 1\,259\,4 \\ \hline 1\,580\,547 \end{array}$$

Example 13

Find the exact value of $\dfrac{2.25 \times 7.5}{4.5}$

$$\begin{aligned} & \frac{2.25 \times 7.5}{4.5} \\ &= \frac{22.5 \times 75}{45} \quad \textit{(multiplying num. and den. by 10)} \\ &= 0.25 \times 15 \quad \textit{(after equal divisions by 9 and 5)} \\ &= 3.75 \end{aligned}$$

Example 14

When the length of a spring is increased by 8% it becomes 351 mm long. What is its original length?

108% of the original length = 351 mm

1% of the original length = $\dfrac{351}{108}$ mm

Original length (100%) = $\dfrac{351}{108} \times 100$ mm

= 325 mm

Example 15

Express 500 g as a percentage of 3.75 kg.

$$3.75 \text{ kg} = 3750 \text{ g}$$

Required percentage = $\dfrac{500}{3750} \times 100\%$

$$= \frac{40}{3}\%$$

$$= 13\frac{1}{3}\%$$

Exercise 16c

1. Arrange the following fractions in order of size from smallest to largest: $\frac{2}{3}, \frac{11}{15}, \frac{7}{10}, \frac{5}{6}, \frac{4}{5}$

2. Evaluate the following.
   (a) $2\frac{2}{3} + 1\frac{1}{2}$ (b) $5\frac{1}{3} - 2\frac{7}{8}$
   (c) $1\frac{5}{8} + \frac{1}{2}$ (d) $1\frac{5}{6} + 2\frac{2}{3} - 3\frac{1}{2}$
   (e) $\frac{3}{4} \times (1\frac{3}{5} \div \frac{1}{6})$ (f) $1\frac{1}{2} \times 6\frac{2}{5} \div \frac{2}{5}$
   (g) $\dfrac{1\frac{2}{3} \times 7}{3\frac{1}{2}}$ (h) $2\frac{1}{2} \div (1\frac{1}{4} \div 3\frac{1}{3})$

3. (a) $\frac{2}{5}$ of the students in a class do history. If 14 students do history, how many students are there in the class?
   (b) In an election there were three candidates; $\frac{2}{3}$ of the electors voted for the first candidate, $\frac{1}{4}$ for the second candidate and the rest for the third. If the third candidate got 3290 votes, how many votes did the winner get?

4. Calculate 129×54 and use the result to write down the value of
   (a) 1.29×5.4 (b) 12.9×0.54
   (c) 12.9×0.0054 (d) 0.129×0.054

5 Calculate the following without using tables.

(a) 22.7×0.38 (b) $8.848 \div 0.28$
(c) $86.13 \div 2.7$ (d) $0.9916 \div 5.36$

6 (a) Find the exact value of

$$2.55 \times 6.3 - \frac{7.5}{1.25}$$

(b) Write your answer correct to 1 decimal place.
[CXC (adapted) June 87]

7 Express the following quantities in terms of the units shown in brackets.

(a) 405 cm   (km)
(b) 158.7 m   (km)
(c) 905 g   (kg)
(d) 2.4 kg   (g)
(e) 7.03 km   (m)
(f) 1.305 ml   (l)

8 Round off the following numbers to the degree of accuracy shown in brackets.

(a) 3.7846   (3 d.p.)
(b) 75.0794   (2 d.p.)
(c) 144   (2 s.f.)
(d) 9.84   (1 s.f.)
(e) 34.625   (2 s.f.)
(f) 34.625   (2 d.p.)

9 (a) What percentage of 2 is 5?
(b) The original area of a farm is 250 ha. The farmer sells 15% of his land. What area is left?

10 The length round a running track should be 400 m. The actual length is found to be 401.2 m. Calculate the percentage error in the length of the track.

11 A bicycle manufacturer requires wheel spokes to measure 260 mm, with a tolerance (acceptable error) of ±0.5%. Calculate the acceptable range of length for the wheel spokes.

12 A boy spends 57% of his pocket money. If the amount that he spends is $1.40 more than he has left, how much money had he originally?

13 A woman's salary was $2800 in 2004 and was 15% more in 2005. If she paid a tax of $12\frac{1}{2}\%$ of her salary, how much tax did she pay in 2005?

14 A factory produced 3456 radios in 2003 and 2880 in 2004.

(a) Calculate the percentage decrease in production from 2003 to 2004.
(b) How many radios were produced in 2005 if there was a 15% increase over 2004?

15 A man's body-mass increases by 15%. He then goes on a diet and reduces his new body-mass by 15%. is his final mass greater or less than the original, and by how much?

16 The total cost of a camera consists of a basic price plus a tax of 12%. Given that the total cost is $210, calculate the basic price of the camera.

17 Table 16.1 shows certain details of an aeroplane's flight from Trinidad to Kingston (Jamaica) with intermediate stops in Barbados and St Kitts.

Notes:
1   Local time is the time in the given country.
2   The time in Kingston is one hour *behind* the time in the other islands.

Table 16.1

		Local time	Time spent travelling	Distance between airports	Average speed in km/h
Dep.	Trinidad	08:25	a	350 km	b
Arr.	Barbados	09:00			
Dep.	Barbados	10:00	1 h 10 min	780 km	670
Arr.	St. Kitts	11:10			
Dep.	St. Kitts	11:30	c	d	680
Arr.	Kingston	12:45			

(a) Calculate the values of a, b, c and d in Table 16.1.
(b) Excluding the times the aeroplane spent on the ground, calculate the average speed for the journey from Trinidad to Kingston.
[CXC Jan 89]

Ratio, rate

A **ratio** is a numerical way of comparing quantities of the same kind. The quantities should be expressed in the same units. For example, the ratio of \$2 to 40c is $5:1$, whether working in cents, $200:40 = 5:1$, or in dollars, $2:\frac{2}{5} = 5:1$.

Example 16

Which ratio if greater, $3:4$ or $6:11$?

Express each ratio in the form $n:1$.
$3:4 = \frac{3}{4}:1 = 0.75:1$
$6:11 = \frac{6}{11}:1 = 0.545...:1$
The ratio $3:4$ is greater.

Example 17

Decrease \$73.35 in the ratio $5:9$.

Express the ratio $5:9$ as the fraction $\frac{5}{9}$.
Since the money is decreased, multiply \$73.35 by $\frac{5}{9}$.
Required amount $= \$73.35 \times \frac{5}{9}$
$\qquad\qquad\qquad = \$8.15 \times 5$
$\qquad\qquad\qquad = \$40.75$

Example 18

If 9 men paint a building in 21 days, how long would 7 men take?

7 men take more time than 9 men.
The number of men is *decreased* in the ratio $7:9$.
Hence the time taken is *increased* in the ratio $9:7$. Time is to be found, so time comes last in each line of working.
$\qquad$ 9 men take 21 days
$\qquad$ 7 men take $21 \times \frac{9}{7}$ days $= 27$ days.

This example illustrates **inverse** ratio.

Example 19

9 notebooks cost \$12.15. How many can be bought for \$17.55?

Money is *increased* in the ratio $17.55:12.15$. The number of books will be *increased* in the same ratio. The number of books is to be found, so this comes last in each line of working.

\$12.15 is the cost of 9 notebooks
\$17.55 is the cost of $9 \times \dfrac{17.55}{12.15}$ notebooks
$= \dfrac{1755}{135}$ notebooks $= 13$ notebooks

This is an example of **direct** ratio.

Quantities of different kinds may be connected in the form of a **rate**. For example, km/h, g/cm^3 and number of people/km^2 are all examples of rates.

Note that km/h, m/s, g/cm^3 may be written as kmh^{-1}, ms^{-1}, gcm^{-3} respectively.

Example 20

A car travels 132 km in 1 h 15 min. Calculate the speed of the car.

1 h 15 min $= 1\frac{1}{4}$ h
In $1\frac{1}{4}$ h the car travels 132 km.
In 1 h the car travels $\dfrac{132}{1\frac{1}{4}}$ km
$\qquad = \dfrac{132 \times 4}{5}$ km
$\qquad = \dfrac{528}{5}$ km $= 105.6$ km
The speed of the car is 105.6 km/h.

Example 21

A wooden cube has a mass of 50.48 g. If one edge of the cube measures 4 cm, calculate the density of the wood correct to 2 s.f.

Volume of cube $= (4 \times 4 \times 4)\,\text{cm}^3 = 64\,\text{cm}^3$
Density of wood $= \dfrac{50.48}{64}$ g/cm$^3 = \dfrac{6.31}{8}$ g/cm^3
$\qquad\qquad\qquad = 0.788\,75$ g/cm^3
$\qquad\qquad\qquad = 0.79$ g/cm^3 to 2 s.f.

Example 22

The population density of a village is 520 people/km^2. If the village has an area of about 3.3 km^2, find its population to the nearest 100 people.

The population density is 520 people/km^2.
i.e. 1 km^2 contains 520 people
Hence 3.3 km^2 contains 520×3.3 people $=$ 1716 people
The village contains 1700 people (to the nearest 100 people).

Exercise 16d

1. Write the following as ratios in their simplest form.
   (a) $16:20$
   (b) 150c to $1
   (c) $3\frac{1}{3}:8$
   (d) $40\,\text{cm}:1\,\text{m}\,20\,\text{cm}$

2. Express each of the following ratios in the form $1:n$.
   (a) $6:9$
   (b) $3:7$
   (c) $16:12$
   (d) $3.5:0.7$

3. In each case, find out which one of the two ratios is greater.
   (a) $17:8$ or $15:6$
   (b) $\$1.70:\$2$ or $\$3:\$4.80$
   (c) $1.5\,\text{g}:2\,\text{kg}$ or $0.5\,\text{kg}:600\,\text{kg}$

4. The cost price, $750, of a bicycle is reduced by $225.
   (a) Find the ratio of the new price to the old price.
   (b) Express the price reduction as a rate of cents in the dollar.

5. A car goes 60 km in 48 min. Find the speed of the car in km/h.

6. A shop sells oranges at 6 for $4. A trader sells the same kind of oranges at 8 for $4.80. Which price is cheaper, and by how much per orange?

7. A map is drawn on a scale of 1 cm to 5 km.
   (a) Write the scale as a ratio in the form $1:n$.
   (b) What distance does 2.8 cm on the map represent?

8. A town has an area of 80 hectares and a population of 2500 people. Calculate the population density of the town in people/ha correct to the nearest whole person.

9. The density of aluminium is $2.7\,\text{g/cm}^3$. Calculate the volume of a piece of aluminium of mass 13.5 g.

10. The telegraph poles along a road are 20 m apart and a car travels from the first to the fifteenth in $10\frac{1}{2}$ seconds. Calculate the speed of the car in km/h.

11. A train 180 m long travels through a station at 60 km/h. Find how many seconds it takes for the train to pass a man who is standing on the station platform.

12. The ages of a mother and daughter are in the ratio $8:3$. If the daughter's age now is 12, what will be the ratio of their ages in 4 years' time?

13. A person's income is increased in the ratio $47:40$. Find the increase per cent.

14. A tankful of water lasts 15 weeks if 3 litres a day are used. How long will the tankful last if 10 litres a day are used?

15. A train normally travels between two stations at v km/h. If its average speed is increased in the ratio $m:n$, will it take more or less time? In what ratio is the time changed?

Example 23

Three people share 30 eggs in the ratio $1:2:3$. How many eggs does each get?

$1 + 2 + 3 = 6$

1st person gets $\frac{1}{6}$ of 30 eggs $= 5$ eggs

2nd person gets $\frac{2}{6}$ of 30 eggs $= 10$ eggs

3rd person gets $\frac{3}{6}$ of 30 eggs $= 15$ eggs

Check: $5 + 10 + 15 = 30$

Example 24

Mark, Betty and Gina share $180 so that for every $1 that Mark gets, Betty gets 50c, and for every $2 that Betty gets, Gina gets $3. Find Betty's share.

Assuming that Mark has 1 share, then Betty gets a $\frac{1}{2}$ share. Gina gets $1\frac{1}{2}$ times as much as Betty.

Hence Gina's share $= \frac{1}{2} \times 1\frac{1}{2} = \frac{3}{4}$.

They share the money in the ratio

$$1:\frac{1}{2}:\frac{3}{4} = 4:2:3$$
$$4 + 2 + 3 = 9$$
$$\text{Betty's share} = \frac{2}{9} \text{ of } \$180$$
$$= \$40$$

Example 25

Three numbers d, m, n are in the ratio $3:6:4$.

Find the value of $\dfrac{4d - m}{m + 2n}$.

Since $d:m:n = 3:6:4$,
let $d = 3k$, $m = 6k$, $n = 4k$.

Hence $\dfrac{4d - m}{m + 2n} = \dfrac{12k - 6k}{6k + 8k} = \dfrac{6k}{14k} = \dfrac{3}{7}$

Example 26

A and B are partners in a business. A invests $10 000 for one year and B invests $25 000 for nine months. How should they share the first year's profit of $4600?

A's investment $= \$10\,000$ for 12 months
$ = 120\,000$ dollar-months
B's investment $= \$25\,000$ for 9 months
$ = 225\,000$ dollar-months
Ratio of investments, $A:B = 120\,000:225\,000$
$ = 8:15$
$ 8 + 15 = 23$
A should get $\frac{8}{23}$ of $\$4600 = \$1600$
B should get $\frac{15}{23}$ of $\$4600 = \$3000$

Exercise 16e

1. 3 children divide $1.05 between them in the ratio $6:7:8$. What is the size of the largest share?

2. In a lottery, $28 845 is divided between the 1st, 2nd and 3rd prizewinners in the ratio $4:3:2$. How much does the 3rd prizewinner get?

3. $90 is divided between Martin, Bob and Sophie so that Sophie's share is $\frac{3}{8}$ of Bob's and Bob's share is twice that of Martin. How much does Sophie receive?

4. If $a:b:c = 5:2:3$, evaluate
   (a) $\dfrac{a - 2b}{3b - c}$
   (b) $\dfrac{a + b + c}{5a}$
   (c) $a - b:b + c$
   (d) $4a - b:a + 2b - c$

5. Find the value of the ratio $m:n$ if
   (a) $4m = 7n$
   (b) $5m - n = 2m + 5n$

6. 10 kg of coffee costing $36 per kg is mixed with 5 kg costing $48 per kg. What is the cost of the mixture per kg?

7. A shopkeeper has 5 shirts priced at $46.40 each, 3 at $44.00 each and 4 at $32.00 each. He mixes the shirts up and sells them all at $40 each. How much does he gain or lose altogether by doing this? What is his average gain or loss per shirt?

8. Teas costing $17.20 and $15.40 per kg are mixed together in the ratio $2:1$. What is the value of the mixture per kg?

9. Copper sulphate is made from
   32 parts of copper,
   16 parts of sulphur,
   32 parts of oxygen,
   45 parts of water.

   Find the mass of copper, sulphur, oxygen and water in 4.5 kg of copper sulphate.

10. A motorist averages 80 km/h for the first 60 km of a journey and 96 km/h for the next 120 km. What is her average speed for the whole journey?

11. Fred invests $15 000 in a business. Four months later Damon invests $36 000 in the business. At the end of the first year the profit is $7865. How much should Fred and Damon get if the profit is shared according to the amount and duration of investment?

12. Wines at $9 and $10 per litre are mixed with a third wine in the ratio $2:3:4$. If the mixture costs $10.40 a litre, what is the cost of the third wine?

Indices

The following **laws of indices** are true for all non-zero values of a, b and x.

1. $x^a \times b^b = x^{a + b}$
2. $x^a \div x^b = x^{a - b}$
3. $x^0 = 1$
4. $x^{-a} = \dfrac{1}{x^a}$
5. $(x^a)^b = x^{ab}$

Example 27

Table 16.2

Simplify	Working	Result
(a) $4^3 \div 4^5$	$= 4^{3-5} = 4^{-2} = \dfrac{1}{4^2}$	$= \dfrac{1}{16}$
(b) $2^6 \times 2^{-6}$	$= 2^{6+(-6)} = 2^{6-6} = 2^0$	$= 1$
(c) $(2^3)^{-2}$	$= 2^{3 \times -2} = 2^{-6} = \dfrac{1}{2^6}$	$= \dfrac{1}{64}$

Example 28

Simplify (a) $9x^{-3} \times 2x^5$, (b) $(2d^3)^2$, (c) $2a^3 \div a^4$.

(a) $9x^{-3} \times 2x^5 = 9 \times 2 \times x^{-3} \times x^5$
$= 18 \times x^{-3+5} \quad = 18x^2$

(b) $(2d^3)^2 = 2^2 \times (d^3)^2$
$= 4d^6$

(c) $2a^3 \div a^4 = 2(a^3 \div a^4) = 2a^{3-4}$
$= 2a^{-1} = \dfrac{2}{a}$

Exercise 16f

Simplify the following expressions.

1 $3^8 \times 3^3$

2 $2^{-3} \div 2^5$

3 $5^3 \times 5^{-1}$

4 $3^6 \div 3^2$

5 $(2^3)^4$

6 $(5^2)^{-4}$

7 $7^3 \times 7^{-3}$

8 $3^{-5} \div 3^{-3}$

9 $2a^2 \times 5a$

10 $(2a)^2 \times 5a$

11 $2a^2 \times (5a)^3$

12 $3a^{-2} \times 5a$

13 $(2a)^{-2} \times 5a$

14 $2a^{-2} \times (5a)^3$

15 2^{-4}

Standard form

The number $A \times 10^n$ is said to be in **standard form** or in **scientific notation**, if n is a positive or negative integer and $1 \leqslant A < 10$. For example 5.3×10^6 and 9×10^{-2} are in standard form.

Example 29

Find the value of $\dfrac{1.26 \times 10^3}{7 \times 10^{-1}}$, expressing your answer in standard form.

$$\frac{1.26 \times 10^3}{7 \times 10^{-1}} = \frac{1.26}{7} \times \frac{10^3}{10^{-1}}$$
$$= 0.18 \times 10^{3-(-1)}$$
$$= 0.18 \times 10^4$$
$$= (1.8 \times 10^{-1}) \times 10^4$$
$$= 1.8 \times 10^{-1+4} = 1.8 \times 10^3$$

Exercise 16g

1 Express the following numbers in standard form.
(a) 950
(b) 9500
(c) 0.95
(d) 0.0095
(e) 23
(f) 0.000 23
(g) 0.023
(h) 23 000

2 Express the following numbers in ordinary form.
(a) 2.6×10^4
(b) 7.01×10^2
(c) 4.55×10^{-2}
(d) 8×10^{-5}
(e) 3.9×10^6
(f) 6.02×10^3
(g) 1×10^{-3}
(h) 8.7×10^{-1}

3 Express $8 \times 10^{-3} + 5 \times 10^{-2} + 2 \times 10^{-1}$
(a) as a decimal fraction, (b) in standard form.

4 Find the value of each of the following, expressing your answers in standard form.
(a) $(6.2 \times 10^{-3}) + (5.08 \times 10^{-2})$
(b) $(7.3 \times 10^5) - (7.9 \times 10^4)$
(c) $(4.2 \times 10^5) \times (5 \times 10^2)$
(d) $(3.87 \times 10^{-2}) \div (9 \times 10^{-6})$
(e) $\dfrac{2.97 \times 10^4}{1.1 \times 10^{-4}}$

5 (a) Express in standard form
(i) 260, (ii) 0.013
(b) Evaluate $260 \div 0.013$ giving your answer in standard form.

6 Given $m = 9.7 \times 10^4$ and $n = 8.3 \times 10^3$, evaluate (a) $m + n$, (b) $m - n$, giving your answers in standard form.

7 Given that $p = 2.4 \times 10^3$ and $q = 6 \times 10^{-2}$, calculate (a) pq, (b) $\dfrac{p}{q}$, expressing each of your answers in standard form.

8 If $r = 3.6 \times 10^3$ express (a) r^2, (b) $\sqrt{r}$ as numbers in standard form.

9 Given that $p = 3 \times 10^5$ and $q = 6 \times 10^2$, find the value of (a) pq, (b) $p + q$, (c) $\dfrac{p}{q}$, expressing each of your answers in standard form.

10 The pages of a dictionary are numbered from 1 to 1632. The dictionary is 6.4 cm thick (neglecting covers).
   (a) How many thicknesses of paper make 1632 numbered pages?
   (b) Estimate the thickness of 1 page. Give your answer in metres in standard form correct to 1 significant figure.

Revision test (Chapter 16)

Select the correct answer and write down your choice.

1 Which of the following is an irrational number?
   A 0   B $\frac{3}{4}$ C $\sqrt{3}$ D $\sqrt{4}$

2 Given that n is a number greater than 1, which of the following conditions is *not* possible?
   A n is a prime number and a perfect square
   B n is an integer and a rational number
   C n is a prime number and an integer
   D n is an integer and a perfect square

3 The LCM of 10, 15, 35 is
   A 5   B 70   C 210   D 420

4 The HCF of 42, 70, 105 is
   A 5   B 7   C 21   D 35

5 The missing term in the sequence 1, 2, 4, ..., 11, 16 is
   A 6   B 7   C 8   D 9

6 Of the following, the largest number is
   A $11\,111_2$ B 122_8
   C 100_{10} D 61_{16}

7 The product of $\frac{2}{3}$ and its reciprocal is
   A -1 B $-\frac{4}{9}$ C 1   D $\frac{3}{2}$

8 $\frac{3}{5} + \frac{1}{4} =$
   A $\frac{17}{20}$ B $\frac{4}{9}$ C $\frac{3}{9}$ D $\frac{3}{20}$

9 The largest fraction in the set $\{\frac{5}{8}, \frac{3}{5}, \frac{5}{7}, \frac{9}{14}\}$ is
   A $\frac{3}{5}$ B $\frac{5}{8}$ C $\frac{9}{14}$ D $\frac{5}{7}$

10 $(\frac{2}{3} + \frac{3}{4}) \times \frac{4}{5} =$
   A $\frac{1}{3}$ B $\frac{4}{7}$ C $\frac{17}{15}$ D $\frac{19}{15}$

11 $5\frac{1}{2} \div 3\frac{1}{7} =$
   A $\frac{4}{7}$ B $1\frac{2}{7}$ C $1\frac{2}{3}$ D $1\frac{3}{4}$

12 $\dfrac{0.09 \times 0.3}{0.06 + 0.04} =$
   A 0.027   B 0.27   C 2.70   D 27.0

13 Correct to 2 s.f., $0.060\,76 =$
   A 0.060 7   B 0.060 8
   C 0.061   D 0.608

14 The ratio of a pole's shadow to a boy's shadow is $3:2$. If the length of the pole is 210 cm, the boy's height, in cm, is
   A 84   B 126   C 140   D 315

15 There are 24 answers correct out of total of 30 questions. The percentage correct is
   A 20   B 25   C 75   D 80

16 If 30% of a number is 12, the number is
   A 30   B 36   C 40   D 48

17 In standard form, $3.2 \times 10^2 + 2.5 \times 10 =$
   A 3.45×10 B 3.45×10^2
   C 34.5×10 D 34.5×10^2

18 A cyclist covers 9 km in 25 min. Her speed, in m/s, is
   A 5   B 6   C 10   D 16

19 $\sqrt{150}$ may be written as
   A $5\sqrt{6}$ B $15\sqrt{10}$
   C $25\sqrt{6}$ D 75

20 If n is an even number, which of the following is an odd number?
   A $n + 2$ B $n + 3$
   C $3n$ D n^3

Chapter 17

Consumer arithmetic

Profit, loss, discount

Example 1

The cost price of a dress is $180. The dress is sold at a loss of 15%. How much money does the saleslady lose?

Cost price of dress (100%) = $180

Loss = 15% of $180 = $$\left(\frac{15}{100} \times 180\right)$

$$= \$27$$

Example 2

A shopkeeper pays $85 for a radio. He sells it at a profit of 30% on the cost price. Calculate the selling price.

Cost price of radio (100%) = $85
Profit = 30% of cost price

Selling price of radio = 130% of $85

$$= \$\left(\frac{130}{100} \times 85\right)$$

$$= \$\left(\frac{13 \times 17}{2}\right) = \$110.50$$

Example 3

An article costing $60 is sold for $75. Express the profit as a percentage of (a) the cost price, (b) the selling price.

Profit = $(75 − 60) = $15
(a) Profit as percentage of cost price

$$= \left(\frac{15}{60} \times 100\%\right)$$

$$= 25\%$$

(b) Profit as percentage of selling price

$$= \left(\frac{15}{75} \times 100\%\right)$$

$$= 20\%$$

Example 4

When a refrigerator is sold for $558 the profit is 24%. What should be the selling price to make a profit of 28%?

124% of cost price = $558

128% of cost price = $$\$558 \times \frac{128}{124}$$

$$= \$558 \times \frac{32}{31}$$

$$= \$18 \times 32$$

$$= \$576$$

Notice that it was not necessary to find the cost price.

Example 5

Carl sells a second-hand car to Tim and makes a profit of 10%. Tim then sells the car to Anna for $2508, making a loss of 5%. How much did Carl pay for the car?

Method 1

Tim's loss = 5% of his cost price
Tim's selling price = 95% of his cost price

$$= \$2508$$

Tim's cost price (100%) = $$\$\left(\frac{100}{95} \times 2508\right)$$

Tim's cost price = Carl's selling price
Carl's profit = 10% of his cost price
Carl's selling price = 110% of his cost price

$$= \$\frac{100}{95} \times 2508$$

Carl's cost price (100%) = $$\$\left(\frac{100}{110} \times \frac{100}{95} \times 2508\right)$$

$$= \$\left(\frac{200}{19} \times 228\right)$$

$$= \$2400$$

Method 2

Carl paid $\frac{100}{110}$ of what Tim paid.

Tim paid $\frac{100}{95}$ of what Anna paid.

Hence, Carl paid $\frac{100}{110}$ of $\frac{100}{95}$ of what Anna paid.

Hence, Carl paid $\frac{100}{110} \times \frac{100}{95} \times \$2508$

$$= \frac{10}{11} \times \frac{20}{19} \times \$2508$$

$$= 10 \times 20 \times \$12$$

$$= \$2400$$

Example 6

In a sale a discount of $12\frac{1}{2}\%$ is given on a pair of shoes marked at $240. Find the selling price.

Marked price = $240

Discount = $12\frac{1}{2}\%$ of $240

Selling price = $87\frac{1}{2}\%$ of $240

$$= \$\left(\frac{175}{200} \times 240\right) = \$210$$

Example 7

A trader sells rice in bags. 30 bags of rice costing $50 per bag are mixed with 40 bags of another rice costing $58.75 per bag. If the mixture is sold at $66 per bag, find the percentage gain.

Cost of first 30 bags = $50 × 30 = $1500

Cost of other 40 bags = $58.75 × 40 = $2350

Total cost of 70 bags = $1500 + $2350 = $3850

Average cost of 1 bag $= \dfrac{\$3850}{70} = \$55$

Gain in cash on 1 bag = $66 − $55 = $11

Gain % $= \frac{11}{55} \times 100 = 20\%$

Exercise 17a

① Calculate the selling price of each item in Table 17.1

Table 17.1

	Cost price ($)	Profit (%)	Loss (%)
(a)	20		10
(b)	45		20
(c)	40	$12\frac{1}{2}$	
(d)	510	30	
(e)	16.20	15	

② For each item in Table 17.2, calculate the profit or loss as a percentage of the cost price.

Table 17.2

	Cost price ($)	Selling price ($)
(a)	12	16
(b)	68	51
(c)	315	210
(d)	400	519
(e)	32	19.20

③ Calculate the cost price of each item in Table 17.3

Table 17.3

	Selling price ($)	Profit (%)	Loss (%)
(a)	18	20	
(b)	22.50		10
(c)	760		5
(d)	325.80	$12\frac{1}{2}$	
(e)	420.55	45	

④ A man loses 12% by selling a watch for $341. Find the cost price of the watch.

⑤ A trader makes a profit of $12\frac{1}{2}\%$ by selling some goods for $94.50. Find her cash profit.

⑥ During a sale there is a discount of 15% on the items shown in Table 17.4.

Table 17.4

Items	Marked price ($)
bed	2250
chest	1950
tables	900

Calculate (a) the sale price of each item, (b) the total savings made by a shopper.

⑦ When a car is sold for $37 400 the profit is 10%. Find the selling price to make a profit of 18%.

⑧ Concentrated acid costing 90c per litre is diluted with water in the ratio 1 : 5 by volume. The diluted acid is sold at 24c per litre. What is the percentage profit?

⑨ Sacks of rice costing $62 and $55 per sack are mixed together in the ratio 5 : 2. Find the selling price of the mixture per sack in order to make a profit of 35%.

⑩ (a) The regular price of a shirt is $40. During a sale, a discount of 20% is given. Calculate the amount that a customer pays for the shirt.
(b) When selling at the discount price, the salesman makes a profit of 60%.

Calculate the cost of the shirt to the salesman.

(c) Calculate the percentage profit the salesman made when a shirt was sold for $40.

(d) After the sale, the salesman bought 100 shirts at $25 each and sold them to make a profit of 56% on his buying price. Calculate the total profit he made on this set of shirts.

[CXC June 88]　　　　(10 marks)

⑪ A merchant sold a pen for $5.35, thereby making a profit of 7% on the cost to him. Calculate

(a) the cost price of the pen to the merchant

(b) the selling price the merchant should request in order to make a 15% profit.

[CXC June 92]　　　　(5 marks)

⑫ The marked price of a freezer is $3000.00. There is a discount of 15% for cash payment. To obtain the freezer on hire-purchase, a deposit of $615.00 and 18 monthly instalments of $157.50 each are required. Calculate

(a) the cash price

(b) the total amount paid if bought on hire-purchase

(c) the difference between the cash price and the hire-purchase price as a percentage of the marked price.

[CXC June 89]　　　　(8 marks)

Simple interest, compound interest, investments, depreciation

Example 8

Find the simple interest paid on a loan of $5000 for $2\frac{1}{2}$ years at 13% per annum.

Interest on $5000 for 1 year $= \$\left(\dfrac{13}{100} \times 5000\right)$

Interest for $2\frac{1}{2}$ years $= \$\left(\dfrac{13}{100} \times 5000 \times \dfrac{5}{2}\right)$

$= \$1625$

Example 9

Find the rate of simple interest in per cent per annum at which $142 will amount to $295.36 in 12 years.

Amount = principal + interest
$295.36 = \$142 + \text{interest}$
Interest $= \$295.36 - \$142 = \$153.36$

But $I = \dfrac{PRT}{100}$, where I is the interest, P is the principal, R is the rate per cent per annum and T is the time in years.

Hence $\$153.36 = \$\dfrac{142 \times R \times 12}{100}$

$R = \dfrac{100 \times 153.36}{142 \times 12}\%$

$= \dfrac{1278}{142}\% = \dfrac{639}{71}\% = 9\%$

Example 10

Find the total interest to be paid on a loan of $5000 for 3 years at 9% per annum, if interest is calculated at the end of each year and added to the principal for the next year.

Interest at the end of the 1st year

$= \$\left(5000 \times \dfrac{9}{100}\right)$

$= \$450$

Principal for 2nd year $= \$(5000 + 450)$

$= \$5450$

Interest for 2nd year $= \$\left(5450 \times \dfrac{9}{100}\right)$

$= \$490.50$

Principal for 3rd year $= \$(5450 + 490.50)$

$= \$5940.50$

Interest for 3rd year $= \$\left(5940.50 \times \dfrac{9}{100}\right)$

$= \$534.65$

Amount at end of 3rd year

$= \$(5940.50 + 534.65)$

$= \$6475.15$

∴　Total interest $= \$(6475.15 - 5000)$

$= \$1475.15$

Check: Total interest $= \$(450 + 490.50$

$+ 534.65)$

$= \$1475.15$

Exercise 17b

1. Calculate the simple interest for the values given in Table 17.6.

   Table 17.6

	P ($)	T	R (% p.a.)
(a)	1500	3 yrs	7
(b)	300	6 mths	$12\frac{1}{2}$
(c)	725	9 mths	9
(d)	215	$1\frac{1}{2}$ yrs	15
(e)	2300	7 yrs	$8\frac{3}{4}$

2. Find the simple interest on $126 for 6 years at 7%.

3. Find the time in which $168.40 will earn $29.47 of interest at 5% per annum.

4. If $206.40 amounts to $237.36 in 2 years, find the rate of simple interest per annum.

5. Calculate the missing values in Table 17.7.

   Table 17.7

	P ($)	R (% p.a.)	T (yrs)	I ($)
(a)	50	4		6
(b)	60		3	9
(c)		6	2	120
(d)		$7\frac{1}{2}$	3	72
(e)	1600		$2\frac{1}{2}$	510
(f)	900	11		330

6. Find the time in which $108.33 will amount to $123.50 at 8% per annum.

7. (a) Mary borrowed $3000 from her Credit Union. The Credit Union charges interest at the rate of 20% per annum on the loan balance at the end of *each* year. Mary paid $200 per month on her account.
   Calculate
   (i) the amount she repaid during the first year
   (ii) the interest charged at the end of the first year
   (iii) the amount owed at the beginning of the second year

   (iv) the fewest number of payments required to pay off the loan and interest during the second year.
   (7 marks)
   (b) Calculate the rate of interest per annum if $405 is the simple interest gained on $2700 in *one* year.
   [CXC June 92] (3 marks)

8. A woman invests $5000 in government bonds for 5 years at 8% simple interest.
   (a) Calculate the total interest she receives for the five years.
   (b) Calculate the sum that must be invested in bonds to obtain a total interest of $3600 in five years.
   [CXC June 89] (7 marks)

9. Interest is added to a loan at the end of each year at the rate of 15% per annum. Calculate the total amount due on $1250 at the end of 3 years.

10. Machinery is bought for $9000. The value of the machinery at the end of each year is estimated at 15% less than the year before. Calculate the estimated value of the machinery at the end of 2 years.

11. (a) Calculate the simple interest on $2000 at $12\frac{1}{2}$% p.a. at the end of 18 months.
    (b) If interest is calculated at the end of every 6 months and added to the deposit, calculate the total deposit, including interest, at the end of the 18 months.

12. The cash price of a car is $25 400. A purchaser agrees to pay a deposit of 25% and the balance plus $1350 interest in 3 equal instalments. Evaluate the cost of one instalment.

Bills, taxes, rates

Example 11

The rates for electricity are as follows:

the first 300 units cost 17 cents per unit,
the next 900 units cost 15 cents per unit,
the next 1500 units cost 10 cents per unit,
the remainder costs 4 cents per unit.

A Government tax of 5% is added to the charges.

Calculate the total bill when 664 units are used.

Charge on 1st 300 units = 300 × 17c
= $51.00

Charge on remaining
364 units* = 364 × 15c
= $54.60

Charges on units used = $105.60

$$\text{Government tax} = \$\left(\frac{5}{100} \times 105.60\right)$$
= $5.28

Total bill = $(105.60 + 5.28)
= $110.88

*Note that rates also apply to parts of stated ranges.

Example 12

The charges for telephone services are as follows.
1   Monthly rental: $34.00
2   Local calls (transaction code 9): charge of 23 cents per call unit.
3   Overseas calls
    International Direct Dialling (transaction code 8):
    Caribbean $2.05 per minute
    UK $5.50 per minute
    USA $6.00 per minute
    Operator-assisted (transaction code 5): additional charge equal to the basic rate for one minute.
4   Government tax (transaction code 10): tax equal to 10% of total charges for all calls.

```
 *  |  TRANSACTION DETAILS            | TIME    | DUR.
 5  | 8028684688  VERMONT, USA        |11:04 AM | 25
 8  | 9853966     JAMAICA             | 8:31 AM |  7
 8  | 9823589     JAMAICA             | 8:42 AM | 18
 8  | 9834988     JAMAICA             | 9:00 AM |  1
 8  | 4290876     BARBADOS            | 8:11 PM |  3
 8  | 4290876     BARBADOS            | 6:41 AM |  1
 8  | 6038984983  N HAMPSHIRE, USA    | 9:06 AM |  4
 8  | 9293868     JAMAICA             | 7:00 AM | 13
 9  |           C/U LESS CR/U=CH/U
    | PVT-71925   116   20   96
10  | GOVT TAX   .
    | C/U:CALL UNITS
    | CR/U:CREDIT UNITS(NO CHARGE)
    | CH/U:CHARGEABLE UNITS

 *  TRANSACTION CODES 5:Operator assisted
                      8:IDD
                      9:Local calls
```

Fig. 17.1

Fig. 17.1 shows the transaction details on a telephone bill for overseas calls and local calls. Calculate
(a) the charges for overseas calls,
(b) the charges for local calls,
(c) the Government tax,
(d) the total amount due.

(a) Duration of IDD overseas calls (USA)
     = 4 min
     Charges for IDD overseas calls (USA)
     = $(6.00 × 4) = $24
     Duration of operator-assisted overseas calls
     = 25 min
     Charges for operator-assisted overseas calls (USA)
     = $(6.00 × 25) + $6.00 = $156
     Duration of IDD overseas calls (Caribbean)
     = (7 + 18 + 1 + 3 + 1 + 13) min = 43 min
     Charges for IDD overseas calls (Caribbean)
     = $(2.05 × 43) = $88.15
     Total charges for overseas calls
     = $(24 + 156 + 88.15) = $268.15

(b) Charges for 96 chargeable local call units
     = $(96 × 0.23)
     = $22.08

(c) Total charges for all calls = $(268.15 + 22.08)
     = $290.23
     Government tax = 10% of $290.23
     = $29.02

(d) Total amount due
     = $(34.00 + 290.23 + 29.02)
     = $353.25

Example 13

Tables 17.8 and 17.9 are used to calculate income tax.

Table 17.8

Tax-free allowances (per year)	Amount ($)
personal	2500
spouse	1300
each child under 18 yrs	800
other dependent relative	400

Table 17.9

Annual taxable income	Tax rate (%)
first £2000	5
next $5000	10
next $7500	15
next $10 000	25
next $20 000	40
> $44 500	50

A married man with two children under 18 years earns an annual salary of $24 000. If he has no other income, calculate (a) the annual income tax due, (b) the amount of his take-home salary.

(a) Tax-free allowances
     = $(2500 + 1300 + 2 × 800)
     = $5400
     Taxable income
     = $(24 000 − 5400)
     = $18 600
     Income tax due
     = $\$\left(\dfrac{5}{100} \times 2000 + \dfrac{10}{100} \times 5000\right.$
     $\left. + \dfrac{15}{100} \times 7500 + \dfrac{25}{100} \times (18\,600 - 14\,500)\right)$
     = $(100 + 500 + 1125 + 1025)
     = $2750

(b) Amount of take-home salary
     = $(24 000 − 2750)
     = $21 250

Example 14

Calculate the value of EC$510 in J$ if US$1 = EC$2.04 and US$1 = J$35.30.

EC$2.04 = J$35.30 (= US$1)

EC$1 = J$$\dfrac{35.30}{2.04}$

EC$510 = J$$\dfrac{35.30}{2.04} \times 510$ = J$$\left(\dfrac{3530}{204} \times 510\right)$

= J$(1765 × 5) = J$8825

Exercise 17c

1 Using the rates given in Example 11, find the electricity bill to be paid when the number of units used is (a) 1234, (b) 3280.

2 Using the charges given in Example 12, calculate the total bill, including tax, for telephone service in a month when there are 210 call units (including 20 credit units) for local calls and the charges for overseas calls amount to $322.40.

3 The water rates for a property are 13% of the value of the property. The sewerage tax is a fixed amount of $77. Find the total amount due for water and sewerage taxes on a property valued at $3500.

4 An unmarried woman earns $1200 per month. Using Tables 17.8 and 17.9, calculate the amount that she pays monthly for income tax.

5 If TT$1 = Guy$16.50 and TT$1 = EC$0.35, change Guy$500 to EC$. Give your answer to the nearest cent.

6 The tax rate on a property is 4%. Calculate
   (a) the tax on a property valued at $8000,
   (b) the value of a property if the tax paid is $720.

7 In addition to the rates and tax listed in Example 11 for the electricity supply, there is a fuel charge at the rate of 1.45 cents per unit used. On a bill, the previous meter reading is 74 680, the present reading is 75 421. Find the total amount due including tax.

8 A woman must change US$275 to BD$. If she only knows that N$1 = US$0.19 and N$1 = BD$0.39, calculate the equivalent BD$ she expects to receive.

9 In November 1985, Mr Rock took TT$800 in Caricom Travellers' Cheques to a bank in Antigua to exchange for Eastern Caribbean currency. Half of this amount was in $100 cheques and the other half was in $50 cheques. He was given the information shown below:

TT$1.00 = EC$1.125

$\frac{1}{2}$% tax is charged on the total foreign transaction.
EC$0.25 stamp duty is charged for each cheque.

Calculate, in Eastern Caribbean currency,

(a) the tax Mr Rock had to pay,

(b) the amount Mr Rock received for TT$800 after paying tax and stamp duty.

[CXC June 87]

10 The import tax on cars is 60%. Calculate

(a) the tax paid on a car valued at $70 000,

(b) the cost of a car before tax if the final cost of the car is $45 000.

11 (a) The customs duty on imported vehicles is 30% of the imported price.

(i) Calculate the customs duty on a car for which the imported price is $8500.

(ii) Calculate the imported price of a bus for which the amount paid, including customs duty, is $15 600.

(b) Charges for electricity in a Caricom country are made up of a fixed fuel charge of 45 cents per unit and an energy charge computed under three schemes as follows:

Scheme A   Homes   15 cents per unit
Scheme B   Schools   20 cents per unit
Scheme C   Business places   30 cents per unit

The meter reading of a certain business place reads as follows:

Meter reading (units) Present	Meter reading (units) Previous	Units used	Scheme	Energy charge ($)	Fuel charge ($)
39 421	18 368		C		

Calculate

(i) the number of units used,

(ii) the energy charge,

(iii) the fuel charge,

(iv) the amount the business place had to pay for the electricity it used.

[CXC June 87]

12 In addition to the tax-free allowances given in Table 17.8, annual allowances are given as follows:

(i) 75% of all medical bills,

(ii) $6000 for the upkeep of each child studying at a tertiary level institution abroad.

A woman's annual income is $70 000. She claims a personal allowance, allowances for a child under 18 years old and another child studying at university abroad. In addition she submits medical bills totalling $960. Using Tables 17.8 and 17.9, calculate

(a) the total tax-free allowances,

(b) the annual income tax due,

(c) the amount of income tax due as a percentage of her gross income.

13 In addition to the rates and charges listed in Example 12, the rates are reduced by 50% for overseas calls made between 6 pm and 6 am. Note that the additional charge for operator-assisted calls remains at the basic rate for 1 minute. Table 17.10 shows the details of overseas calls made in one months from a Caribbean country.

Table 17.10

*	Transaction details	Time	Duration (mins)
5	USA	5:40 am	12
5	UK	3:26 pm	7
5	Barbados	8:00 am	4
8	St Lucia	11:42 am	3
8	Jamaica	7:11 pm	21
8	USA	2:25 pm	4
8	Antigua	6:13 pm	2
8	USA	9:41 pm	3

*Transaction codes

Calculate the total charges due for overseas calls in the month.

14 Income tax is calculated as follows.

(i) Mortgage payments are deducted from gross income.

(ii) A tax surcharge of 5% is deducted from remaining income.

(iii) Tax-free allowances as in Table 17.8 are then deducted.

(iv) The remaining income after all deductions above is the taxable income.

(v) Tax rates, as shown in Table 17.11, are then applied.

Table 17.11

Annual taxable income ($)	Tax rate (%)
0–$7000	10
next $5000	15
next $10 000	20
next $20 000	25
> $42 000	40

Calculate the total tax paid on a gross income of

(a) $25 000 of a married man with three children aged 7, 11 and 15 years who pays an annual mortgage of $10 800.

(b) $90 000 of a single woman who has a dependent relative and pays an annual mortgage of $30 000.

15 A mechanic's annual gross salary is $14 400. He contributes 2% of his salary to a medical scheme and his company contributes 3%. His contribution to the medical scheme is a non-taxable allowance. Other non-taxable allowances and the income tax rates on taxable income are given in Table 17.12.

Table 17.12

Non-taxable allowances	Income tax rates on annual taxable income
$75 per month for National Insurance	20% on first $4000 25% on remainder
$3000 per annum for Personal Allowance	

Calculate for the mechanic:

(a) the monthly amount the company contributes to his medical scheme;

(b) the total amount of his annual salary that is not taxed;

(c) his annual taxable income;

(d) the tax he pays monthly.

(Answer to the nearest cent.)

[CXC June 86]

Revision test (Chapter 17)

Select the correct answer and write down your choice.

1 The selling price of a radio is $50 not including sales tax of 5%. The sales tax equals

   A $2.25   B $2.38
   C $2.50   D $2.63

2 A shopkeeper pays $32 for an item which she sells for $40. The profit percentage on the cost price is

   A   3.2%   B   8.0%
   C   20%   D   25%

3 Calculate the simple interest earned on $7000, at 9% per annum for 6 months.

   A $46.67   B $105
   C $315   D $378

4 A discount of 20% is given on a marked price of $125. The price actually paid is

   A $85   B $100
   C $105   D $110

5 A fan is bought for $325 and sold at a loss of 15%. The selling price is

   A $240   B $276.25
   C $282.65   D $310

6 $20 000 earns $2400 of simple interest in $1\frac{1}{2}$ years. The rate per cent per annum is

   A   6   B   8
   C   12   D   18

7 A collector made a profit of 15% when he sold a stamp for $37.50. The price he had paid was

   A $22.50   B $28.59
   C $31.87   D $32.61

8 If £1 = TT$10.60 and US$1 = TT$6.30, US$1 =

   A $\$\frac{1.0}{4.3}$ B $\$\frac{4.3}{1.0}$
   C $\$\frac{6.3}{10.6}$ D $\$\frac{10.6}{6.3}$

9 What sum was borrowed if the simple interest came to $1000 after 2 years at 8% per annum?

   A $2500   B $4000

   C $6250   D $16 000

10 If $5000 at $6\frac{1}{4}$% per annum simple interest earns $625 in t years $t =$

   A $\frac{1}{2}$ B $1\frac{1}{2}$

   C 2   D $5\frac{1}{3}$

11 A woman spends $3000 to furnish a room for rental. She wishes to make a profit of $12\frac{1}{2}$% at the end of 1 year. Calculate the monthly rental.

   A $194.44   B $218.75

   C $222.22   D $281.25

12 A car which cost $180 000 when new depreciates each year by 15% of its value at the beginning of the year. Find its value after 2 years.

   A $153 000   B $136 100

   C $133 050   D $130 050

13 Rates for electricity are as follows:

first 300 units at 12 cents per unit,
next 500 units at 18 cents per unit,
next 1000 units at 25 cents per unit.

A householder uses 920 units. The total electricity bill amounts to

   A $110.40   B $147.60

   C $156.00   D $230.00

14 Income tax is charged at 25% on taxable income after allowances are deducted. Annual allowances amount to $6000 and there is a surcharge of 3% on gross incomes above $100 000. The tax due on a gross annual income of $70 000 is

   A $16 000   B $16 975

   C $17 500   D $19 600

15 Compound interest is charged semi-annually on a loan of $1000 at 14% per annum. The amount due at the end of the first year is

   A $1074.90   B $1105.00

   C $1140.00   D $1144.90

16 An article is marked down from $18 to $13.50. The discount percentage is

   A 25%   B $29\frac{1}{6}$%

   C $33\frac{1}{3}$%   D $58\frac{1}{3}$%

Use the following information in questions 17 and 18.

The cash price of a machine is $3812. To buy with hire purchase there is an initial payment of 30%. Interest of $500 is added to the balance which is then paid in ten equal monthly instalments.

17 The initial payment is

   A $614.36   B $1143.60

   C $1270.66   D $1633.71

18 Each monthly instalment is

   A $301.84   B $316.84

   C $381.20   D $452.63

19 $P may be invested at compound interest which is calculated annually. The amount earned

(i) at 3% for 2 yrs is $S,

(ii) at 2% for 3 yrs is $T.

Which one of the following statements is correct?

   A $S > T$ B $S < T$

   C $S = T$ D S and T cannot be compared

20 A man marks an item to make a profit of 45%. He gives a discount of 10% for cash. The percentage profit made on a cash sale is

   A 30.5%   B 35.0%

   C 40.5%   D 55.0%

In mathematics a collection of things is called a set. Refer to the list in Table 6.1 on page 64 for the symbols of set language and an explanation of their meanings.

Example 1

Describe the following sets in words and state the number of elements in each set.
(a) {Monday, Tuesday, Wednesday, Thursday, Friday, Saturday, Sunday}
(b) {2, 3, 5, 7, 11}

(a) {days of the week}, 7
(b) {prime numbers less than 12}, 5

Example 2

A = {multiples of 2 less than 10}
B = {odd numbers less than 8}
C = {prime numbers less than 9}
List the elements of each set and state the number of elements in each set.

A = {2, 4, 6, 8}, n(A) = 4;
B = {1, 3, 5, 7}, n(B) = 4;
C = {2, 3, 5, 7}, n(C) = 4

Example 3

P = {letters of the word **connotation**}
Q = {letters of the word **action**}
List the elements of the sets P and Q.

P = {c, o, n, a, t, i}
Q = {a, c, t, i, o, n}

In example 2, sets A, B and C have the same number of elements. Sets that have the same number of elements are called **equivalent** sets.

In example 3, sets P and Q have the same (identical) elements. Sets that have the same (identical) elements are **equal** sets, P = Q. Remember the order in which the elements are written does not matter.

Example 4

If U = {1, 2, 3, 6, 9, 18}, X = {2, 6, 18} and Y = {1, 3, 6}, list the elements of (a) X′ ∩ Y, (b) X ∪ Y′, (c) (X ∪ Y)′.
Remember that X′ is the **complement** of X, i.e. those elements in U which are *not* in X.

  U = {1, 2, 3, 6, 9, 18}
  X = {2, 6, 18}; X′ = {1, 3, 9}
  Y = {1, 3, 6}; Y′ = {2, 9, 18}
(a) X′ ∩ Y = {1, 3, 9} ∩ {1, 3, 6}
        = {1, 3}
(b) X ∪ Y′ = {2, 6, 18} ∪ {2, 9, 18}
        = {2, 6, 9, 18}
(c) (X ∪ Y)′ is the complement of X ∪ Y.
   X ∪ Y = {1, 2, 3, 6, 18}
   (X ∪ Y)′ = {9}

Example 5

There are 38 students in Form 4. 34 of them eat lunch at school and then some of them play games, 12 of them play games in the courtyard and 16 in the classroom. Draw a Venn diagram to show this information, indicating the number of members in each set.
Let U = {students in Form 4}
  L = {students who eat lunch at school}
  Y = {students who play games in the courtyard}
  C = {students who play games in the classroom}
The Venn diagram is shown in Fig. 18.1.

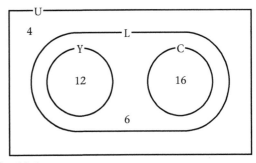

Fig. 18.1

Example 6

A survey taken among 100 households revealed that 8 households keep dogs, cats and fish, 61 households keep dogs, 32 households keep dogs only, 10 households keep fish and dogs only, and 3 keep no pets at all. The Venn diagram in Fig. 18.2 shows the results of the survey, where

F = {households that keep fish},
D = {households that keep dogs},
and C = {households that keep cats}.

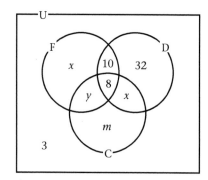

Fig. 18.2

(a) Find the value of x.

(b) If 35 households keep fish, find how many households keep (i) fish and cats only, (ii) two types of pets only, (iii) cats only.

(a)
$$n(D) = 61$$
$$10 + 8 + 32 + x = 61$$
$$x = 61 - 50$$
$$x = 11$$

(b) (i) Let y be the number which keep fish and cats only.
Then $n(F \cap C) - n(F \cap C \cap D) = y$
$$y + 11 + 10 + 8 = 35$$
$$y = 35 - 29$$
$$y = 6$$

(ii) Number keeping two pets only
$$= 10 + 11 + 6$$
$$= 27$$

(iii) Let number keeping cats only be m.
Then $61 + (35 - 18) + m + 3 = 100$
$$m = 100 - 81$$
$$m = 19$$

Example 7

Draw a Venn diagram to show that
(a) all squares are rectangles.
(b) all rectangles are polygons.

Let U {polygons}, S = {squares}, R = {rectangles}.

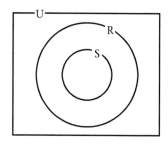

Fig. 18.3

Exercise 18a

① Given U = {r, e, v, o, l, t, i, n, g},
A = {l, i, o, n} and B = {t, i, g, e, r}, list the members of the following sets.
(a) A ∪ B (b) A ∩ B
(c) A′ (d) B′
(e) A′ ∪ B′ (f) A′ ∩ B′
(g) (A ∪ B)′ (h) (A ∩ B)′
(i) A′ ∩ B

② For each of the following, make a copy of Fig. 18.4 then shade the given set.
(a) (P ∪ Q)′ (b) P′ ∪ Q′
(c) (P ∩ Q)′ (d) P′ ∩ Q′
(e) P′ ∪ Q (f) P ∩ Q′

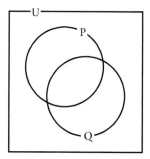

Fig. 18.4

③ For each of the following, make a copy of Fig. 18.5 then shade the given set.

(a) $(A \cup B) \cap C$ (b) $(A \cap B) \cup C$
(c) $A \cup (B \cap C')$ (d) $A \cap (B \cup C')$
(e) $A \cap (B \cup C)'$ (f) $(A \cap C') \cup (B \cap C)$

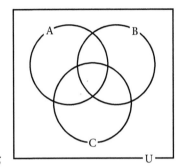

Fig. 18.5

④ If $U = \{1, 2, 3, ..., 10\}$, list the members of the following subsets of U.

(a) $\{x: x \leqslant 2\}$
(b) $\{x: x$ is a multiple of 5$\}$
(c) $\{x: 3 < x < 8\}$
(d) $\{x: 3x - 3 = 3\}$
(e) $\{x: x$ is a factor of 360$\}$

⑤ In a survey, 60 people were asked if they had a radio or a television set. Thirty people said they had a television set and 45 said they had a radio.

(a) Draw a Venn diagram to show this information.
(b) How many had both radio and television?
(c) How many had only a television set?
(d) How many had only a radio?

⑥ Fig. 18.6 shows two intersecting sets P and Q with the number of elements in each region as shown.

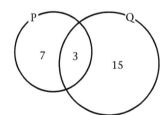

Fig. 18.6

Find (a) n(P), (b) n(Q), (c) n(P ∩ Q),
(d) n(P ∪ Q).

⑦ Let U = multiples of 2,
    A = multiples of 4,
    B = multiples of 8.

Draw a Venn diagram to show that all multiples of 4 are multiples of 2, and not all multiples of 4 are multiples of 8.

⑧ Let P = {boys who wear glasses},
    Q = {boys who ride bicycles},
    R = {boys who like games}.

Draw diagrams to illustrate the following statements.

(a) All boys who ride bicycles like games.
(b) Some boys who wear glasses do not like games.
(c) No boys who wear glasses ride bicycles.

⑨ Let U = {books}, A = {algebra books}, B = {books with brown covers}. Show, by shading on a Venn diagram, the set of books with brown covers which are not algebra books.

⑩ A company employs 79 people, 52 of whom are men. 38 people, including all the women, are clerical staff. Draw a suitable Venn diagram to show this information. Hence or otherwise find the number of men who are clerical staff.

⑪ The numbers of elements of the subsets of the Venn diagram in Fig. 18.7 are as shown.

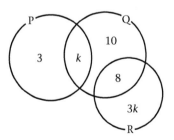

Fig. 18.7

If U = P ∪ Q ∪ R and n(U) = 33, find
(a) k
(b) n(P ∪ R)
(c) n(P ∩ R)
(d) n(R′ ∩ Q)

Revision test (Chapter 18)

1 Which of the diagrams in Fig. 18.8 shows the relation P ⊂ Q?

A

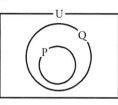

B

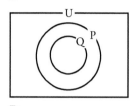

C

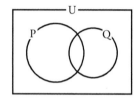

D

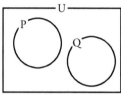

Fig. 18.8

2

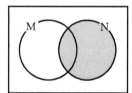

Fig. 18.9

The shaded part of Fig. 18.9 represents

A M ∪ N   B M ∩ N
C N   D N′

Use the following information in questions 3 and 4.

F = {2, 3, 4, 5}
G = {2, 3, 7}
H = {7, 8, 9, 10}

3 n(F ∪ G) =

A 0   B 2   C 3   D 5

4 Which of the following is an empty set?

A (F ∩ G)   B (F ∩ H)
C (G ∪ H)   D (G ∩ H)

5 If U = {odd numbers less than 20},
   P = {1, 5, 9, 13}
and Q = {5, 13}, the Venn diagram representing this data is given by which diagram in Fig. 18.10?

A

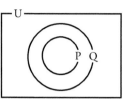

B

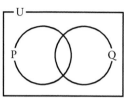

C

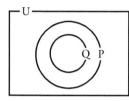

D

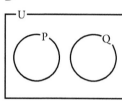

Fig. 18.10

6

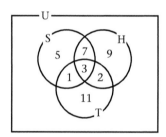

Fig. 18.11

The Venn diagram in Fig. 18.11 illustrates
U = {girls in Form 4},
S = {girls who play soccer},
H = {girls who play hockey},
T = {girls who play tennis}.
n(H ∪ T) =

A 2   B 10   C 13   D 33

7 P = {letters in the word **fruitful**}.
Then n(P) =

A 8   B 7   C 6   D 5

8 If U = {1, 2, 3, 4, 5, 6, 7, 8}
   V = {even numbers}
   W = {odd numbers}
Which of the following statements is *not* true?

A   V and W are disjoint sets
B   V and W are equivalent sets
C   V and W are equal sets
D   V ⊄ W

9 If U = {letters of the alphabet}
  P = {letters of the word **resist**}
  Q = {letters of the word **rest**}

Which of the following statements is *not* true?

A (P ∩ Q) = Q   B P ∪ Q = P
C Q ⊂ P   D P ⊂ Q

10 In a class of 50 students, 30 like English, 25 like History and 10 like both subjects. How many students like neither History nor English?

A 5   B 10   C 20   D 25

Use the Venn diagram in Fig. 18.12 to answer questions 11 and 12.

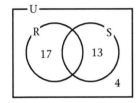

Fig. 18.12

11 If n(R ∪ S) = 40 then n(R ∩ S) =

A 4   B 6   C 10   D 30

12 The number of elements in the universal set in Fig. 18.12 is

A 48   B 44   C 40   D 36

13 Given P ⊂ Q, Q ⊂ R, P ∩ R = P which of the diagrams in Fig. 18.13 **best** illustrates this information?

A 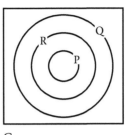   B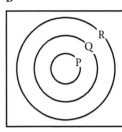

C 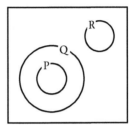   D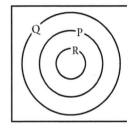

Fig. 18.13

14 In a class, 20 students passed a mathematics test, 15 passed a science test, 8 passed both tests and 3 students passed neither mathematics nor science. The number of students in the class is

A 46   B 43   C 30   D 27

15 In a group of 30 children, 20 enjoy swimming, 15 enjoy riding, and 3 of them enjoy neither swimming nor riding. Which of the following statements is true?

A 8 children enjoy both swimming and riding
B 3 children enjoy both swimming and riding
C 8 children enjoy neither swimming nor riding
D 11 children enjoy neither swimming nor riding

Chapter 19

Relations, functions and graphs

Relations

A relation is a connection between elements of
the same set or between elements of two sets.

Example 1

A relation on each of the following sets is
defined as stated.

(a) {Gemma, Indra, Charles, Arthur, Ken,
George}
rule: is the sister of

(b) {1, 2, 3, 4, 8, 27, 64}
rule: is the cube of

(c) {10, 12, 14, 16, 18, 20, 5, 6, 7, 8}
rule: is divisible by

These may be shown in arrow diagrams, Fig. 19.1
and Fig. 19.2.

(a)
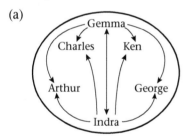

(b) 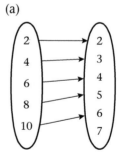 (c)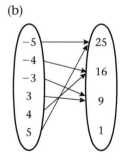

Fig. 19.1

(a)
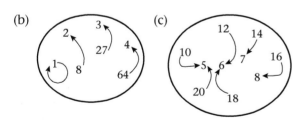

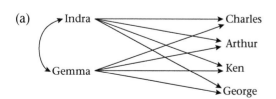

(b)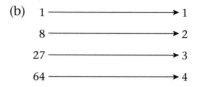

Fig. 19.2

When the connection is between the elements
of two sets, the elements of the first set form the
domain and the elements of the second set are
the **range**.

Example 2

For each relation shown in Fig. 19.3, write down
(i) the domain, (ii) the range.

(a)   (b)
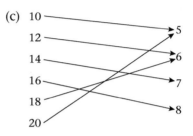

Fig. 19.3

(a) (i) {2, 4, 6, 8, 10}
   (ii) (2, 3, 4, 5, 6)

(b) (i) {−5, −4, −3, 3, 4, 5}
   (ii) {25, 16, 9}

A relation may also be written as ordered pairs
(x, y) where x is an element of the first set, that
is the domain and y is the corresponding
element from the second set that is, the range.

Example 3

Write the relations shown in Fig. 19.3 as a set of ordered pairs.

(a) {(2, 2), (4, 3), (6, 4), (8, 5), (10, 6)}
(b) {(−5, 25), (−4, 16), (−3, 9), (3, 9), (4, 16), (5, 25)}

A relation can also be shown in a table.
Table 19.1 shows the information in Fig. 19.3(b).

Table 19.1

x	−5	−4	−3	3	4	5
y	25	16	9	9	16	25

The relation can also be represented using the Cartesian graph. Fig. 19.4 represents the information in Table 19.1.

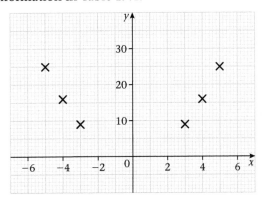

Fig. 19.4

A relation may be one-to-one, one-to-many, many-to-one or many-to-many. Fig. 19.5 shows diagrams of these types of relation.

(a) one-to-one   (b) one-to-many

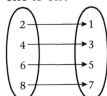 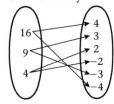

(c) many-to-one   (d) many-to-many

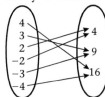 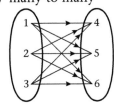

Fig. 19.5

Exercise 19a

1. For each relation shown in Fig. 19.2, write down (i) the domain, (ii) the range.

2. A relation from set A to set B is given by the set of ordered pairs
{(−5, −3), (−4, −2), (−3, −1), (−2, 0), (−1, 1), (0, 2), (1, 3)}
   (a) List the elements of set A and of set B.
   (b) State the relation in words.

3. State the type of relation illustrated in each diagram in Fig. 19.6.

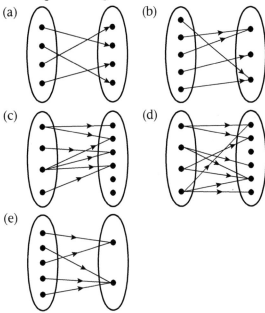

Fig. 19.6

Functions

One-to-one and many-to-one relations are important relations known as **functions**.

A function is a relation between two sets in which each and every element of the first set is mapped (related) to one and only one element in the second set.

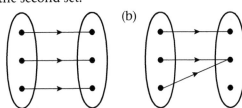

(c) 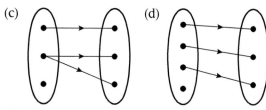 (d)

(b)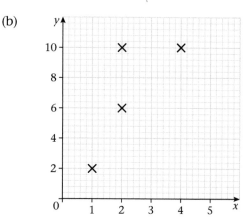

Fig. 19.7

In Fig. 19.7 (a) and (b) represent functions; (c) and (d) do not represent functions. In (c) an element in the first set is related to two elements in the second set, and in (d) every element in the first set is not related to an element in the second set.

A function may also be represented by a set of ordered pairs.

Example 4

Say which of the following sets of ordered pairs represents a function

(i)   P= {(3, 1), (5, 3), (7, 5), (9, 7)}
(ii) Q= {(3, 1), (3, 3), (7, 5), (7, 7)}
(iii) R= {(−1, 4), (0, 5), (1, 6), (2, 7), (3, 8)}

(i) and (iii) represent functions. (ii) is not a function since 3 and 7 are first elements in two different pairs.

Graphs

Caretesian graphs may also be used to illustrate a function.

Example 5

Which of the graphs in Fig. 19.8 does not represent a function?

(a)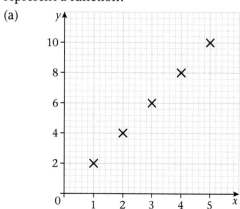

(c)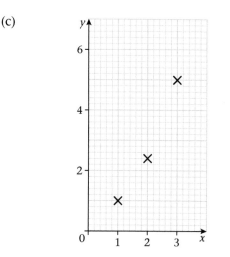

Fig. 19.8

Fig. 19.8(b) is not a function. Can you say why? Discuss why the graph in Fig. 19.9 would not represent a function.

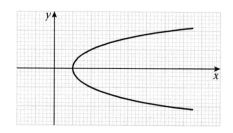

Fig. 19.9

Revision test (Chapter 19)

1 Which of the following relations in Fig. 19.10 is a function?

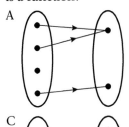
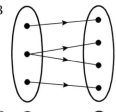
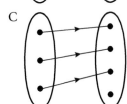
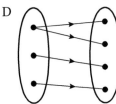

Fig. 19.10

2 Which of the following sets of ordered pairs represents a function?

A {(1, 3), (1, 4), (2, 5), (2, 6)}
B {(2, 2), (2, 5), (3, 6), (4, 7)}
C {(1, 2), (2, 5), (3, 6), (3, 7)}
D {(1, 3), (2, 5), (3, 6), (4, 7)}

3 Which of the following statements is *not* necessarily true?

A A function is a relation.
B A function can be represented by a set of ordered pairs.
C A relation is a function.
D A relation can be represented by an arrow diagram.

4 Which of the following represents a set of ordered pairs for the relation $x \rightarrow 2x - 1$?

A {1, 3, 5, 7}
B {7, 5, 3, 1}
C {(1, 7), (3, 5), (5, 3), (7, 1)}
D {(1, 1), (2, 3), (3, 5), (4, 7)}

5 A relation from set A to set B is given by the set of ordered pairs {(−3, −7), (−2, −5), (−1, −3), (0, −1), (1, 1)}. The relation may be written as

A $x \rightarrow x - 4$ B $x \rightarrow 2x - 1$
C $x \rightarrow 2x + 1$ D $x \rightarrow x + 4$

6 A function may *not* be represented as

A the elements of the domain
B a set of ordered pairs
C an arrow diagram
D a graph of the relation

7 Which of the following relations could represent a function?
(i) one-to-one, (ii) one-to-many, (iii) many-to-one, (iv) many-to-many

A (i) and (ii) only
B (i), (ii) and (iii)
C (i), (ii), (iii) and (iv)
D (i) and (iii) only

8 Which of the following sets of ordered pairs does *not* represent a function?

A {(9, −3), (4, −2), (1, −1), (0, 0), (1, 1), (4, 2), (9, 3)}
B {(−3, 9), (−2, 4), (−1, 1), (0, 0), (1, −1), (2, 4), (3, 9)}
C {(−2, 5), (−1, 2), (0, 1), (1, 2), (2, 5), (3, 10)}
D {(−2, 8), (−1, 2), (0, 0), (1, 2), (2, 8), (3, 18)}

9 Which of the following statements is *always true* for a function?

A Each first element of an ordered pair has only one second element.
B Each element of a set A may be mapped on two elements of a second set B
C All elements of a first set may not be mapped on to elements of a second set.
D The elements of the domain are the same as the elements of the range.

10 Which of the graphs in Fig. 19.11 does not represent a one-to-one relation?

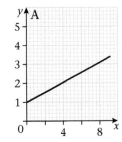
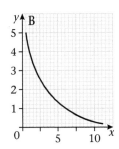

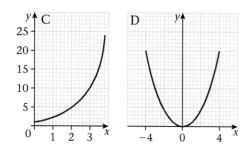

Fig. 19.11

11 Which arrow diagram shows a one-to-one relation?

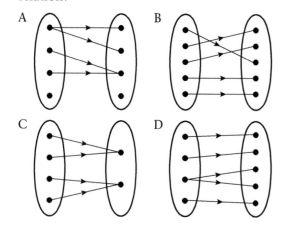

Fig. 19.12

13 Which of the graphs in Fig. 19.13 represents a function?

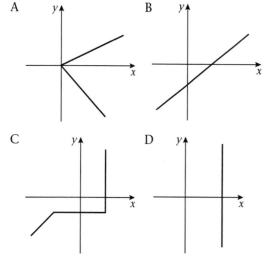

Fig. 19.13

13 Which mapping is represented by the diagram in Fig. 19.14?

   A $x \rightarrow 2x + 2$ B $x \rightarrow x - 2$
   C $x \rightarrow 2x$ D $x \rightarrow x^2$

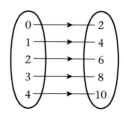

Fig. 19.14

14 Which of the following statements is *always* true?

   A   A relation is a set of ordered pairs.
   B   All relations are functions.
   C   Each element of a domain can be related to only one element in the range.
   D   The domain and the range must have the same number of elements.

15 Given the relation $x \rightarrow x^2 - 1$, this may be represented as

   A $\{(-2, -5), (-1, -3), (0, -1), (1, 1)\}$
   B $\{(-2, 5), (-1, 3), (0, 1), (1, 3)\}$
   C $\{(-2, 3), (-1, 0), (0, -1), (1, 0)\}$
   D $\{(-2, -3), (-1, 0), (0, -1), (1, 1)\}$

16 If $x \rightarrow 2x - 3$, then $-3 \rightarrow$

   A -3 B -9 C 3   D 9

17 If in a given relation between two sets, 3 maps on to 10, which of the following may *not* be true for the same relation:

   A   4 maps on to 13 B   4 maps on to 17
   C   5 maps on to 9   D   5 maps on to 16

18 Given that f means 'is a factor of' so that $3 = f(12)$ reads '3 is a factor of 12', which of the following is *not* true?

   A $5 = f(15)$ B $6 = f(3)$
   C $7 = f(14)$ D $8 = f(48)$

19 Given that m means 'is a multiple of', which of the following is true?

   A $5 = m(15)$ B $6 = m(12)$
   C $7 = m(14)$ D $18 = m(9)$

Algebra

Algebraic expressions, linear equations, linear inequalities in one unknown

Use of letters

Algebra uses letters of the alphabet to represent general numbers. The easiest way of dealing with letters which represent numbers is to imagine first that a letter stands for a particular number.

Example 1

Express p metres in centimetres.

If $1\,\text{m} = 100\,\text{cm}$,
      $2\,\text{m} = 100 \times 2\,\text{cm}$,
and $3\,\text{m} = 100 \times 3\,\text{cm}$,
then $p\,\text{m} = 100 \times p\,\text{cm}$
         $= 100p\,\text{cm}$

In Example 1 notice that $100p$ is short for $100 \times p$. In the expression $100p$, the 100 is called the **coefficient** of p.

Example 2

In a class of p students the average mark is x and in another class of n students the average mark is y. What is the average mark for both classes?

Average mark $= \dfrac{\text{total marks scored}}{\text{number of students}}$
In 1st class, total marks scored $= p \times x = px$
In 2nd class, total marks scored $= n \times y = ny$
For both classes, total marks scored $= px + ny$
         Number of students $= p + n$
             Average mark $= \dfrac{px + ny}{p + n}$

Exercise 20a

1. A girl has 20 cents. If someone gives her 16 cents, how much will she have? If someone gives her x cents how much will she have?

2. A boy is 15 years old. How old will he be in 4 years' time? How old will he be in y years' time?

3. A woman has $20. If she spends $8, how much will she have left? If she spends a instead, how much will she have left?

4. 6 years ago a man was 20 years old. How old is he now? p years ago another man was 23 years old. How old is he now?

5. A lorry carries $8\,\text{t}$ of blocks.
   (a) If it delivers $3\,\text{t}$, what mass of blocks is left on the lorry?
   (b) If it delivers $d\,\text{t}$, what mass remains on the lorry?

6. What number is (a) 5 times as big as 20, (b) 5 times as big as x?

7. How many cents in (a) $3, (b) e?

8. How many hours are there in (a) $120\,\text{min}$, (b) $m\,\text{min}$?

9. A rectangular sheet of paper has an area of $300\,\text{cm}^2$.
   (a) How long is the paper if its breadth is $15\,\text{cm}$?
   (b) How long is the paper if its breadth is $b\,\text{cm}$?

10. Express x dollars and y cents in cents.

11. Express a dollars and b cents in dollars.

12. The perimeter of a rectangle is $14\,\text{m}$ and the length is $x\,\text{m}$. Express the breadth of the rectangle in terms of x.

13. The perimeter of a rectangle is $20\,\text{m}$ and the length is $x\,\text{m}$. Find the area of the rectangle in terms of x.

14. A rectangle has one side equal to $(2x + 4)\,\text{cm}$ and a perimeter of $(6x + 4)\,\text{cm}$. What is the area of the rectangle in terms of x?

15. A car travels $d\,\text{km}$ at an average speed of $u\,\text{km/h}$. How long does it take?

16 A train moves at an average speed of x km/h. How many hours will it take to cover 340 km?

17 A train travels for t hours at a speed of v km/h. How far does it go?

18 Two motor cars start together from the same place and travel in the same direction along a road. If the faster one has an average speed of u km/h and the other v km/h, how far ahead is the faster one after t hours?

Simplification

Certain properties of the basic operations of addition and multiplication have been used both in making calculations simpler and in simplifying algebraic expressions. For example,

$$8a + 2a = 2a + 8a$$

This illustrates the **commutative** property of addition. For any two terms, changing the order of the terms to be added does not change the result of the operation.

Similarly, multiplication is commutative; for example

$$7 \times 2b = 2b \times 7$$
$$\text{and} \quad 3a \times -5b = -5b \times 3a$$

Note that subtraction and division are *not* commutative, that is
$$9a - 4a \neq 4a - 9a$$
$$15a \div 3b \neq 3b \div 15a$$

Grouping terms

Expressions such as $2x$, x, $\frac{1}{3}x$ and $-5x$ are called **terms in x**. Expressions containing a number of terms can be simplified by grouping terms together.

Example 3

Simplify $9a - 5a - 8a + 10a$.

Either, by grouping positive and negative terms together:
$$9a - 5a - 8a + 10a = 9a + 10a - 5a - 8a$$
$$= 19a - 13a$$
$$= 6a$$

or, by treating the terms as directed numbers:
$$9a - 5a - 8a + 10a = 4a - 8a + 10a$$
$$= -4a + 10a$$
$$= 6a$$

This example illustrates the **associative** property of addition. An operation is associative when the same operation is repeated more than once and the order, in which each one of the operations is done, does not matter. In this example, the addition of directed (positive and negative) numbers is associative.

Collecting like terms

Since expressions such as $2x$, $3x$, $-7x$ are all terms in x they are called **like terms**. $2x$ and $3y$ are **unlike terms**. When simplifying algebraic expressions like terms are grouped together. $3y - 2x$ cannot be simplified any further.

Example 4

Simplify $2x^2 - 3ax + 4x^2 - 6xa - x^2$

$$2x^2 - 3ax + 4x^2 - 6xa - x^2$$
$$= 2x^2 + 4x^2 - x^2 - 3ax - 6ax = 5x^2 - 9ax$$

Notice that $6xa = 6ax$; the order of the letters is not important.

Exercise 20b

Simplify the following expressions.

1 $9x - 2x$ **2** $2a - 9a$

3 $-3y - 30y$ **4** $2y + 5y - 3y$

5 $4x - 8x + 9x$ **6** $9z - 3z - z$

7 $3a - 7a - 2a$ **8** $b + 4b - 12b$

9 $6c - 17c + 3c$ **10** $2a + 5x - 3a$

11 $-3h - 6g + 10g$ **12** $8d - 3 - 7d$

13 $3a - 5a + 11a - 4a$ **14** $7x + 3x - x - 5x$

15 $8k - 4k + 3k - 7k$ **16** $6x - 9x + 2x + 4y$

17 $2a - 3b + 5b - 8a$ **18** $3x + 8y - 5x - y$

19 $2m - 9n - 5m + 4n$

20 $r - 3s - 3t - 4r + 10s + 8t$

21 $6x - 9y + 10x$

22 $2p^2 + 3q - 8q$

23 $7a - 2b - 3a - 8b - a$

24 $8 - 3x^2 + 21 + 18x^2 - 6$

25 $x^2 - xy - 9xy + 9y^2$

26 $5u^2v - 3vu^2 + uv^2 - 5v^2u$

27 $d^2 + 2ad - 2d^2 + 8ad + 3d^2 - 20ad$

28 $2a^2 - a + 6a^2 - 8a^2 - 6a + 5a^2 + 10a$

29 $-3x^2 + 5x^2 + 6x^2 + 4x^3 + 2x^4 - x^2$

30 $m^2 - 2mn + 3n^2 + 5n^2 + 8mn - 4n^2 - 7mn$

Brackets

If any quantity multiplies the terms inside a bracket, *every* term inside the bracket must be multiplied by that quantity when the bracket is removed.

For example, $3(2a + 3b) = 6a + 9b$
and $2p(5q - 3r) = 10pq - 6pr$

These are called the **distributive** properties.

In general, $a(x + y) = ax + ay$
and $a(x - y) = ax - ay$

Example 5

Remove the brackets from $4(3x - 5y + z)$.

$4(3x - 5y + z) = 4 \times 3x - 4 \times 5y + 4 \times z$
$= 12x - 20y + 4z$

Example 6

Remove the brackets from the expression $-3(2x - 5y + 6z)$.

$-3(2x - 5y + 6z)$
$= (-3) \times (2x) + (-3) \times (-5y) + (-3) \times (6z)$
$= -6x + 15y - 18z$

Notice that every term inside the bracket is multiplied by the quantity outside the bracket.

Example 7

Remove brackets and simplify $3 - (a - \overline{5 - 6a})$.

Note: the line over the terms $5 - 6a$ acts like a bracket.

$3 - (a - \overline{5 - 6a}) = 3 - (a - 5 + 6a)$
$= 3 - (7a - 5)$
$= 3 - 7a + 5$
$= 8 - 7a$

In example 7, notice that the inner bracket is removed *first*. Notice also that a negative sign outside a bracket is equivalent to -1 outside the bracket, i.e. $-(5 - 6a) = -1(5 - 6a)$.

Example 8

Remove brackets and simplify
$3x(2 - x) - 5(6 + x - 2x^2)$

$3x(2 - x) - 5(6 + x - 2x^2)$
$= 6x - 3x^2 - 30 - 5x + 10x^2$
$= 10x^2 - 3x^2 + 6x - 5x - 30$
$= 7x^2 + x - 30$

Example 9

Remove brackets and simplify
$x(x - 2y) - 5y(3x - 6y)$.

$x(x - 2y) - 5y(3x - 6y)$
$= x^2 - 2xy - 15xy + 30y^2$
$= x^2 - 17xy + 30y^2$

Example 10

Simplify   (a) $3 \times 2a - 5$ (b) $3(2a - 5)$.

(a) $3 \times 2a - 5 = 6a - 5$
(b) $3(2a - 5) = 3 \times 2a - 3 \times 5$
$= 6a - 15$

Exercise 20c

Remove brackets from the following and simplify where possible.

1 $3(a + b - 2)$ 2 $-2(3m - n + 4)$

3 $2a + 3(a + 3b)$ 4 $5(2y - x) + 6x$

5 $\frac{1}{2}(2u - 8) + 5$ 6 $3c - \frac{2}{3}(12c - 3d)$

7 $4 \times 3m - 2$ 8 $4 \times (3m - 2)$

9 $4u - \overline{7u - 3v}$ 10 $5d - \overline{e + 2d}$

11 $2(a - 3b) + 3(a + b)$

12 $5m + 5(2n - m) - 8n$

13 $2(h + 5k) + 5(h + 2k)$

14 $5(2u - 3v + 4w) - 7(u - 2v + 3w)$

15 $4(r - 3s - 3t) - 2(2r - 5s - 4t)$

16 $6x - (3 - \overline{x - 8})$

17 $2a - (4a - \overline{5a + 7})$

⑱ $2x(3x - 1) + 7(x^2 + 3)$

⑲ $6(a^2 - 2a - 1) - 3a(2a - 5)$

⑳ $4x(3x - 2y) + 5y(x + 3y) - 7xy$

Expanding algebraic expressions

The expression $(x + 2)(x - 5)$ means
$(x + 2) \times (x - 5)$. The product of $(x + 2)$ and
$(x - 5)$ is found by multiplying each term in the
first bracket by each term in the second bracket.
Read the following examples carefully.

Example 11

Find the product of $(x + 2)$ and $(x - 5)$.

$$(x + 2)(x - 5) = x(x + 2) - 5(x + 2)$$
$$= x^2 + 2x - 5x - 10$$
$$= x^2 - 3x - 10$$

We say that $(x + 2)(x - 5)$ is **expanded** to
$x^2 - 3x - 10$. Notice that the terms in x, that is,
$+2x$ and $-5x$, are added together in the final
line of working in Example 11.

Example 12

Expand $(2c - 3m)(c - 4m)$.

$$(2c - 3m)(c - 4m) = c(2c - 3m) - 4m(2c - 3m)$$
$$= 2c^2 - 3cm - 8cm + 12m^2$$
$$= 2c^2 - 11cm + 12m^2$$

Example 13

Expand $(3a + 2)^2$.

$$(3a + 2)^2 = (3a + 2)(3a + 2)$$
$$= 3a(3a + 2) + 2(3a + 2)$$
$$= 9a^2 + 6a + 6a + 4$$
$$= 9a^2 + 12a + 4$$

Exercise 20d

Expand each expression. Arrange the working as
in Examples 11–13.

① $(a + 2)(a + 3)$

② $(c + 6)(c - 1)$

③ $(e - 3)(e + 2)$

④ $(d - 6)(d + 3)$

⑤ $(x - 1)(x - 2)$

⑥ $(a + 3)^2$

⑦ $(b - 5)^2$

⑧ $(m + 4)(m - 4)$

⑨ $(n + 5)(n - 4)$

⑩ $(d + 3)(d - 7)$

⑪ $(b - 5)(b + 6)$

⑫ $(p - 3)(p - 5)$

⑬ $(q - 3)(q + 3)$

⑭ $(u - 9)(u + 5)$

⑮ $(v - 4)(v - 9)$

⑯ $(2a + 1)(a + 3)$

⑰ $(b + 4)(3b + 2)$

⑱ $(2c - 5)(c - 3)$

⑲ $(d - 9)(2d + 3)$

⑳ $(2x + 1)^2$

㉑ $(5x + 2)(2x - 3)$

㉒ $(3y - 5)(2y + 1)$

㉓ $(m + 4n)^2$

㉔ $(u + 2v)(u + 3v)$

㉕ $(3d - 2e)(3d + 2e)$

㉖ $(3b + 2c)(2b - c)$

㉗ $(2s - 5t)(3s + t)$

㉘ $(2c - 3d)^2$

㉙ $(4m - n)(3m - 3n)$

㉚ $(2c - 9e)(4c + 5e)$

Direct expansion of products

With practice, it is not necessary to write down
all the working as in Exercise 20d. The product
of an expansion can be written down directly.

For example, in the product
$$(c - 3)(c - 4) = c^2 - 7c + 12$$

notice that

1. the first term in the expansion is the
   product of the first two terms in the
   brackets:
   $$(c - 3)(c - 4) = c^2 - 7c + 12$$

2. the last term in the expansion is the product
   of the last two terms in the brackets:
   $$(c - 3)(c - 4) = c^2 - 7c \mathbf{+ 12}$$

3. the middle term in the expansion is found
   by adding the products of the inner and
   outer pairs of terms:
   $$(c - 3)(c - 4) = c^2 - \mathbf{7c} + 12$$

   that is $(-3c) + (-4c) = -7c$

Fig. 20.1 shows these steps in one diagram:

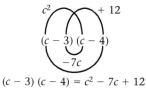

$$(c - 3)(c - 4) = c^2 - 7c + 12$$

Fig. 20.1

Example 14

Expand $(d - 2)(d + 5)$ directly.

$(d - 2)(d + 5) = d^2 + 3d - 10$

In the answer,
d^2 is the product of d and d,
$+3d$ is the result of adding $-2 \times d = -2d$ to $d \times +5 = +5d$,
-10 is the product of -2 and $+5$.

Example 15

Expand $(x + 7)^2$.

$$(x + 7)^2 = (x + 7)(x + 7)$$
$$= x^2 + 14x + 49$$

Example 16

Expand $(10 + x)(10 - x)$ without showing any working.

$(10 + x)(10 - x) = 100 - x^2$

Notice in Example 16 that the middle term reduces to zero:
$+10x + (-10x) = 0$.

Exercise 20e

Expand the following without showing any working.

① $(a + 1)(a + 2)$ ② $(a + 2)(a + 3)$

③ $(a + 3)(a + 4)$ ④ $(b + 1)(b - 2)$

⑤ $(b + 2)(b - 3)$ ⑥ $(b + 3)(b - 4)$

⑦ $(c - 3)(c - 4)$ ⑧ $(d + 7)(d + 1)$

⑨ $(e + 2)(e + 9)$ ⑩ $(f - 5)(f - 4)$

⑪ $(x - 7)(x - 1)$ ⑫ $(y - 2)(y - 9)$

⑬ $(h + 6)^2$ ⑭ $(k - 5)^2$

⑮ $(z + 2)(z - 9)$ ⑯ $(a + 4)(a + 6)$

⑰ $(a - 4)(a - 6)$ ⑱ $(a - 4)(a + 6)$

⑲ $(a + 4)(a - 6)$ ⑳ $(b + 6)(b - 3)$

㉑ $(c - 1)(c - 2)$ ㉒ $(m - 1)^2$

㉓ $(n + 1)^2$ ㉔ $(f + 9)(f + 11)$

㉕ $(e - 3)(e - 5)$ ㉖ $(d - 2)(d + 10)$

㉗ $(h + 3)(h - 8)$ ㉘ $(a + 3)^2$

㉙ $(a - 3)^2$ ㉚ $(a + 3)(a - 3)$

㉛ $(b - 5)(b + 5)$ ㉜ $(c + 7)(c - 7)$

Coefficients of terms

The **coefficient** of an algebraic term is the number which multiplies the unknown. For example,

in $3x^2$, the coefficient of x^2 is 3
in $-2y$, the coefficient of y is -2
in $\frac{2}{3}d$, the coefficient of d is $\frac{2}{3}$

Example 17

Find the coefficient of x in the expansion of $(x + 9)(x + 3)$.

It is not necessary to expand the expression fully. The middle term is the term in x.
Middle term $= (+9x) + (+3x) = +12x$
The coefficient of x in the expansion is $+12$.

Example 18

Find the coefficient of ab in the expansion of $(5a + 2b)(4a - 3b)$.

The terms containing ab are
$$2b \times 4a = 8ab$$
and $5a \times (-3b) = -15ab$.
These add to give $-7ab$. The coefficient of ab in the expansion is -7.

With practice, it should be possible to find coefficients without any written work.

Exercise 20f

① Find the coefficient of d in the expansion of
(a) $(d + 2)(d + 7)$ (b) $(d - 4)(d + 6)$
(c) $(d - 3)(d - 1)$ (d) $(d + 8)(d - 3)$
(e) $(d + 7)^2$ (f) $(d - 5)^2$

② Find the coefficient of u in the expansion of
(a) $(u + 2)(2u + 3)$ (b) $(u - 4)(3u + 5)$
(c) $(2u - 5)(3u + 5)$ (d) $(4u - 5)(2u - 7)$
(e) $(3u - 4)^2$

③ Find the coefficient of ab in the expansion of
(a) $(3a + b)(a + 2b)$ (b) $(a - b)(3a - 2b)$
(c) $(4a + 3b)(5a - 3b)$
(d) $(5a + 2b)(5a - 2b)$
(e) $(a - 3b)^2$

Example 19

Expand the following, simplifying where possible.
(a) $(r - x)(r + y)$ (b) $(x - 5)(x + 8)$
(c) $(2a + 3b)(5a - 2b)$ (d) $(m + 4n)^2$

(a) $(r - x)(r + y) = r(r - x) + y(r - x)$
$\qquad\qquad = r^2 - rx + ry - xy$

(b) $(x - 5)(x + 8) = x(x - 5) + 8(x - 5)$
$\qquad\qquad = x^2 - 5x + 8x - 40$
$\qquad\qquad = x^2 + 3x - 40$

(c) $(2a + 3b)(5a - 2b)$
$\qquad = 5a(2a + 3b) - 2b(2a + 3b)$
$\qquad = 10a^2 + 15ab - 4ab - 6b^2$
$\qquad = 10a^2 + 11ab - 6b^2$

(d) $(m + 4n)^2 = (m + 4n)(m + 4n)$
$\qquad\qquad = m(m + 4n) + 4n(m + 4n)$
$\qquad\qquad = m^2 + 4mn + 4mn + 16n^2$
$\qquad\qquad = m^2 + 8mn + 16n^2$

In Example 19(d), the expression $(m + 4n)^2$ is called a **perfect square**. The expansions in parts (b), (c) and (d) of Example 19 were simplified by collecting the like terms together.

Exercise 20g

Expand the following, collecting like terms where possible.

① $(p + q)(r + s)$ ② $(6 + y)(5 + x)$

③ $(a - b)(c - d)$ ④ $(2x + 1)(3x - 1)$

⑤ $(2n + 3)(2n + 5)$ ⑥ $(4m - 9)(2m - 3)$

⑦ $(5x - 2)(x + 4)$ ⑧ $(a + 5)(3a - 2)$

⑨ $(x - 8)(x + 9)$ ⑩ $(3h + 4)(5h - 2)$

⑪ $(b - 5)(b + 5)$ ⑫ $(c + 7)(c - 7)$

⑬ $(3t - 2)^2$ ⑭ $(2x - 5y)^2$

⑮ $(5x - y)(2x - 3y)$ ⑯ $(4a - 1)(5a - 2)$

⑰ $(5a + 2d)(3a - 8d)$ ⑱ $(2d + e)(3d + e)$

⑲ $(6m + n)(5m - n)$ ⑳ $(2k - 7)(4k + 5)$

Example 20

Find the coefficient of n in
$(4n + 3)(n - 5) - (2n - 3)^2$.

Either:
$(4n + 3)(n - 5) - (2n - 3)^2$
$= n(4n + 3) - 5(4n + 3) - [2n(2n - 3) - 3(2n - 3)]$
$= 4n^2 + 3n - 20n - 15 - (4n^2 - 6n - 6n + 9)$
$= 4n^2 - 17n - 15 - (4n^2 - 12n + 9)$
$= 4n^2 - 17n - 15 - 4n^2 + 12n - 9$
$= -5n - 24$

The coefficient of n is -5.

or, by inspection:

For $(4n + 3)(n - 5)$, the term in n is found by adding the products of the inner and outer pairs of terms (Fig. 20.2):

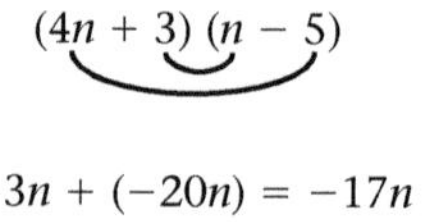

$(4n + 3)\ (n - 5)$

$+3n + (-20n) = -17n$

Fig. 20.2

$(2n - 3)^2 = (2n - 3)(2n - 3)$. Again, the term in n is the sum of the products of the inner and outer terms of the brackets (Fig. 20.3):

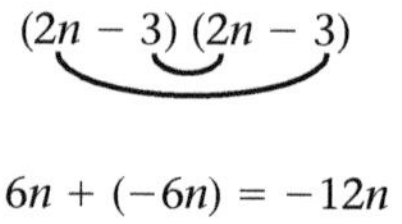

$(2n - 3)\ (2n - 3)$

$-6n + (-6n) = -12n$

Fig. 20.3

For the whole expression,
term in $n = -17n - (-12n)$
$\qquad\quad = -17n + 12n$
$\qquad\quad = -5n$

Coefficient of n is -5.

Fig. 20.4 shows how to expand brackets directly by multiplying the terms inside the brackets in pairs.

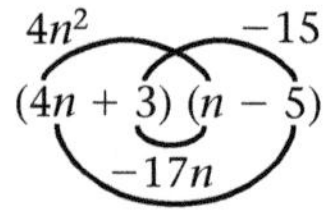

Fig. 20.4

Exercise 20h

1. Either by expanding the brackets of otherwise, write down the coefficient of x in the following.
   (a) $(x + 7)(x + 9)$
   (b) $(x - 3)(x + 8)$
   (c) $(x + 4)(x - 17)$
   (d) $(2x + 5)(4x + 7)$
   (e) $(3x - 1)(5x + 2)$
   (f) $(2x - 8)(2x + 8)$

2. Multiply $(3 + a)$ by $(5 - 2a)$.

3. Simplify the following.
   (a) $(a - 3)(a - 4) + (a - 1)(a + 5)$
   (b) $(m + 6)(m - 2) - (m - 3)(m + 1)$
   (c) $(2x + 3)(x - 2) + 2x(x - 1)$
   (d) $3d(2d + 3) - (3d + 1)(2d + 1)$

4. Simplify $(f + g)^2 + (f - 2g)^2$.

5. Simplify $(3b - 1)^2 - (2b + 1)^2$.

6. Simplify $(2a - 3b)^2 - (a - b)^2$.

7. Simplify $(y + z)^2 - 2yz$.

8. Find the coefficient of (a) x^2, (b) x in $(x + 1)(2x - 3) + (x - 4)^2$.

9. Simplify $(a + b)^2 - (a - b)^2$.

10. Simplify $(3n + t)(n + 3t) - 3(n - 2t)(n - 4t)$.

Factorisation

Example 21

Factorise the following.
(a) $12pq - 6qr$
(b) $3x^2 - 6xy + 9x^2y$

(a) $12pq - 6qr = 6q(2p - r)$
(b) $3x^2 - 6xy + 9x^2y = 3x(x - 2y + 3xy)$

Example 22

Factorise $(x - a)(3x + 2a) - (x - a)^2$.

$(x - a)$ is a **common factor** of the two parts of the expression.

$(x - a)(3x + 2a) - (x - a)^2$
$= (x - a)[(3x + 2a) - (x - a)]$
$= (x - a)(3x + 2a - x + a) = (x - a)(2x + 3a)$

Exercise 20i

Factorise the following and simplify further if possible.

1. $9a - 27$

2. $2a + 4b - 6c + 8d$

3. $3r - 8rt$

4. $a^2 + ac$

5. $2m^2 + 8mn$

6. $42x^2 - 28xy$

7. $42a^2b - 51ab^2$

8. $x^3 + 3x^2y - 5xy^2$

9. $\dfrac{12x - 18}{6}$

10. $\dfrac{3x^2 - 15x}{15}$

11. $\dfrac{7y + 14}{5y + 10}$

12. $\dfrac{10y + 4}{15y + 6}$

13. $\dfrac{4x + xy}{2y^2 + 8y}$

14. $3(y - z) + 2w(y - z)$

15. $5m + 5n - m^2 - mn$

Example 23

Simplify $63 \times 29 + 37 \times 29$.

$63 \times 29 + 37 \times 29 = 29(63 + 37)$
$ = 29 \times 100 = 2900$

Exercise 20j

Use factorisation to simplify the following.

1. $18 \times 57 + 18 \times 43$

2. $23 \times 119 - 23 \times 19$

3. $243 \times 4 + 243 \times 6$

4. $28 \times 752 + 28 \times 248$

5. $63 \times 47 - 43 \times 47$

6. $61 \times 127 - 77 \times 61$

Algebraic fractions

Example 24

Simplify $\dfrac{3x - 2}{5} + \dfrac{x + 1}{3}$

$\dfrac{3x - 2}{5} + \dfrac{x + 1}{3} = \dfrac{3(3x - 2) + 5(x + 1)}{5 \times 3}$

$= \dfrac{9x - 6 + 5x + 5}{15}$

$= \dfrac{9x + 5x - 6 + 5}{15} = \dfrac{14x - 1}{15}$

Example 25

Simplify $\dfrac{a - 4}{2a} - \dfrac{9b - 2}{6b} + 1$

$\dfrac{a - 4}{2a} - \dfrac{9b - 2}{6b} + 1$

$= \dfrac{3b(a - 4) - a(9b - 2) + 6ab(1)}{6ab}$

$= \dfrac{3ab - 12b - 9ab + 2a + 6ab}{6ab}$

$= \dfrac{2a - 12b}{6ab} = \dfrac{2(a - 6b)}{6ab} = \dfrac{a - 6b}{3ab}$

Example 26

Simplify $\dfrac{x}{x - 3} - \dfrac{x - 1}{x + 2}$.

$\dfrac{x}{x - 3} - \dfrac{x - 1}{x + 2} = \dfrac{x(x + 2) - (x - 1)(x - 3)}{(x - 3)(x + 2)}$

$= \dfrac{x^2 + 2x - (x^2 - 4x + 3)}{(x - 3)(x + 2)}$

$= \dfrac{x^2 + 2x - x^2 + 4x - 3}{(x - 3)(x + 2)}$

$= \dfrac{6x - 3}{(x - 3)(x + 2)} = \dfrac{3(2x - 1)}{(x - 3)(x + 2)}$

Exercise 20k

Simplify the following.

1. $\dfrac{x - 4}{3} + \dfrac{x + 3}{2}$

2. $\dfrac{b + 5}{7} - \dfrac{2 - b}{5}$

3. $\dfrac{3d + 6}{10} + \dfrac{d + 3}{3}$

4. $\dfrac{2x - 5}{7} - \dfrac{8x - 8}{6}$

5. $\dfrac{2}{3ab} - \dfrac{3}{4bc}$

6. $5 - \dfrac{p - q}{q}$

7. $\dfrac{ab + ac}{ab - ac}$

8. $\dfrac{a^2 + ab}{ab + b^2}$

9. $\dfrac{5xy}{5x^2 - 10xy} + \dfrac{8xy}{2y^2 - xy}$

10. $\dfrac{b}{a^2 - ab} + \dfrac{a + b}{ab}$

11. $\dfrac{x - 5}{6} - \dfrac{6}{x - 5}$

12. $\dfrac{1}{m - 1} + \dfrac{9}{2m + 3} - \dfrac{8}{m + 4}$

Substitution

Example 27

Find the value of $3(x + y)$ if $x = -2$ and $y = 7$.

When $x = -2$ and $y = 7$,

$3(x + y) = 3(-2 + 7) = 3 \times 5 = 15$

Example 28

Evaluate $ut - \frac{1}{2}ft^2$ when $t = 5$, $u = -20$ and $f = 10$.

$ut - \frac{1}{2}ft^2 = (-20) \times 5 - \frac{1}{2}(10)(5)^2$

$= -100 - 125 = -225$

Example 29

Evaluate $\dfrac{ab^2 - c^2}{2bc} + \dfrac{a^2}{2b + c}$ when $a = 2$, $b = -3$ and $c = -2$.

$\dfrac{ab^2 - c^2}{2bc} + \dfrac{a^2}{2b + c}$

$= \dfrac{2(-3)^2 - (-2)^2}{2(-3)(-2)} + \dfrac{2^2}{2(-3) + (-2)}$

$= \dfrac{2 \times 9 - 4}{12} + \dfrac{4}{-6 - 2}$

$= \dfrac{14}{12} + \dfrac{4}{-8} = \dfrac{7}{6} - \dfrac{1}{2} = \dfrac{7}{6} - \dfrac{3}{6} = \dfrac{2}{3}$

Exercise 20l

1. Evaluate $u + at$ if $a = 10$, $u = 4$ and $t = 2$.

2. If $p = 42$ and $r = 3$, find the value of $\dfrac{p - r^2}{r}$.

3. Find $\sqrt{(u^2 + 2as)}$ when $u = 11$, $a = 4$ and $s = 13$.

④ If $x = 3$, $y = -5$ and $z = 2$, what is the value of $\dfrac{x^2(y - z)}{6z + y}$?

⑤ Evaluate $x^2 - 3x + 8$ when
(a) $x = 3$, (b) $x = 0$, (c) $x = -1$, (d) $x = -4$.

⑥ If $y = 3x^2 - 5x - 2$, find y when
(a) $x = 0$, (b) $x = 1$, (c) $x = 2$, (d) $x = 3$.

⑦ If $V = k\dfrac{R}{T}$, find k when $V = 14$, $R = 35$ and $T = 45$.

⑧ Evaluate $x^2y + y^2x$ when $x = 5$ and $y = -2$.

⑨ If $a = -4$, $b = 6$ and $c = -3$, find the values of
(a) $5a - 3c + 2b$ (b) $9a^2c^2 - b^2$
(c) $\dfrac{3ac}{b^2 - c^2}$

⑩ If $s = u + 980t$, (a) find s when $u = 0$ and $t = 5$, (b) find t when $s = 10\,000$ and $u = 3000$.

Indices

The laws of indices are summarised in Table 20.1.

Table 20.1

1	$x^a \times x^b = x^{a+b}$
2	$x^a \div x^b = x^{a-b}$
3	$x^0 = 1$
4	$x^{-a} = \dfrac{1}{x^a}$
5	$(x^a)^b = x^{ab}$

Exercise 20m

Simplify the following.

① $a^3 \times a^9$ ② $2^2 \times 2^4$

③ $c^7 \div c$ ④ $\dfrac{9 \times 10^9}{3 \times 10^3}$

⑤ $6 \times z^0$ ⑥ $3x^{-3}$

⑦ $\left(\dfrac{2}{3}\right)^{-1}$ ⑧ $a^{-9} \div b^0$

⑨ $9a^{-5} \times 4a^6$ ⑩ $15 \times 10^4 \div (3 \times 10^{-2})$

⑪ $(4v^3)^2$ ⑫ $(-2b^2)^3$

⑬ $(-u^2)^5$ ⑭ $(-f^4)^5$

⑮ $(x^2y^3)^4$ ⑯ $(-4u^2v)^3$

⑰ $(-a^2m)^4$ ⑱ $-3(de^3)^4$

⑲ $\dfrac{(-x^3)^2}{-x^4}$ ⑳ $\dfrac{(-c)^2 \times c^4}{(-c)^5}$

Binary operations

Example 30

Given that $p * q = p + 3q$.
(a) Evaluate (i) $2 * 3$, (ii) $3 * 2$, (iii) $(2 * 3) * 1$.
(b) Is the operation (i) commutative, (ii) associative?

(a) (i) $2 * 3 = 2 + 3(3) = 11$
(ii) $3 * 2 = 3 + 3(2) = 9$
(iii) $(2 * 3) * 1 = 11 * 1 = 11 + 3(1) = 14$
(b) (i)   Since $2 * 3 \neq 3 * 2$, the operation is not commutative.
(ii) $2 * (3 * 1) = 2 * 6 = 20$
Since $(2 * 3) * 1 \neq 2 * (3 * 1)$, the operation is not associative.

Example 31

Given that $x \otimes y = (x + y)^2 - 3xy$ and $x \otimes 3 = x \otimes 2$, find the value of x.

$$(x + 3)^2 - 3x(3) = (x + 2)^2 - 3x(2)$$
$$x^2 + 6x + 9 - 9x = x^2 + 4x + 4 - 6x$$
$$x^2 - 3x + 9 = x^2 - 2x + 4$$
$$-3x + 2x = 4 - 9$$
$$x = 5$$

Exercise 20n

① Given that $a \square b = 2a - b$, find the value of $5 \square 2$.

② If $r \bigcirc s = \dfrac{1}{3}(r + 4s)$, evaluate $(-3) \bigcirc 9$.

③ If $x \boxtimes y$ means 'square the first number and subtract three times the second', then
(a) calculate the value of (i) $3 \boxtimes 1$, (ii) $2 \boxtimes 7$;
(b) evaluate p if (i) $4 \boxtimes p = 1$, (ii) $4 \boxtimes p = p$.

④ Given that $m * n = m^2 + n^2$,
(a) evaluate
(i) $3 * 4$, (ii) $(3 * 4) * 1$, (iii) $3 * (4 * 1)$.
(b) Is the operation commutative and associative?

⑤ Two binary operations, ⊕ and ⊗ are defined as follows:

⊕ is the remainder when the sum of two terms is divided by 5.

⊗ is the remainder when the product of two terms is divided by 5.

Evaluate

(a) 2 ⊕ 3,　　　　(b) 3 ⊗ 4

(c) (2 ⊕ 3) ⊗ 4,　(d) 2 ⊗ (3 ⊗ 4).

⑥ If $p * q = \dfrac{3p + q^2}{2}$, find the value of x when $x * 1 = 5$

Linear equations

Example 32

Solve the equation $2(x + 3) = -11$.

$$2(x + 3) = -11$$

Remove brackets:

$$2x + 6 = -11$$

Subtract 6 from both sides:

$$2x = -17$$

Divide both sides by 2:

$$x = -8\tfrac{1}{2}$$

Example 33

Solve $\dfrac{8}{t} - 1 = 3$ for t.

$$\frac{8}{t} - 1 \; = 3$$

Multiply both sides by t:

$$8 - t = 3t$$

Add t to both sides:

$$8 = 4t$$

Divide both sides by 4:

$$2 = t \quad \text{or} \quad t = 2$$

Exercise 20o

Solve the following equations for x.

① $3 - 2x = 7$　　　② $10 - 5x + 6 = 0$

③ $x + 13 = 5x - 7$　④ $2x - 13 = 5 + 6x$

⑤ $-3(x - 1) = 9$　　⑥ $2(3x - 1) - 10 = 0$

⑦ Solve $3\dfrac{(x - 2)}{2} - \dfrac{x - 3}{4} = 4$

[CXC June 89]

⑧ $\dfrac{11}{x} - 1\tfrac{1}{2} = \dfrac{5}{x}$　　⑨ $\dfrac{6}{x} + \dfrac{2}{3x} = 2$

⑩ $5(3x + \tfrac{1}{2}) = 2x - 17$

Linear inequalities

Example 34

Represent each of the following inequalities on a number line:　(i) $3x > -6$　(ii) $x + 2 < 5$

(i)　$3x > -6$

$x > \dfrac{-6}{3}$

$x > -2$

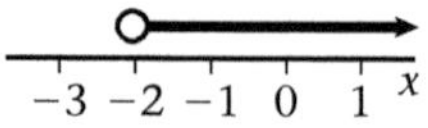

(ii)　$x + 2 < 5$

$x < 5 - 2$

$x < 3$

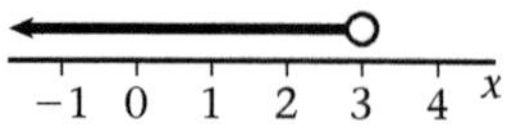

Fig. 20.5　　　　　*Fig. 20.6*

Example 35

Find the range of values of x for which $1 - 6x > 5$.

$$1 - 6x > 5$$

Subtract 1 from both sides:

$$-6x > 4$$

Divide both sides by -6 and *reverse* the inequality:

$$x < -\tfrac{2}{3}$$

Example 36

Solve $\tfrac{3}{4}x - \tfrac{4}{5}x \leqslant -\tfrac{1}{2}$.

$$\tfrac{3}{4}x - \tfrac{4}{5}x \leqslant -\tfrac{1}{2}$$

$$\tfrac{15}{20}x - \tfrac{16}{20}x \leqslant -\tfrac{10}{20}$$

$$-\tfrac{1}{20}x \leqslant -\tfrac{10}{20}$$

Multiply both sides by -20 and *reverse* the inequality.

$$x \geqslant 10$$

Notice in examples 35 and 36 that when both sides of an inequality are multiplied or divided by a negative number, the inequality sign is reversed.

Example 37

Represent the inequality $-6 \leqslant x < 2$ on a number line.

Fig. 20.7 gives two alternative graphs of the inequality.

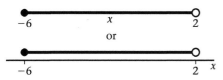

Fig. 20.7

Notice that the value -6 is included in the range. This is shown by a solid dot. The value 2 is *not* included; this is shown by a small circle.

Exercise 20p

① Solve the following inequalities.
  (a) $x + 2 \geqslant 3$ (b) $x - 3 < 1$
  (c) $3 - x \geqslant 1$ (d) $3x \leqslant 12$
  (e) $-5x < 30$ (f) $4x + 20 > 0$

② Solve the following inequalities.
  (a) $5 - 2x \leqslant x - 1$ (b) $8x - 5 \leqslant 6x + 7$
  (c) $p + 2 < 96 - p$ (d) $7x - (5x - 3) \geqslant 9$
  (e) $3x + 4(x - 3) > x - 6$
  (f) $2(x + 4) > 3(x - 1)$
  (g) $\frac{1}{2}(3x - 2) \geqslant x - 6$
  (h) $\dfrac{y + 8}{3} - \dfrac{2y - 4}{7} < 1$
  (i) $\dfrac{x}{2} + \dfrac{3}{4} \leqslant \dfrac{5x}{6} - \dfrac{7}{12}$
  (j) $\dfrac{7}{z + 2} > \dfrac{-2}{5 - 4z}$

③ Make sketch graphs of the inequalities in questions 1 and 2.

④ Sketch graphs of the following inequalities.
  (a) $-2 \leqslant x < 5$ (b) $3 < x \leqslant 11$
  (c) $0 \leqslant x \leqslant 8$ (d) $-7 < x < -4$
  (e) $-4 < x \leqslant 0$ (f) $-9 \leqslant x < 6$

⑤ x is such that $3x + 4 < 7$ and $5x \geqslant 2x - 9$.
  (a) What range of values of x satisfies both inequalities?
  (b) Draw a graph which represents this inequality.

⑥ List the integer values of x which are members of the following sets.
  (a) $\{x: -6\frac{1}{2} < x < -3\frac{2}{3}\}$
  (b) $\{x: -2\frac{1}{4} < x < 2\frac{1}{4}\}$

⑦ If $S = \{x: 3x - 14 < 2 < 2x + 5, x$ is an integer$\}$, list the members of S.

⑧ If $-5 \leqslant x \leqslant 3$ find (a) the smallest, (b) the largest possible values of the following.
  (i) $9 - x$ (ii) $x - 9$
  (iii) $x^2 - 9$ (iv) $9x$

Simultaneous linear equations

$2x - 5y = 16$ is a **linear equation** with two **variables**, x and y. There are many pairs of values of x and y which **satisfy** this equation (i.e. make the equation true). For example, if $x = 13$ and $y = 2$ or if $x = 8$ and $y = 0$ the equation will be true.

Given two equations, such as $2x - 5y = 16$ and $x + 4y = -5$, it is usually possible to find values of x and y which satisfy *both* equations simultaneously (i.e. at the same time).

To solve a pair of **simultaneous linear equations**, either use a graphical method or the method of elimination and substitution.

Example 38

$$2x - 5y = 16 \qquad (1)$$
$$x + 4y = -5 \qquad (2)$$

multiply (2) by 2
$$2x + 8y = -10 \qquad (3)$$

subtract (3) from (1) to eliminate terms in x
$$-13y = 26$$
$$y = -2$$

substitute -2 for y in (2) to find x
$$x + 4(-2) = -5$$
$$x - 8 = -5$$
$$x = 3$$

The solution is $x = 3$ and $y = -2$.

Example 39

Solve the simultaneous equations
$2x - y = 4$, $3x + y = 11$.

Either by substitution:

$$2x - y = 4 \tag{1}$$
$$3x + y = 11 \tag{2}$$

Make y the subject of equation (2):

$$y = 11 - 3x \tag{3}$$

Substitute $11 - 3x$ for y in equation (1):

$$2x - (11 - 3x) = 4$$
$$2x - 11 + 3x = 4$$
$$5x = 15$$
$$x = 3$$

Substitute 3 for x in equation (3):

$$y = 11 - 9 = 2$$

The solution is $x = 3$, $y = 2$.

or by elimination:

$$2x - y = 4 \tag{1}$$
$$3x + y = 11 \tag{2}$$

Add (1) and (2) to eliminate y:

$$5x = 15$$
$$x = 3$$

Substitute 3 for x in (1):

$$6 - y = 4$$
$$y = 2$$

The solution is $x = 3$, $y = 2$.

or by graph:

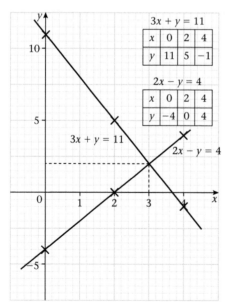

Fig. 20.8

Fig. 20.8 shows the graphs of $2x - y = 4$ and $3x + y = 11$ drawn on the same axes.

The point of intersection of the two lines is (3, 2). The solution of the equations is $x = 3$, $y = 2$.

Example 40

Solve the equations $\frac{2}{3}x - \frac{1}{2}y = 2$, $\frac{3}{4}x - \frac{1}{3}y = 3\frac{1}{6}$.

$$\tfrac{2}{3}x - \tfrac{1}{2}y = 2 \tag{1}$$
$$\tfrac{3}{4}x - \tfrac{1}{3}y = 3\tfrac{1}{6} \tag{2}$$

Simplify the equations by clearing the fractions

(1) ×6: $\quad 4x - 3y = 12 \tag{3}$
(2) ×12: $\quad 9x - 4y = 38 \tag{4}$

Solve in the usual way:

(3) × 4: $\quad 16x - 12y = 48 \tag{5}$
(4) × 3: $\quad 27x - 12y = 114 \tag{6}$
(6) − (5): $\quad 11x = 66$
$$x = 6$$

Substitute 6 for x in (3):

$$24 - 3y = 12$$
$$-3y = -12$$
$$y = 4$$

The solution is $x = 6$, $y = 4$.

Example 41

Solve the equations
$$3x - 2y + 1 = 2x + 5y - 10 = 4x - 3y.$$

Pair the three expressions in any two different ways:

$$3x - 2y + 1 = 4x - 3y$$
$$x - y = 1 \tag{1}$$
$$2x + 5y - 10 = 4x - 3y$$
$$2x - 8y = -10$$
$$x - 4y = -5 \tag{2}$$

Solve equations (1) and (2) in the usual way. This gives $x = 3$ and $y = 2$ as the solution.

Exercise 20q

Solve the following pairs of equations. Use either the substitution, elimination or graphical method, whichever is most suitable.

① $\quad x + y = 6$
$\quad\ \ 2x + 3y = 14$

② $\quad 2x - y = 11$
$\quad\ \ x + 2y = -7$

③ $\quad 3x - 4y = 7$
$\quad\ \ x - 2y = 5$

④ $\quad 3x + 2y = 7$
$\quad\ \ 7x - 3y = 1$

⑤ $5x - 2y = -23$
$3x + 4y = 7$

⑥ $3x - 2y = 4$
$2x - 7y = 31$

⑦ $x - 2y = 1$
$x + 2y = 9$

⑧ $a + 3b = -13$
$2a - 9b = 4$

⑨ $2c + 4d = 17$
$4c - 3d = 1$

⑩ $3x + 4y = -1$
$3x + 8y = 4$

⑪ $5m - 2n = 15$
$3m + 5n = 9$

⑫ $3x + 7y = -8$
$5y = 2x + 15$

⑬ $3a + 4m = 0$
$a = 2m - 5$

⑭ $5d = 4n + 8$
$5n + 10 = 4d$

⑮ $5x + 3y - 2 = 0$
$3x - 32 = 7y$

⑯ $\frac{3}{4}x + \frac{1}{5}y = 4$
$\frac{1}{2}x = \frac{3}{5}y - 1$

⑰ $1.2x - 1.1y = 7.9$
$1.8x + 0.7y = 0.1$

⑱ $\frac{d}{4} - \frac{n}{3} = 6$
$\frac{3d}{2} + \frac{5n}{6} = 2$

⑲ $3x - y + 12 = 5x + 2y + 4 = x + y$

⑳ $5(a + b) = 2(a + 3b) + 1$
$3(a + 2b) - 7 = a + 3b + 1$

Word problems

Example 42

The ages of a man and his son add up to 39 years. In 3 years the man will be 4 times as old as his son. Find the ages of the man and his son.

Let the ages of the man and his son be x and y years respectively. From the first sentence:

$$x + y = 39 \qquad (1)$$

From the second sentence:

$$x + 3 = 4(y + 3)$$
$$x + 3 = 4y + 12$$
$$x - 4y = 9 \qquad (2)$$

Solve equations (1) and (2) simultaneously.
Subtract (2) from (1):

$$5y = 30$$
$$y = 6$$

Substitute 6 for y in (1):

$$x = 33$$

The man is 33 years old and the son is 6 years old.

Check: $33 + 6 = 39$ (*1st sentence*)
               $36 = 4 \times 9$ (*2nd sentence*)

Exercise 20r

① A notebook and a pencil cost $2.64. If the notebook costs 66c more than the pencil, find the cost of each.

② A father is 28 years older than his daughter. In 2 years' time he will be 3 times as old as his daughter. Find their present ages.

③ When a man cycles for 1 hour at x km/h and 2 hours at y km/h, he travels 32 km. When he cycles for 2 hours at x km/h and 1 hour at y km/h, he travels 34 km. Find x and y.

④ If I subtract 1 from the numerator of a fraction, the fraction becomes $\frac{1}{2}$. If I add 1 to both the numerator and denominator of the fraction, it becomes $\frac{2}{3}$. What is the fraction? *Hint*: Let the fraction be $\frac{x}{y}$.

⑤ A pack of 7 pens and 10 pencils costs $36. A second pack of 3 pens and 20 pencils costs $39.

   (a) Using x to denote the cost (in dollars) of a pen and y to denote the cost (in dollars) of a pencil,
      (i) write, in terms of x and y, an expression for the cost of the pack of 7 pens and 10 pencils;
      (ii) write a pair of simultaneous equations to represent the cost of both packs of pens and pencils.
   (b) Calculate (i) the cost of a pen, (ii) the cost of a pencil.
   [CXC June 87]

⑥ A table costs $$x$ and a chair costs $$y$. If the price of each is raised by $20, the ratio of their prices becomes $5:2$ respectively. If the price of each is reduced by $5, the ratio becomes $5:1$. Express the ratio $x:y$ in its lowest terms.

⑦ The side of a square is x metres. The length of a rectangle is 5 metres more than the side of the square. The width of the rectangle is 4 metres more than the side of the square.

   (a) Write, in terms of x, expressions for the length and width of the rectangle.

   The area of the rectangle is $47\,\text{m}^2$ more than the area of the square.

   (b) Determine the area of the rectangle.
   [CXC Jan 89]   (6 marks)

8 A girl had $200. She bought a tape, a book and a record. The book cost half as much as the tape. The record cost $8 more than the book. She has $104 left.

(a) Using x to represent the cost (in dollars) of the tape, state in terms of x, expressions for
   (i) the cost of the book,
   (ii) the cost of the record,
   (iii) the total amount of money spent.
             (3 marks)

(b) State an equation for the amount of money left.   (2 marks)

(c) Hence, determine the cost of the record.   (3 marks)

Revision test (Chapter 20)

Select the correct answer and write down your choice.

1 $2(3y - 4) - 3(y - 1) =$
A $3y - 11$ B $3y - 5$
C $2y - 3$ D $2y - 5$

2 $\dfrac{3a}{5} - \dfrac{a}{3} =$

A $\dfrac{a}{4}$ B $\dfrac{4a}{15}$ C $\dfrac{a}{2}$ D a

3 $a(b + d) = ab + ad$ is an example of the
A   associative law   B   commutative law
C   distributive law   D   laws of indices

4 $6x + 6y =$
A $12xy$ B $6(xy)$
C $6x + y$ D $6(x + y)$

5 Given that $m \otimes n = 3m^3 + 2n$, and $2 \otimes n = 4$, then $n =$
A -14 B -11 C -10 D -7

6 $-\dfrac{6a^6b^2}{2a^2b} =$

A $-4a^4b$ B $-3a^4b$
C $3a^3b$ D $-4a^3b$

7 $(2y + 3)(y - 2) =$
A $2y^2 + 5y + 6$ B $2y^2 + y - 6$
C $2y^2 - y - 6$ D $2y^2 - 5y - 6$

8 $V = x^2h$. If $x = 3$ and $h = 4$, then $V =$
A   24   B   36   C   48   D   144

9 If n represents a number, the symbolic expression that represents 'a number which is 5 less than three times the first number' is given by
A $5n - 3$ B $5 - 3n$
C $3n - 5$ D $3 - 5n$

10 Given that $(2x + 3)(3x + k) = 6x^2 - x - 15$, $k =$
A   5   B   3   C -3 D -5

11 Which of the following are factors of $15y^2 - 5y$?
I $5y$ II $3y - 1$ III $3y$ IV $15y - 5$
A   I, II, III only   B   I and II only
C   I, II, IV only   D   II and IV only

12 If $5x + 7 = 2(x - 1)$, then $x =$
A   3   B $\frac{9}{7}$ C   1   D -3

13 Which one of the following is *not* a member of the solution set of the inequality $3x \leqslant 6$?
A -1 B   0   C $\frac{1}{2}$ D   3

14 4 more than twice a number is 18. The number is
A   11   B   7   C   5   D   4

15 Given that $\dfrac{3a + 1}{4} = \dfrac{5}{2}$, then $a =$
A $\frac{4}{3}$ B   2   C   3   D $\frac{7}{2}$

16 Given that $3x + 2 > x + 8$, then
A $x > -\frac{3}{2}$ B $x > \frac{3}{2}$
C $x > 3$ D $x > 5$

17 Given that $2x + y = 10$ and $y = 3x$, $\{x, y\} =$
A $\{2, 6\}$ B $\{-2, -6\}$
C $\{2, -1\}$ D $\{-2, 14\}$

18 A cycle costs $800. A girl has $350 and saves $100 a week. What is the least number of weeks required so that she has enough money to buy the cycle?
A   3   B   4   C   5   D   8

19 The largest possible value of y in the solution set of $2y + 1 \leqslant 7$ is
A   2   B   3   C   4   D   6

20 If $2x - \dfrac{3 - x}{5} = \dfrac{x + 4}{3} + \dfrac{9}{5}$, then $x =$
A -2 B   1   C   2   D   4

Measurement
Solids and plane shapes, approximation and estimation

Plane shapes

Perimeter

The **perimeter** of a plane shape is the distance round the edge of the shape. The perimeters of shapes are found by measurement. In some cases there are formulae which enable perimeters to be calculated:

Perimeter of rectangle in Fig. 21.1 = $2(l + b)$

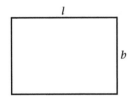

Fig. 21.1

Perimeter of circle in Fig. 21.2 = $2\pi r$.

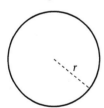

Fig. 21.2

The perimeter of a circle is often called the **circumference**.

Length of arc in Fig. 22.3 = $\dfrac{\theta}{360} \times 2\pi r$

Perimeter of sector = $\left(\dfrac{\theta}{360} \times 2\pi r\right) + 2r$

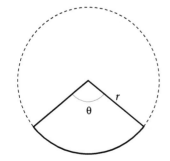

Fig. 21.3

Example 1

Find the perimeter of a sector of a circle of radius 3.5 cm, the angle of the sector being 120°.

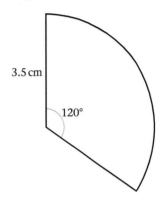

Fig. 21.4

Perimeter = length of arc + 2r

Length of arc = $\left(\dfrac{120}{360} \times 2\pi \times 3.5\right)$ cm

$\qquad\qquad = \left(\dfrac{120}{360} \times 2 \times \dfrac{22}{7} \times \dfrac{7}{2}\right)$ cm

$\qquad\qquad = \dfrac{22}{3}$ cm

$\qquad\qquad = 7.3$ cm

Perimeter = $(7.3 + [2 \times 3.5])$ cm = 14.3 cm

Exercise 21a

Take the value of π to be $3\frac{1}{7}$, unless told otherwise.

1   Use the value 3.14 for π to calculate the circumference of a circle of diameter 20 cm.

2   The minute hand of a wall-clock is 10.5 cm long. How far does its tip travel in 24 hours?

3   Through what angle does the minute hand of a clock move in 25 min? If the minute hand is 6.3 cm long, how far does its tip move in 25 min?

4   How many revolutions does a bicycle wheel of diameter 70 cm make in travelling 110 m?

⑤ An arc subtends an angle of 72° at the centre of a circle of radius 17.5 cm. Find the length of the arc.

⑥ Two circles have circumferences of 10π cm and 12π cm. What is the difference in their radii?

⑦ A piece of thread was wound tightly round a cylinder for 20 complete turns. The thread was found to be 3.96 m long. Calculate the diameter of the cylinder in cm.

⑧ The figure ABCDEF in Fig. 21.5, *not drawn to scale*, represents an athletic track. ABDE is a rectangle and BCD and AFE are semicircles. ED = 78 m and BD, the diameter, is 14 m.

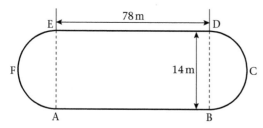

Fig. 21.5

Calculate
(a) the total distance around the track
(b) the number of times an athlete must run around the track to complete 1500 metres. (5 marks)
[CXC June 92]

⑨ Fig. 21.6 is a sketch of a doorway. The arc at the top subtends an angle of 60° at the centre of a circle of radius 75 cm.

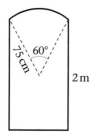

Fig. 21.6
Use the value 3.14 for π to calculate the perimeter of the doorway to the nearest $\frac{1}{2}$ centimetre.

Area

The common units of area are cm², m² and km². The **hectare** (ha) is often used or land measure. 1 ha = 10 000 m². Formulae for the areas of the common plane shapes are given in Fig. 21.7.

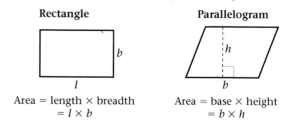

Rectangle

Area = length × breadth
= $l \times b$

Parallelogram

Area = base × height
= $b \times h$

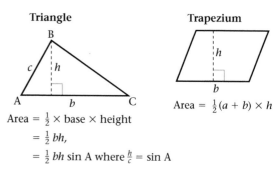

Triangle

Area = $\frac{1}{2}$ × base × height
= $\frac{1}{2} bh$,
= $\frac{1}{2} bh \sin A$ where $\frac{h}{c} = \sin A$

Trapezium

Area = $\frac{1}{2}(a + b) \times h$

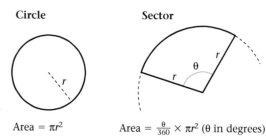

Circle

Area = πr^2

Sector

Area = $\frac{\theta}{360} \times \pi r^2$ (θ in degrees)

Fig. 21.7

Example 2

Calculate the area of △ABC in which AB = 8 cm, AC = 4 cm and $\widehat{BAC}$ = 58°.

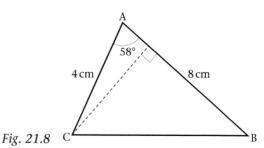

Fig. 21.8

Let base, $b = AB = 8$ cm
and height $= h$ cm

$$\frac{h}{4} = \sin 58°$$
$$\therefore h = 4 \sin 58°$$

Area of $\triangle ABC = \frac{1}{2}bh$
$$= \frac{1}{2} \times 8 \times 4 \sin 58° \, \text{cm}^2$$
$$= 16 \times 0.848 \, \text{cm}^2$$
$$= 13.6 \, \text{cm}^2 \text{ to 3 s.f.}$$

Example 3

Calculate the area of the parallelogram in Fig. 21.9

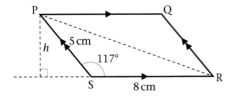

Fig. 21.9

Area of parallelogram
$$= 2 \times \text{area of } \triangle PRS$$
$$= 2 \times (\frac{1}{2} \times 5 \times 8 \sin 117°) \, \text{cm}^2$$
$$= 5 \times 8 \sin 63° \, \text{cm}^2$$
$$= 40 \times 0.891 \, \text{cm}^2$$
$$= 35.6 \, \text{cm}^2 \text{ to 3 s.f.}$$

Example 4

The trapezium in Fig. 21.10 has an area of $456 \, \text{cm}^2$. Calculate the distance between its parallel sides.

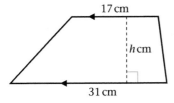

Fig. 21.10

Area of trapezium $= \frac{1}{2}(17 + 31)h \, \text{cm}^2$
$$\frac{1}{2}(17 + 31) \times h = 456$$
$$\frac{1}{2} \times 48 \times h = 456$$
$$24h = 456$$
$$\therefore h = \frac{456}{24} = 19$$

The parallel sides are 19 cm apart.

Example 5

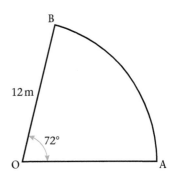

Fig. 21.11

Fig. 21.11 (*not drawn to scale*) represents a flower bed in the shape of a sector of a circle centre O, radius 12 m. Given that $A\widehat{O}B = 72°$, find the area of the flower bed. (Take π to be 3.14.)

Area of sector $= \dfrac{72}{360} \times 3.14 \times 12 \times 12 \, \text{m}^2$
$$= 90.4 \, \text{m}^2 \text{ to 1 d.p.}$$

Example 6

In Fig. 21.12, the chord AB subtends an angle of 120° at the centre of the circle, radius 7 cm. Use the value $\frac{22}{7}$ for π to calculate, to 3 significant figures, the area of the minor segment of the circle.

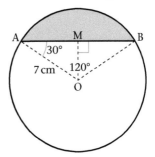

Fig. 21.12

Area of minor segment (shaded)
$$= \text{area of sector } AOB - \text{area of } \triangle AOB$$

Area of sector $AOB = \dfrac{120}{360} \times \pi \times 7^2 \, \text{cm}^2$
$$= \frac{1}{3} \times 22 \times 7 \, \text{cm}^2$$

Area of $\triangle AOB = \frac{1}{2}bh$
$$= AM \times OM$$

$$\frac{AM}{AO} = \cos 30°$$
$$AM = AO \times \cos 30°$$
$$\frac{OM}{AO} = \sin 30°$$
$$OM = AO \times \sin 30°$$

Area of △AOB = 7 cos 30° × 7 sin 30° cm²
$$= (7 \times 7 \times 0.866 \times 0.5)\,\text{cm}^2$$
$$= (7 \times 7 \times 0.433)\,\text{cm}^2$$

Area of (shaded) minor segment
$$= (\tfrac{1}{3} \times 22 \times 7 - 7 \times 7 \times 0.433)\,\text{cm}^2$$
$$= 7(\tfrac{1}{3} \times 22 - 7 \times 0.433)\,\text{cm}^2$$
$$= 7(7.333 - 3.031)\,\text{cm}^2$$
$$= 7 \times 4.302\,\text{cm}^2$$
$$= 30.1\,\text{cm}^2 \text{ to 3 s.f.}$$

Exercise 21b

Use the value $\frac{22}{7}$ for π unless told otherwise.

1. The diagonals of a parallelogram are 6 cm and 8 cm long and they intersect at an angle of 55°. Calculate the area of the parallelogram.

2. Two sides of a triangular field are 120 m and 200 m. If the angle between the sides is 68°, find the area of the field in hectares.

3. Calculate the area shaded in each of the shapes in Fig. 21.13.

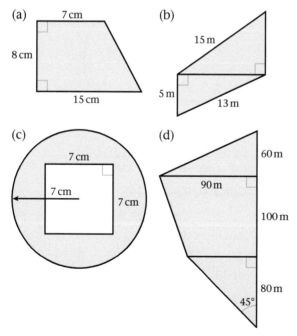

(a) 7 cm, 8 cm, 15 cm

(b) 15 m, 5 m, 13 m

(c) 7 cm, 7 cm, 7 cm

(d) 60 m, 90 m, 100 m, 80 m, 45°

Fig. 21.13

4. Find the area of a circle of radius 35 cm. If a sector of angle 80° is removed from the circle, what area is left?

5. Circular discs of diameter 4 cm are punches out of a sheet of brass of mass 0.84 g/cm². What is the mass of 500 discs?

6. If 350 of the discs in question 5 are punched from a square sheet of brass, 80 cm by 80 cm, what percentage of the sheet is not used?

7.
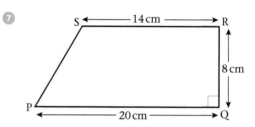

Fig. 21.14

In the trapezium in Fig. 21.14, *not drawn to scale*, PQ = 20 cm, QR = 8 cm, RS = 14 cm and angle PQR = 90°.

(a) Calculate the area of the trapezium.
(b) Calculate the length of PS.   (5 marks)
[CXC June 92]

8. The radius of a circle, centre O, is 4.9 cm. XY is an arc so that angle XÔY is 45°. Calculate to one decimal place
(a) the circumference of the circle
(b) the area of the circle
(c) the area of the *minor* sector XOY.

9. What is the diameter, to the nearest metre, of a circular sports ground of area exactly one hectare?

10. A washer is 4.5 cm in diameter with a central hole of diameter 1.5 cm as shown in Fig. 21.15 (see page 211). Use the value 3.14 for π to calculate the top surface area of the washer correct to 3 s.f.

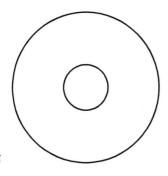

Fig. 21.15

⑪ Fig. 21.16 shows a cross-section of a tunnel in the form of a major segment of a 6 m diameter circle. The path at the base of the tunnel is 3 m wide. Use the value 3.14 for π to find the cross-sectional area of the tunnel correct to 3 s.f.

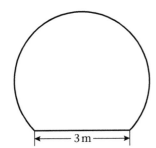

Fig. 21.16 ←— 3 m —→

⑫ Calculate the area of the doorway in Fig. 21.6 on page 209.

Solids

Scale drawing

It is difficult to draw solids on plain paper because solids have **three dimensions**: *length*, *breadth* and *height*, while drawing paper has only two dimensions: *lengths* and *breadth*.

One way of making accurate drawings of solids is by means of an **orthogonal projection**. In an orthogonal projection, the solid is split up into parts and each part is drawn separately.

Freehand sketches of solids

You should always make a freehand sketch of the solid before making an accurate drawing. Fig. 21.17 shows four steps in drawing a cuboid.

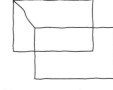

(a) draw two parallel rectangles

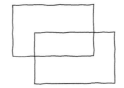

(b) join a pair of corresponding corners

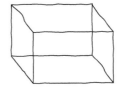

(c) join the other corners in the same way

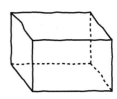

(d) go over the drawing, making hidden lines broken

Fig. 21.17

Exercise 21c (Revision)

Do not use a ruler in this exercise. All drawings should be freehand.

① Use the method of Fig. 21.17 to draw some cuboids. Practise until you can draw a good freehand cuboid.

② Make freehand copies of the cuboids shown in Fig. 21.18.

(a) 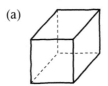 (b)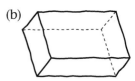

(c) 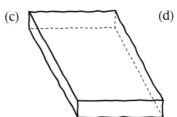 (d)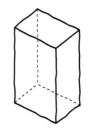

Fig. 21.18

③ Fig. 21.19 shows four steps in drawing a square-based pyramid.

Use the method of Fig. 21.19 to draw some square-based pyramids. Practise until you can draw a good pyramid.

(a) draw a parallelogram

(b) find the centre of the parallelogram

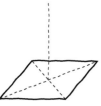

(c) mark a point vertically above the centre

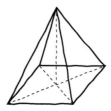

(d) join the point to each corner of the paralellogram

Fig. 21.19

④ Make some copies of Fig. 21.20.

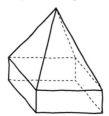

Fig. 21.20

⑤ Fig. 21.21 shows four steps in sketching a cylinder.

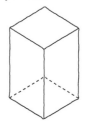

(a) sketch a cuboid such that its top and bottom faces are rhombuses

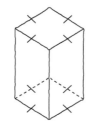

(b) mark out the mid-points of the top and bottom edges

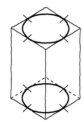

(c) draw ellipses to touch the mid-points of the top and bottom edges

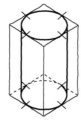

(d) complete the cyclinder as shown

Fig. 21.21

Use the method of Fig. 21.21 to draw some cylinders. Practise until you can draw cylinders without drawing the surrounding cuboids.

⑥ Make freehand copies of the cylinders in Fig. 21.22.

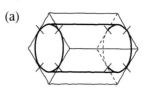

(a)　　　　　　　　　　　　　(b)

Fig. 21.22

⑦ Fig. 21.23 shows three steps in drawing a cone.

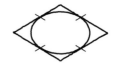

(a) draw an ellipse inside a rhombus

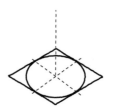

(b) draw a vertical line passing through the centre of the rhombus

(c) complete the cone

Fig. 21.23

Use the method of Fig. 21.23 to draw some cones. Practise until you can draw cones without drawing the starting rhombus.

⑧ Make freehand copies of the solids in Fig. 21.24.

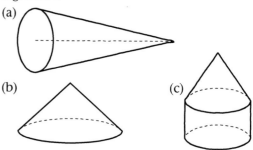

(a)

(b)　　　　　　　　　　(c)

Fig. 21.24

9. Fig. 21.25 shows a method of sketching a triangular prism.

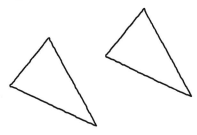

(a) draw a pair of equal triangles so that their corresponding sides are parallel to each other

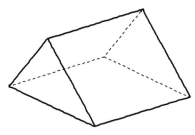

(b) joining corresponding vertices

Fig. 21.25

Use the method of Fig. 21.25 to sketch some triangular prisms.

10. Make freehand copies of the solids in Fig. 21.26.

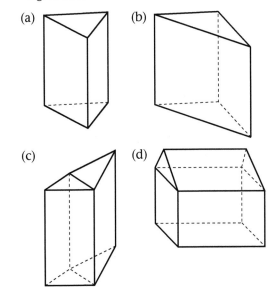

Fig. 21.26

Surface area and volume

Fig. 21.27 contains a summary of the formulae for surface area and volume of common solids.

Cuboid

volume = lbh
surface area = $2(lb + lh + bh)$

Prism

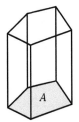

volume = base area × perpendicular height
= $A \times h$

Pyramid

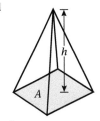

volume = $\frac{1}{3}$ × base area × height
= $\frac{1}{3} Ah$

Cone

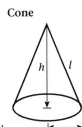

volume = $\frac{1}{3}\pi r^2 h$
curved surface area = πrl
total surface area of a solid cone
= $\pi rl + \pi r^2$
= $\pi r(l + r)$

Cylinder

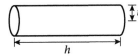

volume = $\pi r^2 h$
curved surface area = $2\pi rh$
total surface area of a solid cylinder
= $2\pi rh + 2\pi r^2$
= $2\pi r(h + r)$

volume = $\frac{4}{3}\pi r^3$
surface area = $4\pi r^2$

Fig. 21.27

Example 7

How many litres of oil does a cylindrical drum 28 cm in diameter and 50 cm deep hold?

Volume of drum $= \pi r^2 h$
$$= \tfrac{22}{7} \times 14^2 \times 50 \, \text{cm}^3$$

Capacity of drum $= \dfrac{22 \times 14^2 \times 50}{7 \times 1000}$ litres
$$= 30.8 \text{ litres}$$

Example 8

Find the capacity in litres of a bucket 24 cm in diameter at the top, 16 cm in diameter at the bottom and 18 cm deep.

The bucket is in the shape of a frustum of a cone. It is necessary to consider the whole cone. Complete the cone as in Fig. 21.28 and let the depth of the extension be x cm.

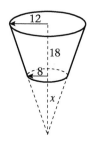

Fig. 21.28

In Fig. 21.28
$$\frac{8}{x} = \frac{12}{x + 18} \qquad \text{(similar } \triangle\text{s)}$$
$$8x + 144 = 12x$$
$$4x = 144$$
$$\therefore \; x = 36$$

Volume of frustum
$$= \tfrac{1}{3}\pi 12^2 \times 54 \, \text{cm}^3 - \tfrac{1}{3}\pi 8^2 \times 36 \, \text{cm}^3$$
$$= \tfrac{1}{3}\pi 4^2 \times 18 \, (3^2 \times 3 - 2^2 \times 2) \, \text{cm}^3$$
$$= \pi \times 16 \times 6 \times 19 \, \text{cm}^3$$
$$= 5730 \, \text{cm}^3$$

Capacity of bucket $\simeq 5.73$ litres

Example 9

A solid is made up of a cylinder with a hemisphere on top as in Fig. 21.29. Calculate (a) the surface area, (b) the volume of the solid.

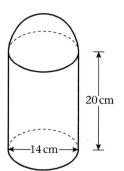

Fig. 21.29

(a) Total surface area
$$= \pi r^2 + 2\pi r h + \tfrac{1}{2}(4\pi r^2)$$
$$= \pi 7^2 + 2\pi \times 7 \times 20 + 2\pi \times 7^2 \, \text{cm}^2$$
$$= 7\pi \, (7 + 40 + 14) \, \text{cm}^2$$
$$= 7 \times \tfrac{22}{7} \times 61 \, \text{cm}^2$$
$$= 1342 \, \text{cm}^2$$

(b) Volume $= \pi r^2 h + \tfrac{1}{2}(\tfrac{4}{3}\pi r^3)$
$$= \pi \times 7^2 \times 20 + \tfrac{2}{3}\pi \times 7^3 \, \text{cm}^3$$
$$= 49\pi \, (20 + \tfrac{2}{3} \times 7) \, \text{cm}^3$$
$$= 49 \times \tfrac{22}{7} \times \tfrac{74}{3} \, \text{cm}^3$$
$$\simeq 3800 \, \text{cm}^3$$

Exercise 21d

Use the value $\tfrac{22}{7}$ or 3.14 for π as appropriate.

1. Water in a 14 mm diameter pipe flow at 2 m/s. How many litres flow along the pipe in 1 min?

2. 154 litres of oil are poured into a cylindrical barrel of diameter 35 cm. Calculate the depth of the oil.

3. 6.6 mm of rain fell onto a rectangular roof 8 m long and 6 m wide. The rainwater drains into a cylindrical barrel of diameter 60 cm. How far does the water level rise in the barrel?

4. The most economical shape for a cylindrical container is one in which the height and diameter are equal. Find the capacity in litres to 2 s.f. of such a tin which is 10 cm high.

5. A cylindrical tin 8 cm in diameter contains water to a depth of 4 cm. If a cylindrical wooden rod 4 cm in diameter and 6 cm long is placed in the tin it floats exactly half submerged. What is the new depth of water?

6. A pyramid 7 cm high stands on a base 12 cm square. Calculate the volume of the pyramid.

7. Calculate (a) the slant height, (b) the curved surface area, (c) the volume, of a cone of height 8 cm and base diameter 12 cm. Leave answers in terms of π.

8. If the cone in question 7 is made of paper and the paper is cut and opened out into a sector of a circle, what is the angle of the sector?

9. A lampshade is in the shape of an open frustum of a cone. Its to and bottom diameters are 10 and 20 cm and its height is 12 cm. Find, in terms of π, the area of material required for the curved surface of the shade.

10. A solid consists of a cone attached to a hemisphere as in Fig. 21.30.

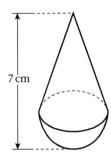

7 cm

Fig. 21.30

Calculate the volume of the solid if the diameter of the hemisphere is 3 cm and the overall height of the object is 7 cm.

11. A rectangular steel pyramid of height 6 cm and base dimensions 11 cm by 16 cm, is melted down and rolled into a cylinder of height 7 cm. Calculate
    (a) the radius of the cylinder in cm;
    (b) the mass of the cylinder in kg, if the density of the steel is 5 g/cm³.
    (*Notes*: Volume of pyramid = $\frac{1}{3}Ah$.
    Volume of cylinder = $\pi r^2 h$.
    Take π to be $\frac{22}{7}$.)
    [CXC June 88]

Similar shapes and similar solids

Areas and volumes

If two similar shapes have corresponding lengths in the ratio $a:b$,
(i) the ratio of their areas is $a^2:b^2$,
(ii) the ratio of their volumes is $a^3:b^3$.

Example 10

A board 1 m long and 80 cm wide costs $4.80. What would be the cost of a similarly shaped board 75 cm long?

Ratio of corresponding lengths $= \dfrac{75\,\text{cm}}{100\,\text{cm}} = \dfrac{3}{4}$.

Ratio of costs = ratio of corresponding areas

$$= \left(\frac{3}{4}\right)^2 = \frac{9}{16}$$

Cost of smaller piece $= \dfrac{9}{16}$ of $4.80

$$= \$2.70$$

Example 11

Two similarly shaped cans hold 2 litres and 6.75 litres respectively. If the smaller can is 16 cm in diameter, what is the diameter of the larger can?

Ratio of volumes $= \dfrac{6.75\,\text{litres}}{2\,\text{litres}} = \dfrac{27}{8} = \left(\dfrac{3}{2}\right)^3$

Ratio of corresponding lengths $= \frac{3}{2}$

Diameter of larger can $= \frac{3}{2}$ of 16 cm = 24 cm

Exercise 21e

1. A photograph is 20 cm long and 15 cm wide. The length of a small print of the photograph is 4 cm. Find (a) the width of the smaller print, (b) the ratio of the areas of the two photographs.

2. The diameter of a sphere is 3 times that of another sphere. How many times greater is its surface area?

3. A box of height 8 cm has a volume of 320 cm³. What is the volume of a similar box of height 6 cm?

4. Two balls have diameters of 10 cm and 6 cm. Find (a) the ratio of their diameters in its

simplest form, (b) the ratio of their surface areas, (c) the ratio of their volumes.

5 A cylinder is 8 cm high and its base diameter is 4 cm. The height of a similar cylinder is 12 cm. (a) Find the diameter of the base of the larger cylinder. (b) What is the ratio of the volume of the larger cylinder to that of the smaller one?

6 The area of a lake is 18 km². It is represented by an area of 2 cm² on the map. (a) What area in km² is represented by 1 cm² on the map? (b) What length does 1 cm on the map represent? (c) What is the ratio of lengths on the map to actual lengths?

7 A school has an area of 3025 m² and it is represented on a plan by an area of 144 cm². Find the actual length of a wall which is shown in the plan by a line 8.4 cm long.

8 Two similar boxes have volumes of 250 cm³ and 54 cm³. What is the ratio of (a) their heights, (b) their surface areas?

9 If an oil drum with a capacity of 25 litres is 60 cm deep, what is the depth of a similar drum which holds 1.6 litres?

Approximation and estimation

Example 12

Calculate the circumference of a circle of radius 4.2 m, giving your answer correct to (a) 3 s.f., (b) 2 d.p. (Use $\pi = 3.14$)

Circumference of circle $= 2 \times 3.14 \times 4.2$ m
$$= 26.376 \text{ m}$$
$$= 26.4 \text{ m to 3 s.f.}$$
$$= 26.38 \text{ m to 2 d.p.}$$

Example 13

Find the exact value of $\dfrac{2.4 \times 0.028}{1.5}$ and give your answer to 2 s.f.

$$\frac{2.4 \times 0.028}{1.5} = \frac{24 \times 0.028}{15}$$
$$= 0.0448$$
$$= 0.045 \text{ to 2 s.f.}$$

Example 14

The sides of a rectangular plate are measured as 8.4 cm and 2.6 cm correct to the nearest mm.

(a) Give a rough estimate of the area of the plate.

(b) Calculate from the measurements, correct to 1 d.p.,
    (i) the area of the plate in cm²,
    (ii) the smallest possible value of the area.

(a) Rough estimate of area $= (8 \times 3)$ cm²
$$= 24 \text{ cm}^2$$

(b) (i) Area of plate $= 8.4 \times 2.6$ cm²
$$= 21.84 \text{ cm}^2$$
$$= 21.8 \text{ cm}^2 \text{ to 1 d.p.}$$
    (ii) Smallest possible area $= 8.35 \times 2.55$ cm²
$$= 21.2925 \text{ cm}^2$$
$$= 21.3 \text{ cm}^2 \text{ to 1 d.p.}$$

Exercise 21f

1 The distance between two towns, C and D, is 11 km. A bus leaves C at 10:55 hrs and arrives at D at 11:09 hrs the same day. Calculate

(a) the time, in minutes, taken for the journey,

(b) the average speed of the bus in kilometres per hour, giving your answer to the nearest whole number.

2 The distance that a man runs in 20 seconds to the nearest second is measured to be 85 m to the nearest metre.

(a) Calculate his average speed to 2 significant figures.

(b) Write down
    (i) the greatest possible distance
    (ii) the least possible distance
    (iii) the greatest possible time
    (iv) the least possible time.

(c) Calculate to 2 significant figures
    (i) his greatest possible speed
    (ii) his least possible speed.

③ The dimensions of a rectangular field, measured to the nearest metre, are given as 8 m long and 5 m wide.
(a) State the range of values for the width of the field.
(b) Calculate
   (i) the maximum possible area
   (ii) the minimum possible length of fencing required to enclose the field.

④ To answer this question use $\pi = \frac{22}{7}$ and $V = \pi r^2 h$, where V = volume, r = radius and h = height of the cylinder.
(a) The internal dimensions of a saucepan shaped like a cylinder are height 20 cm and diameter 35 cm. Calculate, to the nearest whole number,
   (i) the area, in cm², of the bottom of the saucepan
   (ii) the largest volume of liquid, in cm³, which the saucepan can hold.
                                            (5 marks)
(b) The saucepan is filled with water which is then poured into an empty cylindrical pot of radius 21 cm. Calculate, to the nearest cm, the height of the water level above the bottom of the pot.
[CXC Jan 91]   (4 marks)

⑤ (a) An athlete was timed to run the 100 m race in 11.3 s
   (i) Calculate, correct to 3 significant figures, the athlete's average speed, $v\,\text{ms}^{-1}$.
   The length of the 100 m track was in the range $(100 + 1)$ m and the time was in the range (11.3 ± 0.1) s. Calculate
   (ii) the athlete's greatest possible speed
   (iii) the athlete's least possible speed
   (iv) Express the range within which his speed lies in the form $(v - x)\,\text{ms}^{-1}$.
(b) The height of Fig. 21.31, not drawn to scale, is 8.2 cm. LM = 3.8 cm, NO = 7.7 cm and LQ = 5.2 cm. LQ is perpendicular to LM and QP. LQ is parallel to NO. Calculate the area enclosed by the figure LMNOPQ.

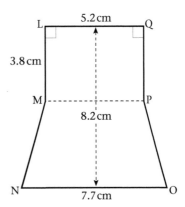

Fig. 21.31

Scale drawing and bearings

Example 15

Fig. 21.32 represents a map of the island of Jetui drawn to a scale of 1 : 25 000.

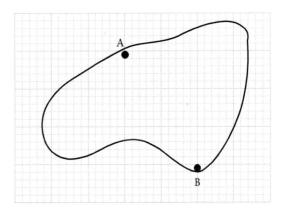

Fig. 21.32

(a) Estimate the actual distance, in kilometres, by a direct route, between the two towns marked A and B.
(b) Estimate in cm² the area of the scale drawing.
(c) Calculate the actual area of the island in km² giving your answer correct to 2 s.f.

(a) Distance between A and B measured by direct route on map ≈ 3.6 cm
Actual distance between A and B
$$\approx \frac{3.6 \times 25\,000}{100 \times 1000}$$
$$\approx 0.9\,\text{km}$$

(b) Estimate of area (counting small squares)
$\simeq 14\,cm^2$

(c) Actual area of island $= \dfrac{14 \times 25\,000^2}{100^2 \times 1000^2}\,km^2$

$$= \dfrac{14 \times 625 \times 10^6}{10^{10}}\,km^2$$

$$= 0.875\,km^2$$

$$= 0.88\,km^2 \text{ to 2 s.f.}$$

Exercise 21g

① The scale used for a map is 1 : 500 000. The distance AB on the map is 4.4 cm. Calculate, in kilometres, the actual distance of A from B.

② In Fig. 21.33, A and D are on opposite sides of a river.

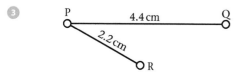

Fig. 21.33

AB = 52 m, BC = 119 m and CD = 86 m. Make a scale drawing and hence find the distance AD.

③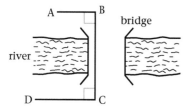

Fig. 21.34

Fig. 21.34, *not drawn to scale*, shows the positions of three towns P, Q and R on a map. On the map the distance between Town P and Town Q is 4.4 cm and the distance on the map between Town P and Town R is 2.2 cm. The actual distance between Town P and Town Q is 11 km. Calculate (a) the scale used on the map, (b) the actual distance between Towns P and Q.

④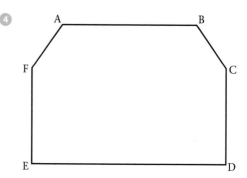

Fig. 21.35

In Fig. 21.35, the figure ABCDEF is an *accurate scale drawing* of a pane of glass where AB and ED represent the top and bottom edges respectively.

(a) Given that CD = 3.0 cm, AB = 5.0 cm and ED = 7.0 cm, calculate in cm² the area of the figure ABCDEF. (4 marks)

(b) Given that the top edge AB of the actual pane of glass measures 2.5 m, calculate the scale used in the drawing. (2 marks)

(c) Using your answer in part (b), calculate for the actual pane of glass:
  (i) the length of the bottom edge in metres,
  (ii) the actual area of the glass in square metres.
  [CXC June 88] (4 marks)

⑤ The course for a 20 km marathon begins at a point B, heads 5 km due East, then 4 km on a bearing of 030°, then 7 km due West, and finally 4 km due South ending at a point F.

(a) Make a scale drawing of the course of the marathon, stating the scale used.
Determine
(b) by measurement
(c) by calculation
  (i) the distance FB
  (ii) the bearing of F from B.

⑥ The distance on a map between the two towns G and H is 4 cm. The actual distance GH is 16 km.

(a) Calculate, in the form 1 : n, the scale used to draw the map.

H is on a bearing of 070° from G.

(b) Using the given scale, draw an accurate diagram to show the positions of G and H.

Another town K is 20 km away from G on a bearing of 340°.

(c) Show the position of K on the drawing.
(d) Find, by measurement on the drawing, the bearing of K from H.
(e) Calculate the distance KH, correct to 2 significant figures.

Revision test (Chapter 21)

Select the correct answer and write down your choice.

① When 0.093 6 is rounded off to 2 s.f. the result is

A 0.094 B 0.09 C 0.93 D 0.94

② 3579 rounded off to 3 s.f. is

A 357.9 B 358 C 3579 D 3580

③ The value of $\dfrac{1.2 \times 10^{-4}}{2.4 \times 10^{-2}}$ written in standard form is

A 0.5×10^{-2} B 0.5×10^{-3}
C 5×10^{-3} D 2×10^{-2}

④ Calculate the value of $\dfrac{63.8}{5.24 \times 0.89}$ to 3 s.f.

A 1.37 B 10.8 C 13.7 D 137

⑤ The circumference, in centimetres, of a circle of diameter 10 cm is

A 10π B 20π C 25π D 100π

⑥ ABCD is a trapezium in which AB∥CD and AB = 2CD. The area of the trapezium is 54 cm² and the height is the same as the length of CD. The height of the trapezium is

A 3 cm B 6 cm C 9 cm D 12 cm

⑦ The area of △ABC is the same as the area of a rectangle ABPQ on the side AB of the triangle. The ratio of the height of △ABC to the length BP of the rectangle is

A 1 : 1   B 1 : 2
C 2 : 1   D 4 : 1

⑧ Given that $\frac{1}{2}bc \sin A = \frac{1}{2}b \times h$ and $\widehat{A} = 30°$, then $h =$

A $2c$ B c
C $\dfrac{c}{2}$ D $\dfrac{2}{c}$

⑨ Two circles have circumferences of 12π cm and 16π cm. The difference between their radii is

A 2 cm   B 4 cm
C 2π cm   D 4π cm

⑩ A map is drawn on a scale of 1 cm : 120 km. The distance between two towns separated on the map by 2.3 cm is

A 2.76 m   B 27.6 m
C 276 km   D 2760 km

⑪ In Fig. 21.36, AOB is a sector of a circle centre O and radius 5 cm.

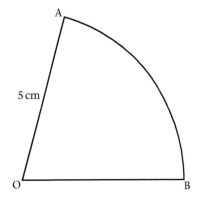

Fig. 21.36

If AOB = 72° the perimeter of the sector, in cm, is

A π B 2π
C $2\pi + 5$ D $2\pi + 10$

⑫ A map is drawn on a scale of 1 : 100 000. What is the actual area, in km², of a region represented on the map by an area of 1.5 cm²?

A 1   B 1.5
C 2.25   D 10

13 Calculate to 2 s.f. the area of a circular region of radius 6.7 cm.

A 14 cm²   B 41 cm²
C 140 cm²   D 141 cm²

14 The ratio of the heights of two similar cuboids is 2 : 3. The ratio of the surface areas of the cuboids is

A 2 : 3   B 4 : 3
C 4 : 9   D 8 : 27

15 The front elevation of a solid is a triangle. Which of the following shapes could the solid possible be?

A cube   B cylinder
C rectangular prism   D cone

16 The height of a solid cylinder is twice the radius of the circular cross-section. The volume of the cylinder is

A $\frac{\pi r^3}{4}$ B $\frac{\pi r^3}{2}$ C πr^3 D $2\pi r^3$

17 The triangular face of a square pyramid has a base of 5 cm. If the height of the pyramid is 12 cm, the volume of the pyramid is

A 100 cm³   B 240 cm³
C 390 cm³   D 78 cm³

18 A solid metal ball is melted down to make smaller balls whose radius is one-third the radius of the larger ball. The number of smaller balls obtained is

A 3   B 6   C 9   D 27

Chapter 22

Geometry

Angles, triangles, polygons and circles, constructions, transformations – congruence, similarity

Angles

Angle is a measure of rotation or turning.
1 revolution = 360 degrees (1 rev = 360°)
1 degree = 60 minutes (1° = 60′)
The names of angles change with their size (Fig. 22.1).

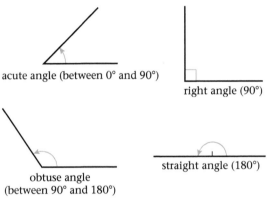

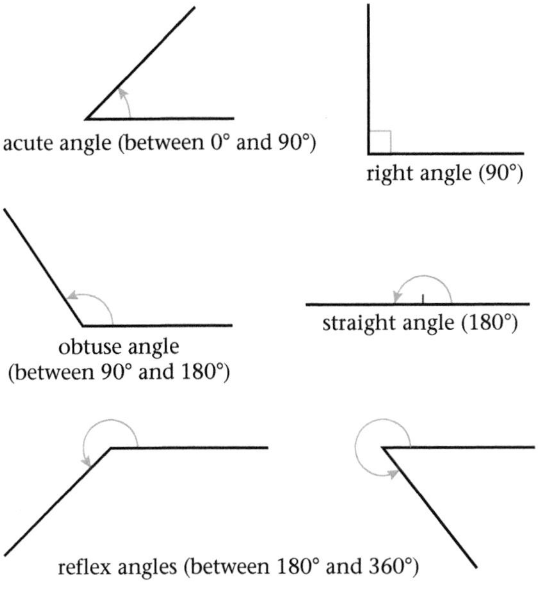

acute angle (between 0° and 90°)
right angle (90°)
obtuse angle
(between 90° and 180°)
straight angle (180°)
reflex angles (between 180° and 360°)

Fig. 22.1

Angles are formed when lines meet or intersect.

Remember the following facts:

Angles at a point add up to 360°. In Fig. 22.2, $a + b + c + d + e = 360°$.

b
a
c
e
d

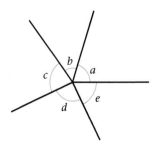

Fig. 22.2

Adjacent angles on a straight line add up to 180°. In Fig. 22.3, $a + b = 180°$.

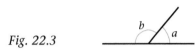

Fig. 22.3

b
a

Vertically opposite angles are equal. In Fig. 22.4, $p = q$ and $r = s$.

r
p
q
s

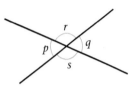

Fig. 22.4

Alternate angles are equal. In Fig. 22.5, $x = y$ and $m = n$.

x
m
n
y

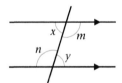

Fig. 22.5

Corresponding angles are equal. In Fig. 22.6, $a = b$ and $p = q$.

a
p
b
q

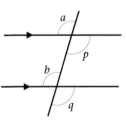

Fig. 22.6

Interior opposite angles add up to 180°. In Fig. 22.7, $c + d = 180° = r + s$.

c
r
d
s

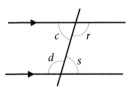

Fig. 22.7

Example 1

Find the angles marked *a*, *b* and *c* in Fig. 22.8.

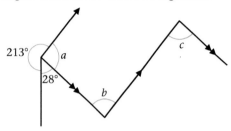

Fig. 22.8

$$a = 360° - (213° + 28°) \quad (angles\ at\ a\ point)$$
$$= 360° - 241°$$
$$= 119°$$

$$b = 180° - a \quad (int.\ opp)$$
$$= 180° - 119°$$
$$= 61°$$

$$c = b \quad (alt.\ angles)$$
$$= 61°$$

Exercise 22a (oral or written)

Find the lettered angles in Fig. 22.9.

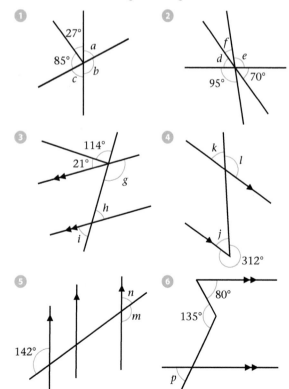

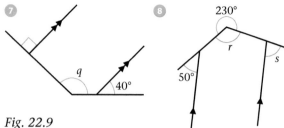

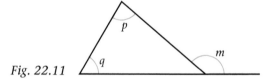

Fig. 22.9

Triangles

Angles in a triangle

The sum of the angles of a triangles is 2 right angles, or 180°. In Fig. 22.10, $\widehat{A} + \widehat{B} + \widehat{C} = 180°$.

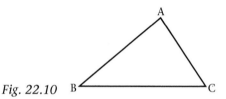

Fig. 22.10

The exterior angle of a triangle equals the sum of the two interior opposite angles. In Fig. 22.11, $m = p + q$.

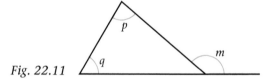

Fig. 22.11

Types of triangle

Fig. 22.12 gives the names and properties of the common types of triangle.

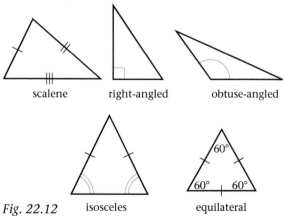

Fig. 22.12

Example 2

Find $\widehat{P}$ in Fig. 22.13.

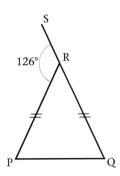

Fig. 22.13

$$\widehat{P} = \widehat{Q} \qquad \text{(base angles of isos. △)}$$
$$\widehat{P} + \widehat{Q} = 126° \qquad \text{(ext. angle of △)}$$
$$\therefore \quad 2\widehat{P} = 126° \qquad (\widehat{P} = \widehat{Q})$$
$$\therefore \quad \widehat{P} = 63°$$

In a right-angled triangle the area of the square on the hypotenuse is equal to the sum of the areas of the squares on the other two sides.

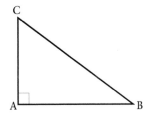

Fig. 22.14

In Fig. 22.14,

$$(BC)^2 = (AB)^2 + (AC)^2$$

Congruent triangles

Two triangles are congruent if they can fit exactly one over the other. Congruent triangles have their corresponding sides and angles equal.

Example 3

Given Fig. 22.15 name the triangle which is congruent to △XYZ, keeping the letters in the correct order.

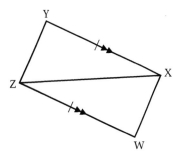

Fig. 22.15

$$\triangle XYZ \equiv \triangle ZWX$$

Reason: XY = ZW   (*given*)

        $Y\widehat{X}Z = W\widehat{Z}X$ (*alt. angles* XY ∥ ZW)

        XZ = ZX   (*common side*)

Exercise 22b

1.   Name and calculate the sizes of the exterior angles shown in Fig. 22.16.

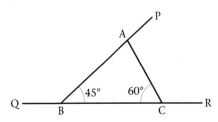

Fig. 22.16

2.   In each of the following, two angles of a triangle are given. Find the third angle and name the type of triangle.
   - (a)   69°, 46°   (b)   38°, 71°
   - (c)   43°, 94°   (d)   60°, 60°
   - (e)   35°, 55°   (f)   58°, 25°

3.   Calculate the sizes of the lettered angles in Fig. 22.17.
   - (a)   (b)

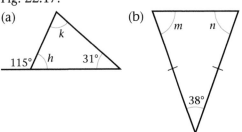

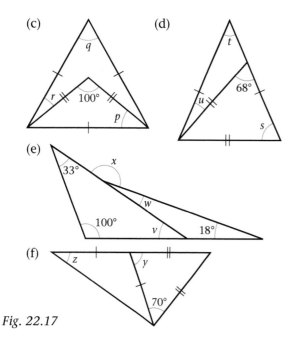

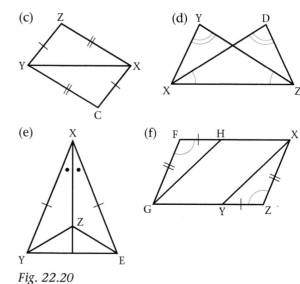

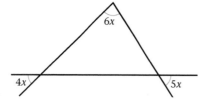

Fig. 22.17

④ Find the value of x in Fig. 22.18.

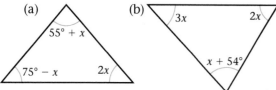

Fig. 22.18

⑤ Find the value of x in each part of Fig. 22.19. Hence state which type of triangle each is.

(a)

55° + x

75° − x 2x

(b)

3x 2x

x + 54°

Fig. 22.19

⑥ In each part of Fig. 22.20, name the triangle which is congruent to △XYZ. Keep the letters in the correct order.

(a)   X

Y   Z   A

(b)   X

Y   Z

B

Fig. 22.20

⑦ In Fig. 22.21, CD∥AB and AD bisects BÂC. If CD̂E = 27° and AĈB = 69°, find BÊD.

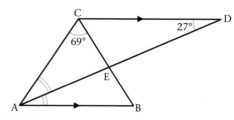

Fig. 22.21

⑧ Calculate the length of the sides marked x in each of the triangles in Fig. 22.22.

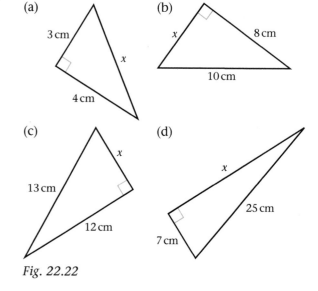

Fig. 22.22

9 In Fig. 22.23, BC∥XY, B$\widehat{X}$Y = 50°, B$\widehat{Y}$X = 28° and AB = BY. Calculate A$\widehat{C}$B.

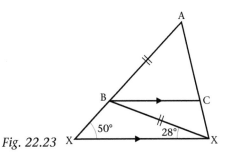

Fig. 22.23

10 Which of the triangles in Fig. 22.24 are congruent? State the case(s) of congruency.

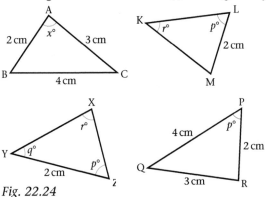

Fig. 22.24

Other polygons

Any plane figure bounded by straight lines is called a **polygon**. A **regular polygon** has all its sides of equal length and all its angles of equal size. Polygons are named after the number of sides they have. Fig. 22.25 gives the names of some common regular polygons.

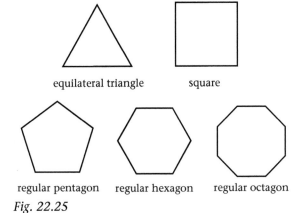

equilateral triangle   square

regular pentagon   regular hexagon   regular octagon

Fig. 22.25

Angles of a polygon

The sum of the angles of an *n*-sided polygon is $(n - 2) \times 180°$ or $(2n - 4)$ right angles.

The sum of the exterior angles of a polygon is 4 right angles, or 360°.

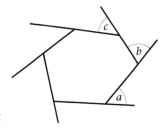

Fig. 22.26

In Fig. 22.26, $a + b + c + \ldots = 360°$.

Example 4

Calculate the interior angles of a regular pentagon.

A pentagon has 5 sides.
The five exterior angles add up to 360°.
Since the pentagon is regular, the exterior angles are equal.

Each exterior angle $= \dfrac{360°}{5} = 72°$

Hence each interior angle
$$= 180° - 72° \qquad \text{(angles on a str. line)}$$
$$= 108°$$

Example 5

Each of the angles of a polygon is 140°. Find the number of sides that the polygon has.

Either:
Let the polygon have *n* sides.
Sum of angles of polygon $= n \times 140° = 140n°$
Also, sum of angles $= (n - 2)180°$
So $(n - 2)180° = 140n°$
$$180n - 360° = 140n$$
$$40n = 360$$
$$\therefore \quad n = \frac{360}{40} = 9$$

or:
Each exterior angle $= 180° - 140° = 40°$
But the sum of the exterior angles $= 360°$
Number of exterior angles $= \dfrac{360°}{40°} = 9$

The polygon has 9 sides.

Properties of quadrilaterals

A **trapezium** is a quadrilateral which has one pair of opposite sides parallel (Fig. 22.27).

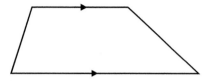

Fig. 22.27

A **parallelogram** is a quadrilateral which has both pairs of opposite sides parallel (Fig. 22.28).

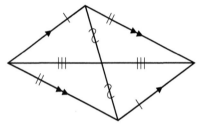

Fig. 22.28

In a parallelogram,
1   the opposite sides are parallel,
2   the opposite sides are equal,
3   the opposite angles are equal,
4   the diagonals bisect one another.

A **rhombus** is a quadrilateral which has all four sides equal (Fig. 22.29).

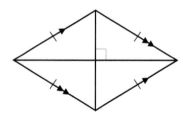

Fig. 22.29

In a rhombus,
1   all four sides are equal,
2   the opposite sides are parallel,
3   the opposite angles are equal,
4   the diagonals bisect one another at right angles,
5   the diagonals bisect the angles.

A **rectangle** is a quadrilateral in which every angle is a right angle (Fig. 22.30).

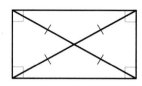

Fig. 22.30

In a rectangle, all of the facts given for a parallelogram are true. In addition:
1   all four angles are right angles,
2   the diagonals are of equal length.

A **square** is a rectangle which has all four sides equal (Fig. 22.31).

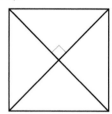

Fig. 22.31

In a square, all of the facts given for a rhombus are true. In addition:
1   all four angles are right angles,
2   the diagonals are of equal length,
3   the diagonals meet the sides at 45°.

Exercise 22c

1   Use the $(2n - 4) \times 90°$ formula to complete Table 22.1.

Table 22.1

Polygon	Sum of interior angles
3 sides triangle	
4 sides quadrilateral	
5 sides pentagon	
6 sides hexagon	
7 sides heptagon	
8 sides octagon	
10 sides decagon	
12 sides dodecagon	

2 Calculate the sizes of the lettered angles in Fig. 22.32.

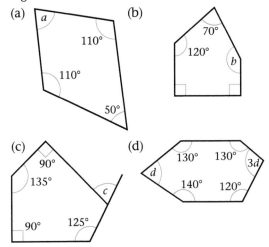

(a), (b)

(c), (d)

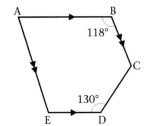

Fig. 22.32

3 Calculate the interior angles of a regular 15-sided polygon.

4 A regular polygon has interior angles of 160°. How many sides has it?

5 In Fig. 22.33, pentagon ABCDE is such that AB∥ED and BC∥AE. If AB̂C = 118° and CD̂E = 130°, calculate BĈD, EÂB and AÊD.

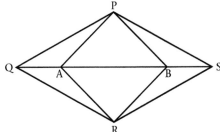

Fig. 22.33

6 In Fig. 22.34, QABS is a diagonal of rhombus PQRS and square PARB.

Fig. 22.34

If PŜR = 68°, what is the size of QP̂A?

7 In Fig. 22.35, ABDE and BCDE are parallelograms. Use the measurements given on the figure to calculate the perimeter of trapezium ABCDE.

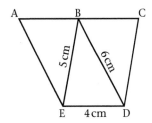

Fig. 22.35

8 ABCD is a trapezium such that AB∥DC. X is a point on CD such that CX = BA. If AB̂C = 102° and DÂX = 47°, calculate AD̂X.

9 Two of the exterior angles of a polygon are 63° each. The remaining exterior angles are each 26°. How many sides has the polygon?

10 The angles of a pentagon are 4x, 5x, 6x, 7x and 8x. Find x and hence state the sizes of the angles of the pentagon.

Circles

Fig. 22.36 gives the names of the lines and regions of a circle.

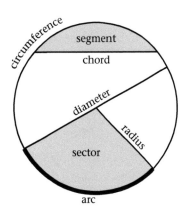

Fig. 22.36

Angle properties of circles

1 The angle which an arc of a circle subtends at the centre of a circle is twice that which it subtends at any point on the remaining part of the circumference. In Fig. 22.37 $\widehat{POQ} = 2 \times \widehat{PRQ}$.

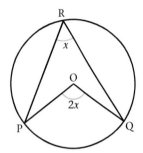

Fig. 22.37

2 The angle in a semicircle is a right angle. In Fig. 22.38, $\widehat{APB} = 90°$.

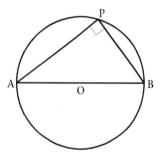

Fig. 22.38

3 Angles in the same segment are equal. In Fig. 22.39, $\widehat{APB} = \widehat{AQB} = \widehat{ARB} = \ldots$

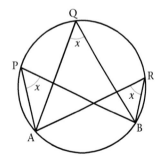

Fig. 22.39

Exercise 22d

In each part of Fig. 22.40, find the value of x. O is the centre of the circle, where given.

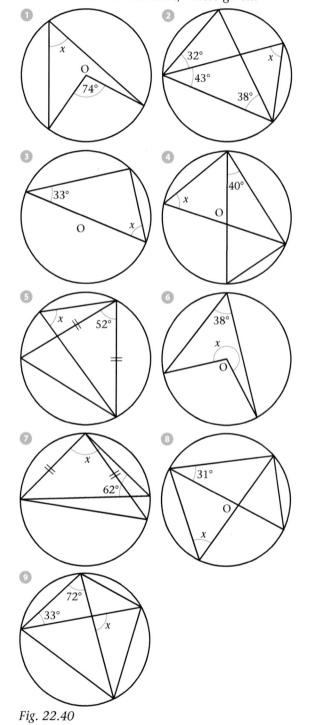

Fig. 22.40

Constructions

Remember the following when making constructions.

1   Make a rough sketch. This helps in anticipating problems associated with the construction.

2   Leave all construction lines visible. Do not rub off anything that contributes to the final result.

3   Use a hard pencil with a sharp point. This enables lines and points to be as fine and accurate as possible.

To bisect a given line segment, AB (Fig. 22.41)

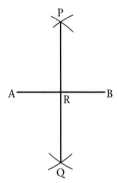

Fig. 22.41

With centres A and B and equal radii, draw arcs to cut each other at P and Q. Join PQ to cut AB at R. R is the mid-point of AB.

To bisect a given angle, AB̂C (Fig. 22.42)

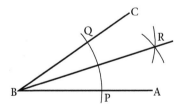

Fig. 22.42

With centre B and any radius, draw an arc to cut BA and BC at P and Q. With centres P and Q and equal radii, draw arcs to cut each other at R. Draw BR. BR is the required bisector.

To construct a line perpendicular to a given straight line, AB, from a point, M, outside the line (Fig. 22.43)

With centre M and any radius, draw an arc to cut AB at P and Q. With centres P and Q and equal radii, draw arcs to cut each other at R. MR is perpendicular to AB.

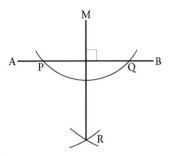

Fig. 22.43

To construct parallel lines, using ruler and set square (Fig. 22.44)

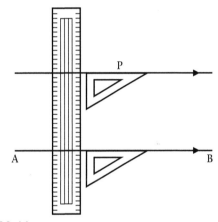

Fig. 22.44

Place a set square with one edge accurately along a given line AB. Place a ruler against one of the other edges of the set square. Holding the ruler firmly, slide the set square along its edge until the edge that was originally along AB passes through the required point P. Use that edge of the set square to draw a line parallel to AB.

To construct an angle of 60° (Fig. 22.45)

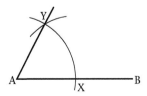

Fig. 22.45

With centre A and any convenient radius draw an arc cutting AB at X. With centre X and the *same* radius, draw an arc to cut the first arc at Y. Join AY to give $B\widehat{A}Y = 60°$.

To construct and angle of 30°
First construct an angle of 60° as above and then bisect it. Further bisections will give angles of 15°, $7\frac{1}{2}°$, etc.

To construct an angle of 45°
First construct a right angle and then bisect it. Further bisections will give angles of $22\frac{1}{2}°$, etc.

To copy a given angle, $A\widehat{B}C$ (Fig. 22.46)

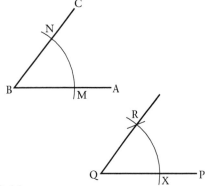

Fig. 22.46

Draw any line PQ. With centre B and any radius draw an arc to cut BA and BC at M and N. With centre Q and the *same* radius, draw an arc to cut QP at X. With centre X and radius MN, draw an arc to cut the arc through X at R. Then $P\widehat{Q}R = A\widehat{B}C$.

Exercise 22e

1. Construct angles of 60°, 30°, 75°, 45°, 120°, $37\frac{1}{2}°$, 135°.

2. Construct an equilateral triangle with sides of length 7.2 cm. Construct the perpendicular from a vertex to the opposite side and measure its length.

3. Draw a triangle ABC in which BC = 6 cm, AB = 4 cm and angle ABC = 50°. State the length of AC. Through C draw CD parallel to BA. If BC is produced to F, state the size of angle DCF.
[CXC June 87]

4. Construct $\triangle PQR$ such that $\widehat{Q} = 90°$, $\widehat{P} = 60°$ and PQ = 8 cm. Draw the bisectors of $\widehat{P}$ and $\widehat{R}$ to intersect at O. Measure OP.

5. Construct an isosceles triangle with the equal sides 9 cm long and the angle between them 45°. Measure the third side.

6. Use ruler and compasses only for this question. All construction lines and arcs must be clearly shown.
   (a) Construct a triangle ABC in which AB = 8 cm, $A\widehat{B}C = 60°$, BC = 6.5 cm.
   (b) Construct the perpendicular from C to AB to meet AB at X.
   (c) Measure and state the lengths AC and CX.

7. Construct the parallelogram ABCD in which BD = 105 mm, DC = 48 mm and $B\widehat{C}D = 30°$. Measure AC.

8. Construct a trapezium PQRS in which PQ∥SR, PQ = 6 cm, PS = 5 cm, SR = 11 cm and QS = 9 cm. Measure QR.

9. Construct a rhombus ABCD so that AC = 6 cm and BD = 8 cm. Measure a side of the rhombus.

10. Construct a triangle with sides of 5 cm, 6 cm, 7 cm. A rhombus with sides of length 4 cm has acute angles each equal to the smallest angle of the triangle. Copy the smallest angle of the triangle and hence construct the rhombus. Measure the longer diagonal of the rhombus.

Transformations

A geometric transformation changes the position, size or shape of an object. Translation, reflection, rotation and enlargement are geometric transformations.

A **translation** is a movement in a straight line. In Fig. 22.47, $\triangle ABC$ is translated in the direction CR to give the $\triangle PQR$. The translation is described in terms of the vector $\overrightarrow{CR}$.

$\triangle ABC$ is translated by vector $\begin{pmatrix} 3 \\ 2 \end{pmatrix}$.

Every point in △ABC moves through the same distance in the same direction. △ABC is **congruent** to △PQR.

Translation is a **congruency transformation**.

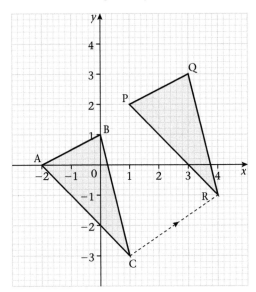

Fig. 22.47

In Fig. 22.48, △PQR is the image of △ABC after a **reflection** in the y-axis. The lines joining the object and image points are perpendicular to the mirror line. △ABC is congruent to △PQR. Note that the orientation of the triangles is different. Reflection is a **congruency transformation**.

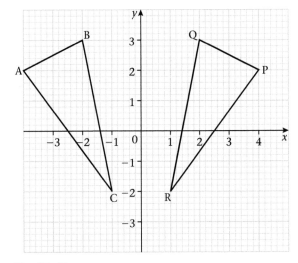

Fig. 22.48

In Fig. 22.49, △PQR is the image of △ABC after a reflection in the line $y = x$ (mirror line).

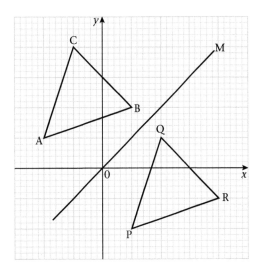

Fig. 22.49

In Fig. 22.50, △PQR is the image of △ABC after a **rotation** of 90° clockwise about the origin. △ABC and △PQR are congruent triangles. Rotation is a **congruency transformation**.

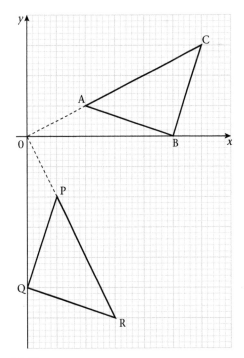

Fig. 22.50

In Fig. 22.51,

(i) △OCD is the enlargement of △OAB about O by a scale factor 2.

(ii) △PQR is the image of △ABC after an enlargement about O, by scale factor 2.

(i)

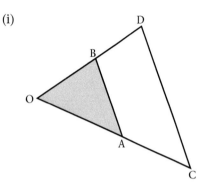

(ii)

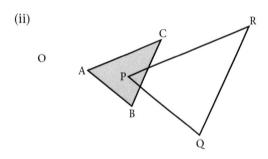

Fig. 22.51

An **enlargement** changes the size of the original shape. The lengths of the sides of the images in Fig. 22.51 (i) and (ii) are all twice the lengths of the corresponding sides of the original shape.

(i) $\dfrac{OD}{OB} = \dfrac{OC}{OA} = \dfrac{CD}{AB} = 2$

Triangles OAB and OCD are similar triangles.

$\dfrac{\text{Area of } \triangle OCD}{\text{Area of } \triangle OAB} = 2^2 = 4$

(ii) $\dfrac{PQ}{AB} = \dfrac{PR}{AC} = \dfrac{QR}{BC} = 2$

Triangles ABC and PQR are similar triangles.

$\dfrac{\text{Area of } \triangle PQR}{\text{Area of } \triangle ABC} = 2^2 = 4$

An enlargement is described by naming the centre of enlargement and the scale factor.

Exercise 22f

① In Fig. 22.52, △PQR is the image of △DEF after a translation.

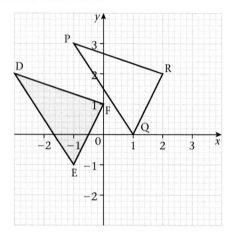

Fig. 22.52

(a) State the coordinates of any two vertices of △DEF.

(b) State the coordinates of the two corresponding vertices of △PQR.

(c) State the translation vector in the form $\begin{pmatrix} x \\ y \end{pmatrix}$.

② (a) Copy Fig. 22.53 and draw the reflection of △ABC in the line $x = -1$.

(b) State the coordinates of the vertices of the image.

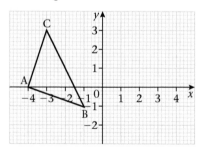

Fig. 22.53

③ Triangle DEF has coordinates D(−2, −2), E(3, 4), F(4, 0). Using graph paper and a scale of 1 cm to 1 unit:

(a) Draw triangle DEF.

(b) Reflect △DEF in the *y*-axis to give △JKL.

(c) State the coordinates of the vertices of △JKL.

④ Write down the coordinates of the following points after each is reflected in (a) the *x*-axis, (b) the *y*-axis.
(i) (2, 1)   (ii) (−2, 3),   (iii) (2, 0),
(iv) (−3, −3),   (v) (0, 5).

⑤ The coordinates of the vertices of △PQR are P(2, 2), Q(6, 3), R(3, 6).
(a) Using a scale of 1 cm to 1 unit, draw △PQR on graph paper.
(b) Draw the image of △PQR after a rotation of 90° clockwise about the origin. Name the image as △P′Q′R′.
(c) Reflect △P′Q′R′ in the *y*-axis to give △P″Q″R″.
(d) What other transformation of △PQR would have produced the final image, △P″Q″R″?

⑥ Triangle ABC has vertices A(2, 1), B(4, 0), C(5, 2).
(a) Using a scale of 1 cm to 1 unit, draw △ABC on graph paper.
(b) Draw the image of △ABC after an enlargement of scale factor 2, with the origin as the centre of enlargement. Name the image, △A′B′C′.
(c) State the coordinates of the vertices of △A′B′C′.
(d) State the value of $\dfrac{\text{area of triangle ABC}}{\text{area of triangle A′B′C′}}$

Revision test (Chapter 22)

Select the correct answer and write down your choice.

① The size of an exterior angle of a regular polygon is 30°. The number of sides of the polygon is
A 3   B 6   C 9   D 12

② Which one of the following is *not* true?
A The sum of the exterior angles of a polygon is 360°.
B A parallelogram is a quadrilateral with one pair of opposite parallel sides.
C The diagonals of a rhombus bisect each other.
D The diagonals of a square meet at right angles.

③ In Fig. 22.54, AB = AC = BC and EX = EC.

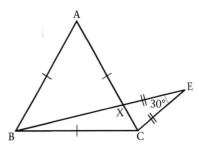

Fig. 22.54

The size of $\widehat{ABE}$ in Fig. 22.54 is
A 15°   B 30°
C 45°   D 60°

④ Which one of the following is always true?
A All isosceles triangles are similar.
B Similar triangles are also congruent.
C Similar triangles are equiangular.
D All right-angles triangles are similar.

⑤ O is the centre of the circle and AB = BC (Fig. 22.55).

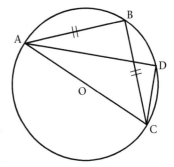

Fig. 22.55

If $\widehat{DAC}$ = 15° then $\widehat{BCD}$ =
A 75°   B 45°
C 30°   D 15°

⑥ Say which of the following could *not* be drawn.
A A triangle with angles 15°, 25°, 140°
B A quadrilateral with angles 35°, 150°, 70°, 105°.
C A polygon the sum of whose angles is 550°
D A rhombus with angles 45°, 45°, 135°, 135°

Mathematics for Caribbean Schools

7 When the point (3, −2) is reflected in the y-axis the coordinates of the image are

A (−3, −2) B (3, 2)
C (3, −2) D (−2, 3)

8 U is the set of all quadrilaterals,
P is the set of quadrilaterals with four right angles,
Q is the set of quadrilaterals with four equal sides,
then P ∩ Q is the set of:

A parallelograms B rectangles
C rhombuses D squares

9 Which of the following best describes a polygon?

A A plane figure with more than 3 sides
B A plane figure with more than 4 sides
D A plane figure bounded by straight lines
E A plane figure with more than 4 right angles

10 In Fig. 22.56, O is the centre of the circle.

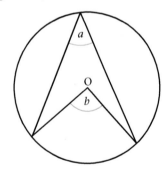

Fig. 22.56

In terms of b, a =

A $\dfrac{b}{2}$ B 2b

C $\dfrac{360 - b}{2}$ D $\dfrac{180 - b}{2}$

11 In Fig. 22.57, △ABC is translated to form the image △A′B′C′.

The vector of translation is

A $\begin{pmatrix} 2 \\ 1 \end{pmatrix}$ B $\begin{pmatrix} 1 \\ 2 \end{pmatrix}$

C $\begin{pmatrix} 2 \\ 2 \end{pmatrix}$ D $\begin{pmatrix} 2 \\ 0 \end{pmatrix}$

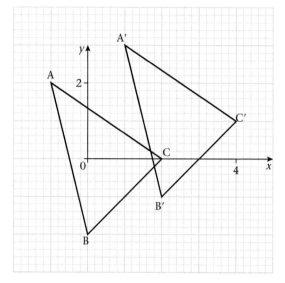

Fig. 22.57

12 The image of the point (−2, 5) under reflection in the y-axis is

A (−2, −5) B (2, 5)
C (2, −5) D (−2, 5)

13 Given that $\dfrac{\text{area } \triangle ABC}{\text{area of } \triangle PQR} = \dfrac{4}{9}$ then PQ : AB is

A 4 : 9 B 9 : 4
C 2 : 3 D 3 : 2

14 The image of (3, −4) under a clockwise rotation of 90° about the origin is

A (4, 3) B (3, 4)
C (−3, −4) D (−4, −3)

15 When a shape is enlarged by a scale factor

A the sides are enlarged by the same scale factor
B the angles are enlarged by the same scale factor
C the size of the shape is enlarged by the same scale factor
D the vertices are moved equal distances.

16 In Fig. 22.58, DE = $\frac{1}{3}$BC.

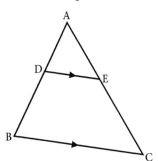

Fig. 22.58

Which of the following is *not* true?

A △ABC is an enlargement of △ADE

B AD = $\frac{1}{3}$ AB

C AE = $\frac{1}{3}$AC

D $\dfrac{\text{Area } \triangle ADE}{\text{Area } \triangle ABC} = \dfrac{1}{3}$

17 Which of the following statements is *not necessarily* true for a parallelogram?

A the opposite sides are parallel

B the opposite sides are equal

C the diagonals bisect one another at right angles

D the opposite angles are equal

18 The least number of triangles into which a polygon of *n* sides can be divided is

A $2n - 4$ B $2n$

C $n - 2$ D n

19 The sum of the angles of a polygon having $2n$ sides is

A $(2n - 4)$ right angles

B $2(2n - 4)$ right angles

C $2(2n - 2)$ right angles

D $(4n - 2)$ right angles

20 The sum of the interior angles of a regular polygon is five times the sum of the exterior angles. The number of sides of the polygon is

A 4   B 5

C 10   D 12

Chapter 23

Statistics and probability

Bar charts, pie charts

Example 1

Table 23.1 shows how an income of $450 is spent.

Table 23.1

Item	Amount ($)
food	150
rent	75
clothing	60
savings	90
other expenses	75

Show this information on a bar chart.

Represent the items by bars of the same width. The height, or length, of each bar is proportional to the amount of money. The bars may be vertical as in (i) or horizontal as in (ii).

(i)

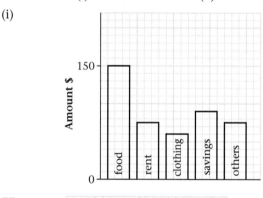

(ii)

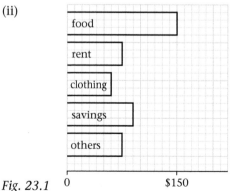

Fig. 23.1

Example 2

The number of students admitted to a university in a particular year is distributed among five faculties as follows: Education, 350; Medicine, 150; Engineering, 200; Law, 100; Arts, 100. Draw a pie chart to represent this information.

Table 23.2 shows how to calculate the angles of the sectors of the pie chart.

Fig. 23.2 is the required pie chart.

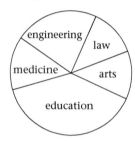

Fig. 23.2

Table 23.2

Faculty	Number of students	Angle of sector in pie chart
education	350	$\frac{350}{900} \times 360° = 140°$
medicine	150	$\frac{150}{900} \times 360° = 60°$
engineering	200	$\frac{200}{900} \times 360° = 80°$
law	100	$\frac{100}{900} \times 360° = 40°$
arts	100	$\frac{100}{900} \times 360° = 140°$
totals	900	360°

Mean, median, mode

Given a set of values, the **mean** is the sum of all the values divided by the number of values. If a set of numbers is arranged in order of size, the middle term is called the **median**. If there is an even number of terms the median is taken as the mean of the two middle terms. Note that the median may not be a value in the distribution.

The number of times any particular value occurs in a set is called its **frequency**. The number which occurs most often, i.e. the value that has the greatest frequency, is called the **mode**.

Example 3

Find (a) the mean, (b) the median, (c) the mode of the following set of numbers: 12, 16, 8, 11, 12, 8, 2, 8, 1, 14.

(a) Sum of the numbers
$$= 12 + 16 + 8 + 11 + 12 + 8 + 2 + 8 + 1 + 14$$
$$= 92$$
Number of numbers = 10
$$\text{Mean} = \frac{92}{10} = 9.2$$

(b) Arrange the numbers in ascending order:
1, 2, 8, 8, 8, 11, 12, 12, 14, 16
$$\text{Median} = \frac{8 + 11}{2} = 9.5$$

(c) The mode is 8.

Example 4

The ages of 15 students in years and months (yr–mo) are 14–5, 15–2, 14–3, 13–9, 14–10, 14–11, 13–8, 15–3, 14–6, 15–6, 15–8, 16–1, 15–4, 14–4, 14–7. Find the average age, that is, the mean age of the students to the nearest month.

The ages range from 13–8 to 16–1. Take 15–0 as a working mean. Make two columns of numbers. The one marked (+) shows all the deviations in months for ages over 15–0; the other, (−), gives the deviations in months for ages under 15–0.

(+)	(−)
2	7
3	9
6	15
8	2
13	1
4	16
	6
	8
	5
36	69
total	33	(i.e. 69 − 36)

The working shows that the total deviation for the 15 students is 33 months less than 15.0. Hence:

$$\text{mean age} = 15\,\text{yr}\,0\,\text{mo} + \left(\frac{-33}{15}\right)\text{mo}$$
$$= 15\,\text{yr}\,0\,\text{mo} - 2.2\,\text{mo}$$
$$= 14\,\text{yr}\,10\,\text{mo to the nearest month}$$

The method in Example 4 is useful when given a large set of numbers of roughly the same size.

Exercise 23a

① A student spent his day as shown in Table 23.3.

Table 23.3

Activity	Time (h)
at lectures	5
reading	6
sleeping	7
sports	2
others	4

Show this data on a bar chart and a pie chart.

② Table 23.4 shows how an income of $200 was spent.

Table 23.4

Item	Amount ($)
food	60
rent	40
clothing	20
savings	40
others	30

Show this data on a bar chart and a pie chart.

③ Table 23.5 shows the numbers of students admitted to a university from 2002 to 2006.

Table 23.5

Year	Number of students
2002	800
2003	1200
2004	1350
2005	1570
2006	2250

(a) Represent this information on a bar chart.
(b) Calculate the mean number of students admitted per year.

4 Table 23.6 shows the numbers of different types of books in a school library.

Table 23.6

Subject	Number of books
mathematics	48
science	110
engineering	54
novels	496
others	372

Draw a pie chart to show this information.

5 Find the mean, median and mode of the following sets of numbers.
(a) 2, 4, 4, 6, 7
(b) 3, 5, 5, 7, 7, 7, 9, 9
(c) 11, 9, 6, 4, 3, 12, 1, 6, 5

6 Use a working mean of 115 to find the mean of the following:

110, 120, 113, 116, 119,
127, 117, 118, 118, 113

7 The ages of 16 students in years and months are as follows:

17–2	17–10	18–2	19–5
18–0	17–11	18–7	19–7
19–3	19–8	16–11	17–9
17–10	17–5	18–5	18–1

Choose a suitable working mean and use it to find the average age of the students.

8 The heights, in cm, of 10 boys are

145	163	159	162	167
149	150	160	170	155

Calculate the mean and median heights.

9 The masses, to the nearest kg, of 15 girls are

45	38	51	44	43
60	55	47	45	42
52	46	41	50	53

Find the mean and median masses.

10 Table 23.7 gives the age distribution of members of a school choir.

Table 23.7

Age	12	13	14	15	16	17
Frequency	2	1	3	6	5	3

(a) How many students are in the choir?
(b) What is the modal age?
(c) Draw a pie chart to show the age distribution.
(d) Calculate the mean age of the choir members.

Grouped data

Histograms

Example 5

Table 23.8 is the frequency distribution of the masses of 40 pupils in a class.

Table 23.8

Mass (kg)	Number of pupils
41–45	3
46–50	7
51–55	12
56–60	10
61–65	6
66–70	2

(a) Draw a histogram of the distribution.
(b) State the modal class.

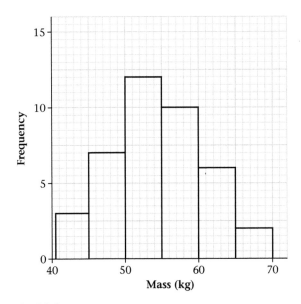

Fig. 23.3

(a) Fig. 23.3 is the required histogram. Notice that the gaps between the bars have been closed by increasing the width of each rectangle by $\frac{1}{2}$ unit on both sides.

(b) The modal class is 51–55. This class has the highest frequency.

Probability

The **probability** of an event happening can be given a numerical value x where

$$x = \frac{\text{number of required outcomes}}{\text{number of possible outcomes}}$$

and $0 \leqslant x \leqslant 1$.

If $x = 1$, then the event is certain to happen. If $x = 0$, then the event cannot happen. $1 - x$ is the probability of the event *not* happening.

Example 6

Table 23.10 shows the numbers of students in each age group in a class.

What is the probability that a student chosen at random from the class is (a) 17 years old, (b) not 17 years old, (c) over 16 years old?

Table 23.10

Age (years)	16	17	18
Number	7	22	13

(a) Probability that the student is 17 years old

$$= \frac{\text{number of 17-year-olds}}{\text{total number of students}}$$

$$= \frac{22}{7 + 22 + 13} = \frac{22}{42} = \frac{11}{21}$$

(b) Probability that the student is *not* 17 years old

$$= 1 - \frac{11}{21} = \frac{10}{21}$$

(c) Probability that the students is over 16 years old

$$= \frac{\text{number of students over 16}}{\text{total number of students}}$$

$$= \frac{22 + 13}{42} = \frac{35}{42} = \frac{5}{6}$$

Spread

The following are marks that Jolan and Joan gained in the same tests in Social Studies over a period of time.

Jolan   15, 15, 16, 17, 17, 18, 18, 18
Joan   11, 12, 15, 16, 18, 18, 19, 20

The median mark for both Jolan and Joan is 17 but Jolan's marks are closer together than Joan's marks. Joan's marks are said to be more widely spread.

The **spread** or **dispersion** of a set of data is a useful measure in statistics. One way of measuring the spread is to use the difference between the highest and lowest observed value.

For Jolan's marks this is $18 - 15 = 3$ marks.
For Joan's marks this is $20 - 11 = 9$ marks.

This difference between the highest and lowest values is called the **range**. Joan's range is 3 marks and Joan's range is 9 marks. The wider the spread of the data, the higher will be the value of the range.

In Mathematics tests, the marks gained were

Jolan   10, 12, 13, 14, 14, 15, 16, 18

Joan   5, 10, 10, 13, 16, 17, 17, 18

Jolan's range in mathematics is 8 marks. Joan's range is 13. Notice that one low mark makes quite a difference in Joan's range.

There are other ways of looking at the spread of the marks. We have seen that the median divides a set of values into two equal parts.

Joan's marks   11, 12, 15, 16, 18, 18, 19, 20

median
17

These marks may be further divided into two equal parts on either side of the median. The further divisions are called quartiles.

11, 12,   15, 16,   18, 18,   19, 20

median 17

lower quartile   upper quartile

The **lower quartile** is $\frac{1}{2}(12 + 15) = 13\frac{1}{2}$

The **upper quartile** is $\frac{1}{2}(18 + 19) = 18\frac{1}{2}$

The difference between the upper quartile and the lower quartile ($18\frac{1}{2} - 13\frac{1}{2}$) is called the **interquartile range**.

The **semi-interquartile range** is half the interquartile range.

For Joan's marks in Social Studies the semi-interquartile range is $2\frac{1}{2}$.

For Jolan's marks in Social Studies

the lower quartile is $\frac{1}{2}(15 + 16) = 15\frac{1}{2}$

the upper quartile is $\frac{1}{2}(18 + 18) = 18$

The interquartile range is (upper quartile − lower quartile) = $(18 - 15\frac{1}{2}) = 2\frac{1}{2}$

The semi-interquartile range is $1\frac{1}{4}$

Range, interquartile range and semi-interquartile range are three measures of the spread (dispersion) of a set of numerical values.

Group work

Discuss the spread of marks that both Jolan and Joan gained in Mathematics.

Example 7

Estimate the interquartile range and the semi-interquartile range for the frequency distribution in Table 23.11.

Table 23.11

Score	0–9	10–19	20–29	30–39	40–49	50–59	60–69	70–79
Frequency	8	17	15	10	25	14	8	3

The total of the frequencies is 100. The three quartiles divide this into 4 parts of 25. There are 25 observations in the first two classes. The lower quartile is at the upper boundary of the second class, that is $19\frac{1}{2}$. There are 25 observations in the top three classes. The upper quartile falls between the top of the fifth class and the bottom of the sixth class, that is $49\frac{1}{2}$.

Interquartile range is $49\frac{1}{2} - 19\frac{1}{2} = 30$

Semi-interquartile range is $\frac{1}{2}(30) = 15$

Exercise 23b

1. Write each set of values in order of size and then find (i) the range, (ii) the semi-interquartile range.

   (a) 15, 24, 2, 3, 12, 18, 19, 20, 14
   (b) 5, 10, 17, 12, 8, 16, 17, 12
   (c) 17, 13, 19, 19, 18, 10, 15, 18, 12, 18, 19, 17

2. Table 23.12 shows the frequency distribution of marks on a test.

Table 23.12

Marks	1–5	6–10	11–15	16–20	21–25
Frequency	17	23	32	34	40

   Estimate (a) the interquartile range, (b) the semi-interquartile range

3. Estimate (a) the interquartile range, (b) the semi-interquartile range of the data in Table 23.11.

Exercise 23c

1 Table 23.13 shows the frequencies, f, of children of age x years in a hospital.

Table 23.13

x	1	2	3	4	5	6	7	8
f	3	4	5	6	7	6	5	4

(a) What is the mode?
(b) How many children are in the hospital?
(c) Calculate the mean age of the children.
(d) Calculate
    (i) the range
    (ii) the interquartile range
    (iii) the semi-interquartile range

2 Table 23.14 shows the distribution of the ages of 30 children in a school choir.

Table 23.14

Age	11	12	13	14	15	16	17
No. of children	3	6	6	6	4	3	2

(a) Calculate the mean age and the median age of this distribution.
(b) Calculate the probability that a child chosen at random is (i) under 15 years old, (ii) at least 15 years old.
[CXC June 88]

3 Table 23.15 shows the length of life of 200 electric light bulbs.

Table 23.15

Length of life (hours)	Number of bulbs
201–300	10
301–400	16
401–500	32
501–600	54
601–700	88

(a) Draw a histogram of this distribution.
(b) Calculate the mean life of the light bulbs.

4 Table 23.16 shows the number of workdays lost through illness among 500 factory employees during a 1-year period.

Table 23.16

Number of days	Number of employees
0–4	250
5–9	158
10–14	33
15–19	29
20–24	15
25–29	10
30–34	5

(a) Draw a histogram of this distribution.
(b) State the modal class.
(c) Calculate the mean number of days lost.

5 The masses, in kg to the nearest kg, of 40 students are as follows:

59   54   51   56   59   61   60   61
59   58   62   61   63   64   58   57
56   60   62   60   61   65   58   57
54   52   62   67   69   49   56   58
60   60   62   58   51   57   70   63

(a) Take class intervals of 46–50, 51–55, …, and make up a table of frequencies.
(b) Draw the corresponding histogram.
(c) Find the median mass.
(d) Calculate the mean mass.

6 The examination marks of 50 students are as follows:

65   58   51   36   23   40   53   59   70   51
46   59   50   67   46   39   61   62   73   60
71   51   47   32   48   40   40   51   58   67
60   69   43   52   37   26   38   50   59   40
44   54   42   47   68   74   45   39   48   55

(a) Make a frequency distribution using class intervals of 21–30, 31–40, …
(b) Estimate (i) the median, (ii) the semi-interquartile range.
(c) Find the percentage of students who got more than 45 marks.

7 Table 23.17 is the frequency distribution of the heights of 40 pupils.

Table 23.17

Height (cm)	No. of pupils
131–140	2
141–150	11
151–160	14
161–170	10
171–180	3

(a) Draw a histogram of this distribution.
(b) State the modal class.
(c) Calculate the mean height of the pupils.

8 Table 23.18 shows the numbers of students in each age group in a class.

Table 23.18

Age (years)	16	17	18	19
Frequency	9	11	11	5

(a) A student is chosen at random from the class. What is the probability that the age of the student is (i) 16 years, (ii) *under* 18 years, (iii) *not* 19 years?
(b) Calculate the average age of the students in years and months.

9 The daily earnings of 50 men are given in Table 23.19.

Table 23.19

Earnings per day ($)	Number of men (frequency)
51–75	1
76–100	4
101–125	17
126–150	15
151–175	11
176–200	2

(a) What is the modal class of daily earnings?
(b) Use mid-values to calculate the mean daily earnings.
(c) One of the men is chosen at random. What is the probability that he earns less than $126 per day?

10 The number of trips a taxi driver made during a period of 60 days is given in Table 23.20.

Table 23.20

Number of trips	0	1	2	3	4	5	6	7
Frequency	8	10	13	16	7	4	0	2

(a) Draw a histogram to illustrate the data.
(b) Calculate, correct to the nearest whole number, the mean number of trips made by the taxi driver per day during the period of 60 days.
(c) Calculate the probability that, on any day chosen at random, the taxi driver made 4 trips.

Revision test (Chapter 23)

Select the correct answer and write down your choice.

Fig. 23.4 is a bar chart showing the numbers of kilometres a group of students walked in a sponsored march. Use Fig. 23.4 to answer questions 1 and 2.

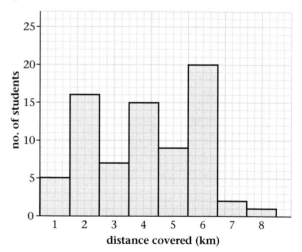

Fig. 23.4

1 The number of students who participated is

    A 20   B 36   C 75   D 305

2 The total distance, in kilometres, covered by all the students is

   A 8   B 20   C 75   D 305

3 The median of 6, 4, 9, 7, 5, 3, 9 is

   A 6   B 7   C 8   D 9

4 Which of the following does *not* represent a statistical average?

   A Mode   B Median
   C Range   D Mean

5 The number of students writing examinations is distributed among five subjects as follows: Mathematics 35, English 20, History 15, Art 10, Geography 10. The size of the angle for English on a pie chart representing this information is

   A 20°   B 40°   C 72°   D 80°

6 The mean of the numbers 0, 2, 5, 9, 3, 11 is

   A 2   B 5   C 6   D 7

7 The mean age of a class of 21 boys is 14 years 3 months. When one boy of age 16 years 1 month joins the class the mean age of the class becomes

   A 14 yr 4 mo   B 15 yr
   C 15 yr 2 mo   D 16 yr

8 Given that each value of a set of observations is doubled, which of the following is true?

   A The mean remains the same.
   B The mean is increased by 2.
   C The mean is decreased by 2.
   D The mean is doubled.

9 The median of the set of numbers 0, 12, 15, 25, 8, 17, 19 is

   A 15   B 16   C 17   D 25

10 The mean weight of 15 boys in a class is 46.5 kg and the mean weight of 21 girls in the same class is 43.3 kg. The mean weight in kg of the class of 36 pupils is

   A 44.9   B 44.6   C 45   D 47

11 Which of the following must be an observed value in given data?

   A mean   B median
   C mode   D range

12 If the smallest value in a set of observations is 15 and the largest is 67, then it is correct to say that

   A the mean is 41.
   B the interquartile range is 52.
   C the semi-interquartile range is 26.
   D the range is 52.

13

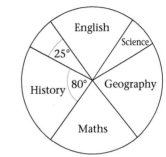

Fig. 23.5

The pie chart in Fig. 23.5 shows the subjects studied by a total of 480 students. The number studying Mathematics is

   A 75   B 100   C 120   D 160

Table 23.21 shows the heights of plants in a sample. Use Table 23.21 to answer questions 12 and 13.

Table 23.21

Height in cm	20	25	30	35	40	45
No. of plants	5	3	1	1	3	2

14 The modal height is

   A 45 cm   B 40 cm
   C 25 cm   D 20 cm

15 If a plant is chosen at random, the probability that it will be of a height greater than 35 cm is

   A $\frac{1}{3}$ B $\frac{1}{5}$ C $\frac{1}{4}$ D $\frac{2}{3}$

16 Which one of the following is *not* true for the set of numbers 0, 2, 3, 3, 5, 8, 9, 10?

   A The median is 4.
   B The mean is 5.
   C The mode is 3.
   D The range is 8.

17 Which one of the following is *not* a measure of spread?

  A   Frequency
  B   Interquartile range
  C   Semi-interquartile range
  D   Range

18 Which one of the following is *not* necessarily true from the histogram in Fig. 23.6?

  A   The highest mark scored is 100.
  B   The modal class is 61–70.
  C   No one scored in class 11–20.
  D   The mode of the data is in class 61–70.

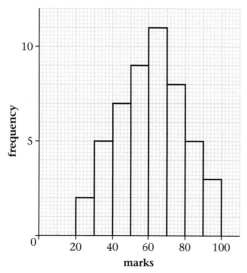

Fig. 23.6

19 Which one of the following statements is *always* true for a set of observations?

  A   The histogram is the same as the frequency polygon.
  B   The frequency polygon *must* be imposed on the histogram.
  C   The mean, median and mode are actual observations.
  D   The area of the columns on a histogram is proportional to the frequency represented.

20 Given that each value of a set of observations is halved, which of the following is true for the mean?

  A   It remains the same.
  B   It is halved.
  C   It is doubled.
  D   It is increased by one-half.

Sine, cosine, tangent

The trigonometric ratios of an angle: sine, cosine and tangent, are defined in terms of the hypotenuse, opposite and adjacent sides of a right-angled triangle as shown in Fig. 24.1, for angle θ.

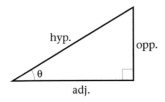

Fig. 24.1

The ratios are defined as

$$\text{sine } \theta = \frac{\text{opp.}}{\text{hyp.}}$$

$$\text{cosine } \theta = \frac{\text{adj.}}{\text{hyp.}}$$

$$\text{tangent } \theta = \frac{\text{opp.}}{\text{adj.}}$$

Notes:

1   sine θ is shortened to sin θ, cosine θ to cos θ, and tangent θ to tan θ.

2   The trigonometric ratios are defined by using a right-angled triangle but the values of the ratio of an angle depend *only* on the size of the angle and not on the shape of the figure.

3   The sine of an angle is equal to the cosine of its complementary angle and the cosine is equal to the sine of its complementary angle, that is
sin (90° − θ) = cos θ
cos (90° − θ) = sin θ

4   When using an electronic calculator to find the value of a trigonometric ratio, use the same degree of accuracy given (decimal places or significant figures) for the other measurement in the problem.

Example 1

Calculate angle *x* in Fig. 24.2.

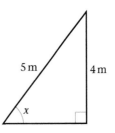

Fig. 24.2

From Fig. 24.2 we have
$$\sin x = \tfrac{4}{5} = 0.800$$
$$x = 53.1° \text{ or } 53°6' \qquad (\textit{from sine tables})$$

Example 2

Calculate the lengths of the sides marked *a* and *b* in Fig. 24.3.

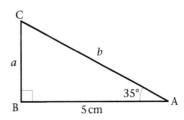

Fig. 24.3

$$\tan 35° = \frac{a}{5}$$
$$a = 5 \times \tan 35°$$
$$= 5 \times 0.700$$
$$= 3.500$$

$$\cos 35° = \frac{5}{b}$$
$$b = \frac{5}{\cos 35°}$$
$$= 6.11$$

working:

No.	Log
5	0.699
cos 35°	$\bar{1}$.913
6.11	0.786

The sides are 3.5 cm and 6.1 cm long (2 s.f.).

Exercise 24a

① The dimensions in Fig. 24.4 are all cm. Find the value of the angle x in each part.

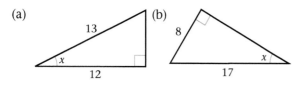

(a)   (b)

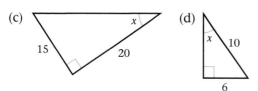

(c)   (d)

(e)

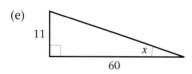

Fig. 24.4

② A radio-mast 35 m high is supported by straight wires attached to its top and to points on the level ground 12 m from its base. Calculate the angle between each wire and the ground.

③ A ladder 8.5 m long leans against a vertical wall so that its upper end is 7.5 m from the ground. Calculate (a) the angle between the ladder and the wall, (b) the distance of the foot of the ladder from the wall.

④ One side of a right-angled triangle is 24 cm long and its hypotenuse is 25 cm long. Calculate (a) the angle between the side and the hypotenuse, (b) the length of the third side of the triangle.

⑤ A cone has a slant height of 29 cm and a circular base of diameter 42 cm. Calculate (a) the vertical angle of the cone, (b) the vertical height of the cone.

⑥ In each part of Fig. 24.5, calculate the value of x correct to 2 s.f.

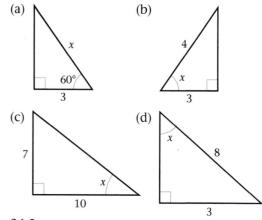

(a)   (b)

(c)   (d)

Fig. 24.5

⑦ A chord of a circle is 10 cm from the centre of the circle. Given that the radius of the circle is 31 cm, calculate
 (a) the angle between the radius and the chord,
 (b) the length of the chord.

⑧ A rectangle is 4.3 cm long and the length of each diagonal is 5.1 cm. Calculate (a) the angle between the long side and the diagonal, (b) the width of the rectangle, giving both answers correct to 3 s.f.

Angle of elevation and angle of depression

Fig. 24.6 shows the **angle of elevation**, e, of the top of a tower, T, from a point A below. The diagram also shows the **angle of depression**, d, of a point B on the ground from a point, P, on the tower.

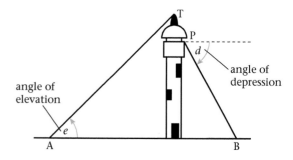

angle of elevation

angle of depression

Fig. 24.6

Example 3

From a window 10 m above level ground, the angle of depression of an object on the ground is 25.4°. Calculate the distance of the object from the foot of the building.

Note that the angle of depression is the angle between the horizontal and the line joining the window and the object (Fig. 24.7).

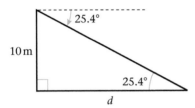

Fig. 24.7

Either, $\tan 25.4° = \dfrac{10}{d}$

$$d = \dfrac{10}{\tan 25.4°} = \dfrac{10}{0.475}$$

$$= 21.05 \qquad \text{(from recip. tables)}$$

or, using the **complement** of 25.4° (see Fig. 24.8),

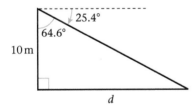

Fig. 24.8

$\tan 64.6° = \dfrac{d}{10}$

$$d = 10 \times \tan 64.6°$$

$$= 10 \times 2.106 = 21.06$$

The object is 21.1 m from the foot of the building (3 s.f.)

Exercise 24b

1. A plane takes off at an angle of 5°10′ to the ground. How high is it when it has moved 2000 m horizontally from its take-off point?

2. The angle of elevation of the top of a vertical mast from a point on level ground 240 m from its foot is 31.5°. How high is the mast?

3. A town B is due north of A. A third town C is 10 km on a bearing 020° from A. If B is on a bearing of 290° from C,
   (a) draw a sketch to show the positions of A, B and C, indicating the bearings given;
   (b) state the size of the angle $A\widehat{C}B$;
   (c) calculate the distances (i) BC, (ii) AB.

4. Calculate the values of a, b and c in Fig. 24.9.

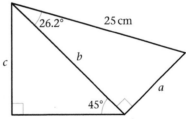

Fig. 24.9

5. A chord of a circle is 5 cm long and subtends an angle of 24.3° in the major segment. Calculate (a) the perpendicular distance of the chord from the centre, (b) the radius of the circle.

6. The angle of elevation of the top of a flag-pole from a point on level ground is 30°. From another point on the ground, 20 m nearer the flag-pole, the angle of elevation is 60°. Calculate the height of the flag-pole.

7. A tripod consists of three legs, each 1.05 m long. The height of the top of the tripod above the ground is 90 cm. What is the inclination of each leg to the horizontal?

8. A man walks 11 km due north from A to B. He then walks 6.5 km due east from B to C. Calculate (a) the bearing of C from A, (b) the distance AC.

9 A plank rests with one end on the ground and the other end on the back of a lorry 1.2 m above the ground. How long is the plank if it is inclined at 21° to the horizontal?

10 In the triangle ABC in Fig. 24.10 (not drawn to scale), AD is perpendicular to BC, BD = 10.0 cm, DC = 6.0 cm and angle ACB = 35°.

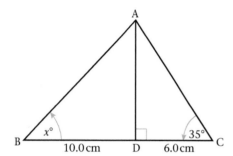

Fig. 24.10

Calculate
(a) the length of AD to the nearest tenth of a centimetre,
(b) the length of AB,
(c) the area of triangle ABC in cm².
[CXC Jan 89]   (6 marks)

Revision test (Chapter 24)

Select the correct answer and write down your choice.

1 In a right-angled triangle, cos α = $\frac{3}{5}$. sin α =

A $\frac{3}{4}$ B $\frac{3}{5}$

C $\frac{4}{5}$ D $\frac{4}{3}$

2 Given that in a right-angled triangle cos θ = $\frac{12}{13}$, then tan θ =

A $\frac{13}{12}$ B $\frac{5}{12}$

C $\frac{5}{13}$ D $\frac{12}{5}$

3 If sin α = $\frac{8}{17}$, cos α = $\frac{15}{17}$, then tan(90° − α) =

A $\frac{8}{17}$ B $\frac{8}{15}$

C $\frac{15}{17}$ D $\frac{15}{8}$

4

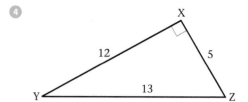

Fig. 24.11

In Fig. 24.11, the correct trigonometric ratio is

A sin XŶZ = $\frac{12}{13}$

B tan XŶZ = $\frac{5}{12}$

C cos XẐY = $\frac{12}{13}$

D tan XẐY = $\frac{5}{12}$

5

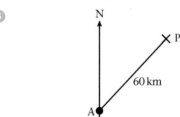

Fig. 24.12

In Fig. 24.12, the bearing of P from A is 40°. if the distance AP is 60 km, the distance that P is north of A is

A $\dfrac{60}{\sin 40°}$ B $\dfrac{60}{\cos 40°}$

C 60 sin 40°   D 60 cos 40°

6 If the angle of elevation of the top of a tower from a point F on the ground is 37°, then the angle of depression of F from the top of the tower is

A   37°   B   53°   C   127°   D   143°

7

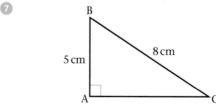

Fig. 24.13

In Fig. 24.13, the ratio $\frac{5}{8}$ is the value of

A   cos AĈB   B   sin AB̂C

C   tan AB̂C   D   sin AĈB

8

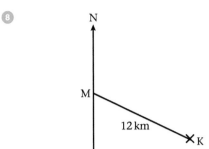

Fig. 24.14

A ship travels from a port M on a bearing 130° to another port K, 12 km from M. How far east of M is K (Fig. 24.14)?

A 12 sin 50°   B 12 sin 40°
C 12 cos 50°   D 12 tan 40°

9

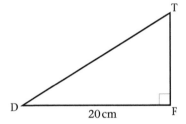

Fig. 24.15

In Fig. 24.15, the angle of elevation of T from D is 26°. If DF = 20 cm, then DT = d cm, where d =

A 20 sin 26°   B $\dfrac{20}{\sin 26°}$

C 20 cos 26°   D $\dfrac{20}{\cos 26°}$

10

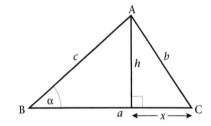

Fig. 24.16

In Fig. 24.16, cos α =

A $\dfrac{h}{c}$ B $\dfrac{a}{c}$

C $\dfrac{a - x}{c}$ D $\dfrac{x - a}{c}$

Certificate-level practice examination

Paper 1 (1½ hours)

*Attempt **all** the questions. Circle the letter of your choice.*

① $\frac{37}{100}$ =

  A 0.000 37   B 0.003 7
  C 0.037   D 0.37

② The HCF of 12, 18, 21 is

  A 3   B 6   C 9   D 12

③ The number, 0.046 78, correct to 3 significant figures, is

  A 0.046   B 0.047
  C 0.046 7   D 0.046 8

④ 1 m² =

  A 10 cm²   B 100 cm²
  C 1000 cm²   D 10 000 cm²

⑤ The mode of the numbers
16, 13, 19, 18, 17, 13, 12, 16, 13 is

  A 12   B 13   C 17   D 19

⑥

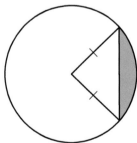

Fig. P1

  The shaded area in Fig. P1 is called

  A an arc   B a chord
  C a sector   D a segment

⑦ A bus left the station at 23:05 h and reached its first stop at 02:55 h. The time taken for the trip was

  A 3 h 10 min   B 3 h 50 min
  C 20 h 10 min   D 20 h 50 min

⑧ 0.5, written as a percentage, is

  A 2%   B 5%   C 20%   D 50%

⑨ The diagram that shows a function in Fig. P2 is

A 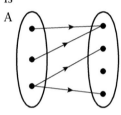   B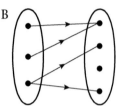

C 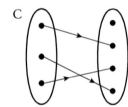   D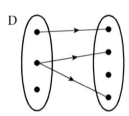

Fig. P2

⑩ $(-0.3)^3$ =

  A −0.027   B 0.27
  C −0.9   D 0.027

⑪ $\frac{3}{2a} + \frac{2}{3b}$ =

  A $\frac{b + a}{ab}$ B $\frac{6a + 4b}{5ab}$

  C $\frac{3b + 2a}{6ab}$ D $\frac{9b + 4a}{6ab}$

⑫ In an election, candidate J gained 5240 votes, and candidate K gained 4061 votes. The ratio of the votes that J got to the total votes cast is

  A $\frac{31}{71}$ B $\frac{40}{71}$ C $\frac{31}{40}$ D $\frac{40}{31}$

⑬ $\sqrt{0.0144}$ =

  A 0.001 2   B 0.012
  C 0.12   D 1.2

⑭ A team scored 2, 6, 3, 2, 4, 5, 4, 6 goals in eight football matches. The median score was

  A 2   B 3   C 4   D 5

15 $10011_2 + 1010_2 =$

A 11021_2 B 11101_2
C 11111_2 D 11121_2

16 The LCM of 12, 18, 21 is

A 3   B 42   C 84   D 252

17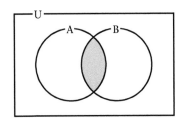

Fig. P3

In Fig. P3, the shaded region represents

A $A \cup B$ B $(A \cup B)'$
C $A \cap B$ D $(A \cap B)'$

18 The total surface area of a solid cylinder of height h and base radius r is

A $2\pi r^2 + 2\pi rh$ B $2\pi r^2 + \pi rh$
C $\pi r^2 + 2\pi rh$ D $\pi r^2 + \pi rh$

19 $4y - 6 =$

A $2y - 3$ B $2(2y - 6)$
C $2(2y - 3)$ D $4(y - 6)$

20 If $f : x \to x^2 + 4$, $f(-3) =$

A -5 B -2 C 10   D 13

21 In a regular pentagon, each exterior angle is equal to

A $18°$ B $36°$ C $54°$ D $72°$

22 The speed of a car is $80 \, \text{km h}^{-1}$. In 1 second, the distance moved is x metres, where $x =$

A $22\frac{2}{9}$ B 48   C $133\frac{1}{3}$ D 288

23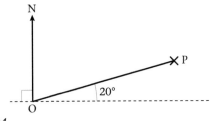

Fig. P4

In Fig. P4, the bearing of P from O is given by

A $020°$ B $070°$ C $110°$ D $290°$

24 $\dfrac{12b^3}{3b^2} =$

A $4b$ B $9b$ C $\dfrac{4}{b}$ D $\dfrac{9}{b}$

25 Simple interest is charged at the rate of 5% per annum. The time in years for $300 to amount to $360 is given by

A $\dfrac{300 \times 60}{5 \times 100}$ B $\dfrac{60 \times 100}{300 \times 5}$

C $\dfrac{300 \times 5}{60 \times 100}$ D $\dfrac{100 \times 5}{60 \times 300}$

Questions 26 and 27 use the following information.

In Fig. P5, AB and CD are two straight lines which intersect at P. The line RP is perpendicular to AB.

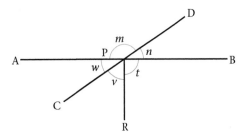

Fig. P5

26 The vertically opposite angles are

A v and w B m and t
C w and n D v and t

27 If $w = 20°$, then $v =$

A $20°$ B $70°$ C $90°$ D $110°$

28 The perimeter of a square is 20 cm. The area of the square in cm^2 is

A 20   B 25   C 40   D 50

29

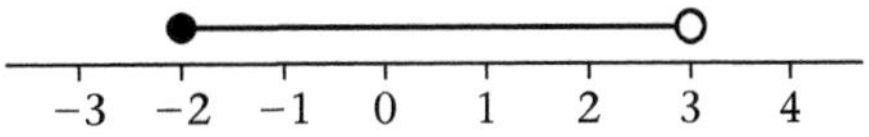

```
       -3  -2  -1   0   1   2   3   4
```

Fig. P6

The inequality shown on the number line in Fig. P6 is

A $-2 \leqslant n < 3$ B $-2 < n \leqslant 3$
C $-2 \geqslant n > 3$ D $-2 > n \geqslant 3$

30 The price of a table is marked at $400. A sales tax of 20% is charged. The total price to be paid is

A $405   B $420

C $460   D $480

31 If $3y + 1 = 2(y - 3)$, then $y =$

A −7   B −4   C −2   D −1

32 The graph of the function $y = 2x - 1$ is shown in Fig. P7 by

A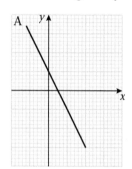
B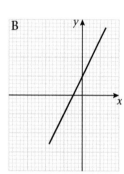

C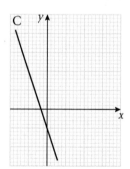
D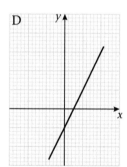

Fig. P7

33 The value of a machine depreciated in 1 year from $30 000 to $24 000. The decrease in value as a percentage of the original price is

A 20%   B 25%   C 75%   D 80%

34 The radius of a circle is 14 cm. The length in cm of the arc of a sector of angle 30° is

A $\frac{30}{180} \times 14\pi$ B $\frac{30}{180} \times 28\pi$

C $\frac{30}{360} \times 14\pi$ D $\frac{30}{360} \times 28\pi$

35 The selling price of an article was $350. If the profit was 40%, the cost price was

A $140   B $210   C $250   D $490

Questions 36 and 37 use the following information.

U = {odd numbers less than 20}

P = {multiples of 3}

Q = {prime numbers between 1 and 20}

R = {factors of 90}

36 $n(Q) =$

A 6   B 7   C 8   D 9

37 $P \cup R =$

A {5}   B {3, 9, 15}

C {1, 3, 5, 9, 15}   D {3, 5, 6, 9, 12, 15, 18}

38 Given that $a = -1$ and $b = 2$, $\dfrac{b + a}{a^2 + ab} =$

A −1   B $-\frac{1}{3}$ C $\frac{1}{3}$ D 1

39 $9.3 \times 10^{-2} \times 1.2 \times 10^3 =$

A 11.16×10^{-6} B 1.116×10^{-5}

C 11.16×10 D 1.116×10^2

40 The gradient of the straight line PQ joining P(1, −3) and Q(4, 7) is

A $\frac{3}{10}$ B $\frac{3}{4}$ C $\frac{4}{3}$ D $\frac{10}{3}$

Questions 41 and 42 use the following information.
Interest is compounded annually on a loan of $60 000 for 3 years at the rate of 12% per annum.

41 The interest for the first year is

A $1200   B $5000

C $7200   D $12 000

42 If $20 000 is paid back at the beginning of the second year, the principal for the second year is

A $41 200   B $45 000

C $47 200   D $52 000

43 $g(x) = 2x - 1$, for the domain −2, −1, 0, 1. The range is

A 3, 1, 0, −1   B 2, 1, 0, 1

C −2, −1, 0, 1   D −5, −3, −1, 1

44 Given that $2a + b = 1$ and $a - b = 2$, $(a, b) =$

A (2, −1)   B (1, −1)

C (0, 1)   D (−1, 1)

45

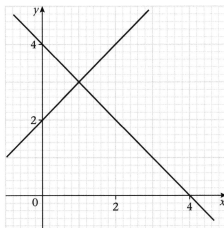

Fig. P8

In Fig. P8, the solution set of the two linear functions $y = f(x)$ and $y = g(x)$ is

A (0, 2) B (0, 4)

C (1, 3) D (4, 0)

46 The correct inequality statement for the numbers -7, 3, -1 is

A $-7 < -1 < 3$ B $-1 < 3 < -7$

C $-1 < -7 < 3$ D $-7 < 3 < -1$

47 In a bag there are 3 red and 6 black buttons. The probability of drawing out 1 black button is

A $\frac{1}{3}$ B $\frac{1}{2}$

C $\frac{2}{3}$ D 2

48 Given that $3x + 1 \geqslant 17$, and that x is an integer, the smallest possible value of x is

A 5 B 6

C 7 D 13

49 The image of the point (3, 1) under a translation is (2, -1). The translation vector is

A $\begin{pmatrix} 1 \\ 2 \end{pmatrix}$ B $\begin{pmatrix} 5 \\ 0 \end{pmatrix}$

C $\begin{pmatrix} -1 \\ -2 \end{pmatrix}$ D $\begin{pmatrix} 5 \\ -2 \end{pmatrix}$

50

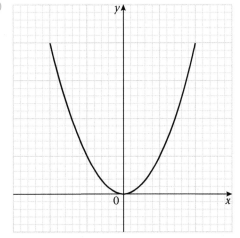

Fig. P9

Fig. P9 is the graph of the function

A $y = 2x$ B $y = -2x$

C $y = x^2$ D $y = -x^2$

Questions 51–53 use the following information.

The results of a 2-hour survey to obtain data on the vehicles using a new road is shown in Fig. P10.

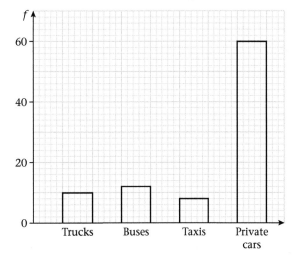

Fig. P10

51 The figure represents the information in a

A pie chart B line graph

C histogram D bar chart

52 The road was used most often by
    A   buses   B   private cars
    C   taxis   D   trucks

53 The total number of vehicles in the survey
    was
    A   4   B   60   C   80   D   90

Questions 54 and 55 use the following information.
In Fig. P11, XY and JK are parallel straight lines
cut by another straight line MPRN.
PR = PS and ∠PRS = 50°.

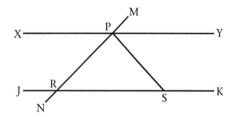

Fig. P11

54 In Fig. P11, ∠MPY =
    A   40°   B   50°   C   130°   D   140°

55 In Fig. P11, ∠RPS =
    A   40°   B   50°   C   80°   D   130°

56 'The square of a number is three more than
    twice the number.'
    If the number is n, the above statement may
    be written as
    A $n^2 = 3 + 2n$ B $n^2 + 3 > 2n$
    C $n^2 > 3 + 2n$ D $n^2 + 3 = 2n$

57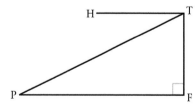

Fig. P12

    In Fig. P12, the angle of elevation of the
    point T from the point P is
    A   ∠TFP   B   ∠FTP
    C   ∠HTP   D   ∠FPT

Questions 58 and 59 use the following information.
Fig. P13 is a triangle PQR, in which ∠PQR = 90°.

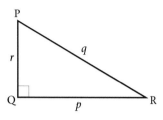

Fig. P13

58 In △PQR,
    A $p^2 = q^2 + r^2$ B $p^2 = q^2 - r^2$
    C $q^2 = r^2 - p^2$ D $q^2 = p^2 - r^2$

59 If $p = 12$ and $q = 13$, $\frac{12}{13} =$
    A   cos $\widehat{PQR}$ B   tan $\widehat{PRQ}$
    C   cos $\widehat{PRQ}$ D   cos $\widehat{QPR}$

60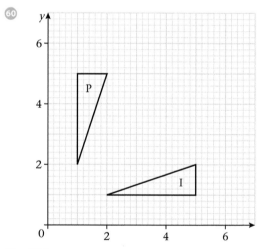

Fig. P14

    In Fig. P14, I is the image of P under a
    A   rotation through 90° anticlockwise about
        (0, 0)
    B   reflection in the line $y = x$
    C   translation by the vector $\begin{pmatrix} -1 \\ 1 \end{pmatrix}$
    D   rotation through 90° clockwise about
        (1, 1)

Paper 2 (2 hours 40 minutes)

1   Attempt **all** the questions.
2   Begin the answer for each **question** on a new page.
3   Full marks may not be awarded unless **necessary** working or explanation is shown with the answer.
4   Mathematical tables, formulae and graph paper are provided.
5   Mathematical instruments and silent electronic calculators may be used for this paper.

①   Evaluate, giving the exact answer

(a) $\dfrac{2\frac{1}{7} + 1\frac{2}{3}}{1\frac{2}{3} - \frac{5}{9}}$

(b) $(3.5)^2 - 7(1.6)$
[8 marks]

②   (a) Write, in its simplest form, 23 cm 7 mm as a percentage of 3 metres.

(b) The scale of a map is $1 : 250\,000$. The distance on the map between two towns J and K measures 2 cm. Calculate, in kilometres, the actual distance JK.
[8 marks]

③   Simplify the following

(a) $3(x + 2y) - 2(2x + y)$

(b) $(p - q)(p + q) - 2q^2$

(c) $\dfrac{3m}{m - 3} + \dfrac{m}{2m + 1}$
[10 marks]

④   (a) Using the exchange rate at a bank, X\$28.50 = US\$1.00 and X\$4.75 = M\$1.00. Calculate

(i)   the amount in US\$ that a tourist receives for X\$5700,

(ii) the exchange rate in M\$ for US\$1.00.

(b) If $c \,\square\, d = 2c - d^2$, find the value of

(i) $3 \,\square\, 1$

(ii) $5 \,\square\, (3 \,\square\, 1)$.
[9 marks]

⑤

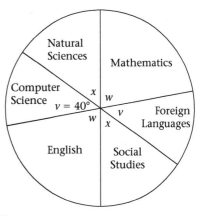

Fig. P15

The pie chart in Fig. P15, *not drawn to scale*, shows how a student spends her time out of class each week on different subjects. The angle w is twice the angle v, and $v = 40°$.

(a) State the size of the angle of the sector representing

(i)   Foreign Languages,

(ii) Mathematics.

(b) Calculate the size of the angle x.

(c) The total time she spends per week is 15 hours.

Calculate

(i)   the time spent on Computer Science as a fraction of the total time per week,

(ii) the time spent doing Computer Science.
[10 marks]

⑥   (a) Solve the following

(i) $4r - 7 = 3r + 2$

(ii) $\dfrac{x - 2}{4} = \dfrac{x + 1}{3}$

(b) (i)   Solve the inequality
$3 - 2x < 6$

(ii) Show your answer on a number line.
[10 marks]

7. PQRS is a trapezium, with PQ parallel to RS, and QR perpendicular to PQ.
   (a) Using ruler and compasses only, construct the trapezium PQRS with PQ = 8 cm, and QR = SR = 4 cm.
   (b) Measure, correct to two significant figures.
      (i) the length of PS,
      (ii) angle SPQ.
   (c) Show by calculation that your answers to (b) are correct.

   [12 marks]

8. A vendor buys oranges at $50.00 per hundred and sells them at $10.50 for a bag of 15.
   (a) Find (i) the cost price of 1 orange,
      (ii) the selling price of 1 orange.
   (b) The vendor bought and sold n oranges and made a profit of $600. Write an equation in n to represent this information.
   (c) Calculate (i) the value of n,
      (ii) the profit as a percentage of the cost price.

   [9 marks]

9. The total charges for the electricity supply to a household include a fixed rental charge of $4.00 per month. The monthly charge, a fuel charge and an exchange rate adjustment charge depend on the number of units of electricity used.

   Table P1 shows the readings and the rates for two months.

   Table P1

Previous reading 014 568	Present reading 015 270
	Rate per kWh
Monthly charge	$15.00
Fuel charge	2.98 cents
Adjustment charge	3.14 cents

   (a) Calculate (i) the number of units used,
      (ii) the total charges.

   (b) In addition, there is a Government tax of 15% of the total charges. Calculate
      (i) the tax,
      (ii) the total amount of the bill to be paid.

   [12 marks]

10. (a) Copy Fig. P16 on graph paper so that 1 cm represents 1 units on each axis. Draw △DEF with D(3, 2), E(6, 2) and F(6, 4).
   (b) △DEF is rotated about the point (1, 2) through an angle of 180° clockwise. Draw, on the same axes, the image △D′E′F′ of △DEF after the rotation.
   (c) △D″E″F″ is the image of △DEF after a transformation T.
      (i) State the coordinates of D″, E″, and F″.
      (ii) Describe fully the transformation T.

   [12 marks]

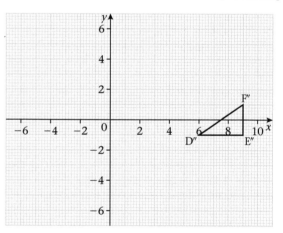

Fig. P16

Basic proficiency examination

Paper 1 (1½ hours)

*Attempt **all** the questions. Circle the letter of your choice.*

1. $\frac{5}{100}$

   A 0.50   B 0.20   C 0.05   D 0.005

2. 0.325 km =

   A 3250 m   B 325 m
   C 32.5 m   D 3.25 m

3. The LCM of 6, 12, 15 is

   A 30   B 60   C 90   D 120

4.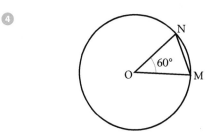

Fig. P17

In Fig. P17, O is the centre of the circle. △OMN is *best* described as

   A equilateral   B isosceles
   C scalene   D right-angled

5. If V = [a, e, i, o, u], then

   A i ∩ V   B i ⊂ V
   C i ∈ V   D i ⊃ V

6. $(x - 2y) - 2(x + 1) + 3y =$

   A $x + y + 1$ B $x - 3y - 2$
   C $-x + y + 2$ D $-x + y - 2$

7. $(27 + 73)15 = 27 \times 15 + 73 \times 15$ is an example of the

   A associative law   B commutative law
   C distributive law   D inverse law

8. The value of the digit 4 in 1425_8 is

   A 4×8 B 4×8^2
   C 4×10^2 D 4×8^3

9. A car travels 150 km for 3 hours. Its average speed in km h is

   A $33\frac{1}{3}$ B 50   C 147   D 450

10. When the point P(1, 4) is reflected in the x-axis, the coordinates of the image are

    A (1, −4)   B (−1, 4)
    C (−1, −4)   D (−4, 1)

11. $110011_2 =$

    A 25_{10} B 36_8 C 52_{10} D 63_8

12. If $3(x + 2) + 2x = -4$, then $x =$

    A −2   B $-\frac{2}{5}$ C $\frac{2}{5}$ D 2

13. The prime factors of 360 are

    A 2, 3, 5   B 3, 5, 9
    C 5, 8, 9   D 10, 20, 30, 40, 60

14. If $m * n = 3m - 2n$, then $4 * 3 =$

    A 1   B 2   C 6   D 7

15. The cost price of a shirt is $40 and the selling price is $50. The percentage profit on the cost price is

    A 50%   B 40%   C 25%   D 20%

16. The median of the numbers
    12, 19, 13, 12, 11, 15, 20, 12, 19, 17, is

    A 12   B 13   C 14   D 15

17.

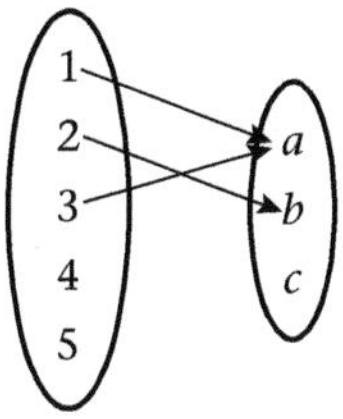

Fig. P18

In Fig. P18, the ordered pairs defining the relation are

   A {(1, *a*), (3, *a*), (2, *b*)}
   B {(2, *b*)}
   C {(1, *a*), (3, *a*)}
   D {(1, *a*), (3, *a*), (5, *a*), (2, *b*), (4, *b*)}

18 X = {4, 8, 12} and Y = {4, 9}. Then X ∪ Y =
A {4}   B {4, 9}
C {4, 8, 12}   D {4, 8, 9, 12}

19 $\dfrac{x - 1}{5} - \dfrac{2x + 1}{3} =$

A $\dfrac{-7x}{8}$ B $\dfrac{-7x - 8}{15}$

C $\dfrac{-7x + 2}{15}$ D $\dfrac{-7x - 2}{2}$

20 Given that $g(x) = x^2 + 1$, $g(-3) =$
A −8   B −4   C 8   D 10

21 0.004 21 written in standard form is
A 4.21×10^{-3} B 4.21×10^{-2}
C 421×10^{-2} D 421×10^{-5}

22

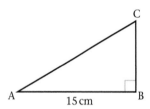

Fig. P19

In Fig. P19, the angle of elevation of C from A is 32°. If AB = 15 cm, then AC = x cm where $x =$

A 15 sin 32°   B $\dfrac{15}{\sin 32°}$

C 15 cos 32°   D $\dfrac{15}{\cos 32°}$

23 Given that $2x = y - 1$
        and $x = 4 - y$,
then $(x, y) =$
A (3, 1)   B (2, 2)   C (1, 3)   D (1, 2)

24

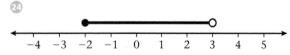

Fig. P20

The set of real numbers defined in Fig. P20 is
A $-2 < x \leqslant 3$ B $-2 > x \geqslant 3$
C $-2 \leqslant x < 3$ D $-2 \geqslant x > 3$

25 60 370 to 2 significant figures is equal to
A 60   B 604   C 60 000   D 60 400

26 The length of the side of a cube is 4 cm. The surface area, in cm², of the cube is
A 24   B 64   C 84   D 96

27 Given that $2.26 \times 3.14 = 7.0964$, then $22.6 \times 0.0314 =$
A 70.964   B 7.0964
C 0.709 64   D 0.070 964

28 $(x - 3)(2x - 1) =$
A $2x^2 + 7x + 3$ B $2x^2 - 7x + 3$
C $2x^2 - 5x - 3$ D $2x^2 - 7x - 3$

29 A mixture contains copper and zinc in the ratio 13 : 7. If the mass of zinc is 182 g, the total mass of the mixture is
A 98 g   B 280 g   C 338 g   D 520 g

30 A plane lands at an airport every 15 minutes. The first plane lands at 05:40 h. At 07:45 h the number of planes that have landed is
A 8   B 9   C 12   D 13

31 $(2x^2)^3 \times 3x =$
A $24x^7$ B $24x^6$ C $18x^6$ D $18x^5$

32

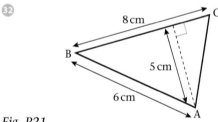

Fig. P21

In Fig. P21, the area △ABC in cm² is
A 15   B 20   C 24   D 40

33 A hall contains 175 people. 12% of them are children and there are 56 men. The number of women in the hall is
A 21   B 35   C 77   D 98

34 If F$1 = J$2.90, then, to the nearest dollar, J$1870 =
A F$5423   B F$3460
C F$1010   D F$645

35 On a map the scale is 1 : 300 000. Two towns are 1.5 cm apart on the map. The actual distance, in km, between the towns is
A 2   B 4.5   C 20   D 45

36

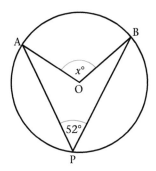

Fig. P22

The size of AÔB (*x*°) in Fig. P22 is

A   104°   B   128°   C   256°   D   308°

Fig. P23 shows the heights of a group of children.
Use Fig. P23 to answer questions 37 and 38.

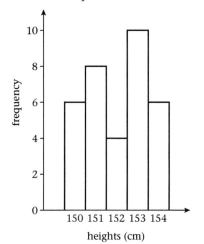

Fig. P23

37   The number of children measured is

A   5   B   30   C   34   D   76

38   The modal height in the group is

A   150 cm   B   152 cm
C   153 cm   D   154 cm

39

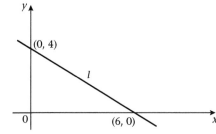

Fig. P24

The gradient of the line *l* in Fig. P24 is

A $-\frac{3}{2}$ B $-\frac{2}{3}$ C $\frac{2}{3}$ D $\frac{3}{2}$

40   The selling price of a radio is $40, not
including a sales tax of $7\frac{1}{2}$%. The total
purchase price paid for the radio is

A $32.50   B $37
C $43   D $47.50

41

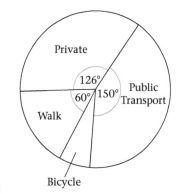

Fig. P25

Fig. P25 is a pie-chart showing the methods
by which 1200 students get to school. The
number of students who ride to school on a
bicycle is

A   15   B   24   C   50   D   80

42

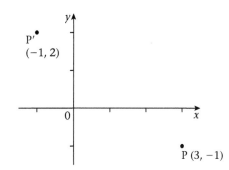

Fig. P26

In Fig. P26, P′ is the image of P under a
translation specified by the vector

A $\begin{pmatrix} 4 \\ 3 \end{pmatrix}$ B $\begin{pmatrix} 4 \\ -3 \end{pmatrix}$

C $\begin{pmatrix} -4 \\ 3 \end{pmatrix}$ D $\begin{pmatrix} -4 \\ -3 \end{pmatrix}$

43   The total amount paid on an investment of
$8000 at the end of 9 months at the rate of
$7\frac{1}{2}$% per annum is

A $13 400   B $12 000
C $8900   D $8450

44 If the perimeter of a square is 52 cm, then the area of the square, in cm², is

   A 169   B 208   C 676   D 2704

45

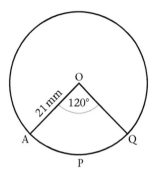

Fig. P27

In Fig. P27, O is the centre of the circle radius 21 mm and $A\widehat{O}Q = 120°$. Taking $\pi = \frac{22}{7}$, the length of the arc APQ is

   A 7 mm   B 22 mm

   C 44 mm   D 132 mm

46 The cash price of a table is $420, the hire purchase price is $684. The deposit is $168 and the balance is paid in 12 equal monthly instalments. Each monthly instalment is

   A $22   B $35   C $43   D $57

47

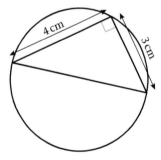

Fig. P28

The diameter of the circle in Fig. P28 is

   A $2\frac{1}{2}$ cm   B 5 cm

   C 7 cm   D 10 cm

48 A worker receives a basic wage of $600 for a 40-hour week. His overtime rate is 120% of the basic hourly rate. He received $780 in one week. The number of overtime hours he worked in that week is

   A 6   B 10   C 12   D 15

49

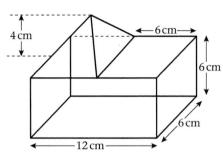

Fig. P29

The total surface area of the solid in Fig. P29, in cm², is

   A 192   B 300   C 360   D 384

50 A packet of 10 pens costs $96 and is sold at $168. During a sale there is a discount of 90 cents per pen. The profit, in dollars, made on a packet of pens during the sale is

   A 63.00 B 71.10 C 78.00 D 87.00

51 The mean score of 30 students in a test is 14. Two students score zero. The mean score of the remaining 28 is

   A 12   B 14   C 15   D 18

52 Given that $\sin A = \frac{5}{13}$ and $\widehat{A}$ is acute, then $\cos A =$

   A $\frac{12}{13}$ B $\frac{8}{13}$ C $\frac{5}{12}$ D $\frac{13}{5}$

53 The selling price of a radio is $690. The profit made on the cost price is 15%. The cost price is

   A $675   B $600

   C $586.50   D $510

54 If $x + 5 < 3x - 1$, then the solution set is

   A $\{x: x > 3\}$ B $\{x: x < -3\}$

   C $\{x: x > 2\}$ D $\{x: x < 1\}$

55

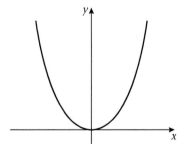

Fig. P30

Fig. P30 represents the graph of the relation

   A $y \propto x$ B $y \propto x^2$ C $y < x$ D $y > x$

56 There are 10 attempts to hit a target. The probability of hitting the target may be equal to

A −1   B $-\frac{3}{10}$ C $\frac{7}{10}$ D 2

57 Given that $f\colon x \to 2x - 3$, the ordered pair that does *not* satisfy the relation is

A (2, 1)   B $(\frac{3}{2}, 0)$
C (0, 3)   D (−1, −5)

58 The statement 'an integer is five more than twice another integer' written in symbols is

A $m - 2n > 5$ B $2n - m > 5$
C $m + 5 = 2n$ D $m - 2n = 5$

59 In a class of 40 students, all students must study either geography or history or both subjects. If 26 study geography and 30 study history, the number that studies both geography and history is

A 4   B 10   C 14   D 16

60 The taxes paid in a country increased by 10% in one year and then by a further 5% in the following year. The total increase for the two years is

A $7\frac{1}{2}$%   B 10%   C 15%   D $15\frac{1}{2}$%

Paper 2 (2 hours 40 minutes)

*1 Attempt **all** the questions.*

2 Begin the answer for each question on a new page.

3 Full marks may not be awarded unless full working or explanation is shown with the answer.

4 Mathematical tables, formulae and graph paper are provided.

*5 Silent electronic calculators may be used **except** in questions where their use is expressly forbidden.*

6 You are advised to use the first 10 minutes of the examination time to read through this paper and to plan your answers. Writing may begin during this 10-minute period.

All steps and calculations must be clearly shown to earn credit for your solutions.

1 Calculate the exact value of

(a) $\left(\frac{3}{4} - \frac{2}{3}\right) \times 1\frac{1}{5}$

(b) $\dfrac{0.6 + 0.75}{0.6 \times 0.75}$ [6 marks]

2 (a) If 1 dollar = 100 cents, express m dollars and n cents in cents.

(b) If $x = 4$ and $y = -1$, find the value of
  (i) $x + 4y$ (ii) $x^2 - y^2$
  (iii) $(x - y)^2$ [7 marks]

3 A woman paid $385 for a radio. This price included a sales tax of 12%. Calculate (a) the basic price before tax, (b) the sales tax.
[6 marks]

4 In a family, the mass of the mother is 59 kg and of the father is 72 kg. The mean mass of the five children is 34 kg. Calculate the mean mass of all seven members of the family.
[7 marks]

5 A triangle has sides of length x cm, $2(x + 1)$ cm and $2x + 3$ cm.

(a) Express the perimeter P of the triangle in terms of x.

(b) Given that $P = 30$, calculate the lengths of the sides of the triangle.

(c) State the size of the largest angle of the triangle, giving a reason.

(d) Calculate the size of the smallest angle of the triangle.   [10 marks]

6 In Fig. P31, ABCD is a trapezium such that $A\widehat{D}B = B\widehat{C}D$, AB = 4 cm and BD = 6 cm.

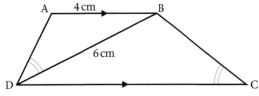

Fig. P31

Using similar triangles, calculate

(a) the length of CD,

(b) the ratio $\dfrac{\text{area of } \triangle \text{ABD}}{\text{area of } \triangle \text{BCD}}$ [9 marks]

7. A salesman receives a basic monthly salary of $3320. He also gets $2\frac{1}{2}\%$ commission on sales. Out of his *total* income he pays 4% as union fees and 13% as income tax. His sales in a month were $35\,200$. Calculate (a) his commission, (b) the union fees, (c) his income tax, (d) his net income.

[12 marks]

8. The total monthly bill y for the use of a machine includes a fixed rental r and a charge that varies with x, the number of times the machine is used. It is given that $y = cx + r$, where c is the charge each time the machine is used. Corresponding values of x and y are given in Table P2.

Table P2

x	y
3	31
8	41
15	55

(a) Using a scale of 1 cm to represent 1 unit on the x-axis and 1 cm to $5 on the y-axis, and the given x and y values, draw a graph to show the relation between x and y. (Show clearly the scale used on each axis.)

(b) From your graph, or otherwise, determine the value of
  (i) the fixed monthly rental,
  (ii) the charge for each time the machine is used,
  (iii) the total monthly bill when the machine is used 11 times.

[14 marks]

9. An answer sheet is provided for this question (Fig. P32).

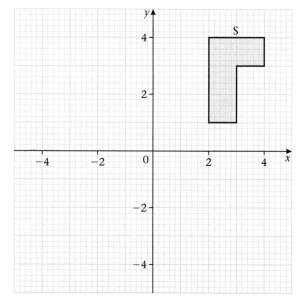

Fig. P32 Answer sheet for question 9

A shape S is drawn on a Cartesian plane. Copy Fig. P32 and draw the image of S under

(a) a translation by vector $\begin{pmatrix} -7 \\ 1 \end{pmatrix}$ (label the image T),

(b) a reflection in the line $y = -x$ (label the image M),

(c) a clockwise rotation of 90° about the point $(0, -2)$ (label the image R).

[12 marks]

10. The following is a list of test scores for a class of 20 students.
94, 72, 76, 53, 52, 60, 84, 65, 63, 63,
63, 78, 87, 55, 59, 66, 81, 61, 52, 57

(a) Calculate the mean score.
(b) Construct a grouped frequency table for this distribution using intervals of 51–60, 61–70, and so on.
(c) Draw a histogram using the table completed in (b).
(d) Find the value of the median score.
(e) Determine the interquartile range.
(f) Find the probability that a student selected at random scored above 80.

[17 marks]

SI units

Mass

The **gram** is the basic unit of mass.

Unit	Abbreviation	Basic unit
1 kilogram	1 kg	1000 g
1 hectogram	1 hg	100 g
1 decagram	1 dag	10 g
1 gram	1 g	1 g
1 decigram	1 dg	0.1 g
1 centigram	1 cg	0.01 g
1 milligram	1 mg	0.001 g

The **tonne** (t) is used for large masses. The most common measures of mass are the milligram, the gram, the kilogram and the tonne.

$1 \text{ g} = 1000 \text{ mg}$
$1 \text{ kg} = 1000 \text{ g} = 1\,000\,000 \text{ mg}$
$1 \text{ t} = 1000 \text{ kg} = 1\,000\,000 \text{ g}$

Length

The **metre** is the basic unit of length.

Unit	Abbreviation	Basic unit
1 kilometre	1 km	1000 m
1 hectometre	1 hm	100 m
1 decametre	1 dam	10 m
1 metre	1 m	1 m
1 decimetre	1 dm	0.1 m
1 centimetre	1 cm	0.01 m
1 millimetre	1 mm	0.001 m

The most common measures are the millimetre, the metre and the kilometre.

$1 \text{ m} = 1000 \text{ mm}$
$1 \text{ km} = 1000 \text{ m} = 1\,000\,000 \text{ mm}$

Time

The **second** is the basic unit of time.

Unit	Abbreviation	Basic unit
1 second	1 s	1 s
1 minute	1 min	60 s
1 hour	1 h	3600 s

Area

The **square metre** is the basic unit of area. Units of area are derived from units of length.

Unit	Abbreviation	Relation to other units of area
square millimetre	mm²	
square centimetre	cm²	$1 \text{ cm}^2 = 100 \text{ mm}^2$
square metre	m²	$1 \text{ m}^2 = 10\,000 \text{ cm}^2$
square kilometre	km²	$1 \text{ km}^2 = 1\,000\,000 \text{ m}^2$
hectare (for land measure)	ha	$1 \text{ ha} = 10\,000 \text{ m}^2$

Volume

The **cubic metre** is the basic unit of volume. Units of volume are derived from units of length.

Unit	Abbreviation	Relation to other units of volume
cubic millimetre	mm³	
cubic centimetre	cm³	$1 \text{ cm}^3 = 1000 \text{ mm}^3$
cubic metre	m³	$1 \text{ m}^3 = 1\,000\,000 \text{ cm}^3$

Capacity

The **litre** is the basic unit of capacity. 1 litre takes up the same space as 1000 cm^3.

Unit	Abbreviation	Relation to other units of capacity	Relation to units of volume
millilitre	$m\ell$		$1\,m\ell = 1 \text{ cm}^3$
litre	ℓ	$1\ell = 1000\,m\ell$	$1\ell = 1000 \text{ cm}^3$
kilolitre	$k\ell$	$1\,k\ell = 1000\,\ell$	$1\,k\ell = 1 \text{ m}^3$

Mensuration formulae

Plane shapes	Perimeter	Area
Square side s	$4s$	s^2
Rectangle length l, breadth b	$2\,(l + b)$	lb
Triangle base b, height h		$\frac{1}{2}bh$
Parallelogram base b, height h		bh
Trapezium height h, parallels a and b		$\frac{1}{2}(a + b)\,h$
Circle radius r	$2\pi r$	πr^2
Sector of circle radius r, angle $\theta°$	$2r + \frac{\theta}{360}\,2\pi r$	$\frac{\theta}{360}\,\pi r^2$

Solid shapes	Surface area	Volume
Cube edge s	$6s^2$	s^3
Cuboid length l, breadth b, height h	$2(lb + bh + lh)$	lbh
Prism height h, base area A		Ah
Cylinder radius r, height h	$2\pi rh + 2\pi r^2$	πr^2h
Cone radius r, slant height l, height h	$\pi rl + \pi r^2$	$\frac{1}{3}\pi r^2h$
Sphere radius r	$4\pi r^2$	$\frac{4}{3}\pi r^3$

Trigonometric formulae

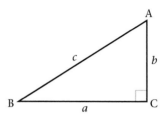

Fig. T1

In the right-angled triangle shown in Fig. T1,
$c^2 = a^2 + b^2$ (*Pythagoras' theorem*)

$$\tan B = \frac{b}{a} \qquad \tan A = \frac{a}{b}$$

$$\sin B = \frac{b}{c} \qquad \sin A = \frac{a}{c}$$

$$\cos B = \frac{a}{c} \qquad \cos A = \frac{b}{c}$$

Divisibility tests

Any whole number is exactly divisible by

2	if its last digit is even
3	if the sum of its digits is divisible by 3
4	if its last two digits form a number divisible by 4
5	if its last digit is 5 or 0
6	if its last digit is even and the sum of its digits is divisible by 3
8	if its last three digits form a number divisible by 8
9	if the sum of its digits is divisible by 9
10	if its last digit is 0

Symbols

Symbol	Meaning
=	is equal to
≠	is not equal to
≃	is approximately equal to
≡	is identical or congruent to
⇒	leads to
∴	therefore
∝	is proportional to
>	is greater than
<	is less than
⩾	is greater than or equal to
⩽	is less than or equal to
°	degree (angle)
°C	degrees Celsius (temperature)

Symbol	Meaning
A, B, C	points (geometry)
AB	the line through points A and B, *or* the distance between points A and B

Symbol	Meaning
△ABC	triangle ABC
$A\widehat{B}C$	angle ABC
⊥	is perpendicular to
∥	is parallel to
π	pi
%	per cent
$A = \{p, q, r\}$	A is the set p, q, r
$B = \{1, 2, 3, \ldots\}$	B is the infinite set 1, 2, 3, etc.
$C = \{x : x$ is an integer$\}$	Set builder notation. C is the set of numbers x such that x is an integer
$n(A)$	number of elements in set A
∈	is an element of
∉	is not an element of
A′	complement of A
{ } or ∅	the empty set
U	the universal set
A ⊂ B	A is a subset of B
A ⊃ B	A contains B
⊄, ⊅	negations of ⊂ and ⊃
A ∪ B	union of A and B
A ∩ B	intersection of A and B

$x \rightarrow \log x$

x	0	1	2	3	4	5	6	7	8	9
55	.740	741	742	743	744	744	745	746	747	747
56	.748	749	750	751	751	752	753	754	754	755
57	.756	757	757	758	759	760	760	761	762	763
58	.763	764	765	766	766	767	768	769	769	770
59	.771	772	772	773	774	775	775	776	777	777
60	.778	779	780	780	781	782	782	783	784	785
61	.785	786	787	787	788	789	790	790	791	792
62	.792	793	794	794	795	796	797	797	798	799
63	.799	800	801	801	802	803	803	804	805	806
64	.806	807	808	808	809	810	810	811	812	812
65	.813	814	814	815	816	816	817	818	818	819
66	.820	820	821	822	822	823	823	824	825	825
67	.826	827	827	828	829	829	830	831	831	832
68	.833	833	834	834	835	836	836	837	838	838
69	.839	839	840	841	841	842	843	843	844	844
70	.845	846	846	847	848	848	849	849	850	851
71	.851	852	852	853	854	854	855	856	856	857
72	.857	858	859	859	860	860	861	862	862	863
73	.863	864	865	865	866	866	867	867	868	869
74	.869	870	870	871	872	872	873	873	874	874
75	.875	876	876	877	877	878	879	879	880	880
76	.881	881	882	883	883	884	884	885	885	886
77	.886	887	888	888	889	889	890	890	891	892
78	.892	893	893	894	894	895	895	896	897	897
79	.898	898	899	899	900	900	901	901	902	903
80	.903	904	904	905	905	906	906	907	907	908
81	.908	909	910	910	911	911	912	912	913	913
82	.914	914	915	915	916	916	917	918	918	919
83	.919	920	920	921	921	922	922	923	923	924
84	.924	925	925	926	926	927	927	928	928	929
85	.929	930	930	931	931	932	932	933	933	934
86	.934	935	936	936	937	937	938	938	939	939
87	.940	940	941	941	942	942	943	943	943	944
88	.944	945	945	946	946	947	947	948	948	949
89	.949	950	950	951	952	952	953	953	954	954
90	.954	955	955	956	956	957	957	958	958	959
91	.959	960	960	960	961	961	962	962	963	963
92	.964	964	965	965	966	966	967	967	968	968
93	.968	969	969	970	970	971	971	972	972	973
94	.973	974	974	975	975	976	976	976	977	977
95	.978	978	979	979	980	980	980	981	981	982
96	.982	983	983	984	984	985	985	985	986	986
97	.987	987	988	988	989	989	989	990	990	991
98	.991	992	992	993	993	993	994	994	995	995
99	.996	996	997	997	997	998	998	999	999	1.000

Logarithms

x	0	1	2	3	4	5	6	7	8	9
10	.000	004	009	013	017	021	025	029	033	037
11	.041	045	049	053	057	061	064	068	072	076
12	.079	083	086	090	093	097	100	104	107	111
13	.114	117	121	124	127	130	134	137	140	143
14	.146	149	152	155	158	161	164	167	170	173
15	.176	179	182	185	188	190	193	196	199	201
16	.204	207	210	212	215	217	220	223	225	228
17	.230	233	236	238	241	243	246	248	250	253
18	.255	258	260	262	265	267	270	272	274	276
19	.279	281	283	286	288	290	292	294	297	299
20	.301	303	305	307	310	312	314	316	318	320
21	.322	324	326	328	330	332	334	336	338	340
22	.342	344	346	348	350	352	354	356	358	360
23	.362	364	365	367	369	371	373	375	377	378
24	.380	382	384	386	387	389	391	393	394	396
25	.398	400	401	403	405	407	408	410	412	413
26	.415	417	418	420	422	423	425	427	428	430
27	.431	433	435	436	438	439	441	442	444	446
28	.447	449	450	452	453	455	456	458	459	461
29	.462	464	465	467	468	470	471	473	474	476
30	.477	479	480	481	483	484	486	487	489	490
31	.491	493	494	496	497	498	500	501	502	504
32	.505	507	508	509	511	512	513	515	516	517
33	.519	520	521	522	524	525	526	528	529	530
34	.531	533	534	535	537	538	539	540	542	543
35	.544	545	547	548	549	550	551	553	554	555
36	.556	558	559	560	561	562	563	565	566	567
37	.568	569	571	572	573	574	575	576	577	579
38	.580	581	582	583	584	585	587	588	589	590
39	.591	592	593	594	595	597	598	599	600	601
40	.602	603	604	605	606	607	609	610	611	612
41	.613	614	615	616	617	618	619	620	621	622
42	.623	624	625	626	627	628	629	630	631	632
43	.633	634	635	636	637	638	639	640	641	642
44	.643	644	645	646	647	648	649	650	651	652
45	.653	654	655	656	657	658	659	660	661	662
46	.663	664	665	666	667	667	668	669	670	671
47	.672	673	674	675	676	677	677	679	679	680
48	.681	682	683	684	685	686	687	688	688	689
49	.690	691	692	693	694	695	695	696	697	698
50	.699	700	701	702	702	703	704	705	706	707
51	.708	708	709	710	711	712	713	713	714	715
52	.716	717	718	719	719	720	721	722	723	723
53	.724	725	726	727	728	728	729	730	731	732
54	.732	733	734	735	736	736	737	738	739	740

$x \rightarrow 10^x$

x	0	1	2	3	4	5	6	7	8	9
.50	316	317	318	318	319	320	321	321	322	323
.51	324	324	325	326	327	327	328	329	330	330
.52	331	332	333	333	334	335	336	337	337	338
.53	339	340	340	341	342	343	344	344	345	346
.54	347	348	348	349	350	351	352	352	353	354
.55	355	356	356	357	358	359	360	361	361	362
.56	363	364	365	366	366	367	368	369	370	371
.57	372	372	373	374	375	376	377	378	378	379
.58	380	381	382	383	384	385	385	386	387	388
.59	389	390	391	392	393	394	394	395	396	397
.60	398	399	400	401	402	403	404	405	406	406
.61	407	408	409	410	411	412	413	414	415	416
.62	417	418	419	420	421	422	423	424	425	426
.63	427	428	429	430	431	432	433	434	435	436
.64	437	438	439	440	441	442	443	444	445	446
.65	447	448	449	450	451	452	453	454	455	456
.66	457	458	459	460	461	462	463	465	466	467
.67	468	469	470	471	472	473	474	475	476	478
.68	479	480	481	482	483	484	485	486	488	489
.69	490	491	492	493	494	495	497	498	499	500
.70	501	502	504	505	506	507	508	509	511	512
.71	513	514	515	516	518	519	520	521	522	524
.72	525	526	527	528	530	531	532	533	535	536
.73	537	538	540	541	542	543	545	546	547	548
.74	550	551	552	553	555	556	557	558	560	561
.75	562	564	565	566	568	569	570	571	573	574
.76	575	577	578	579	581	582	583	585	586	587
.77	589	590	592	593	594	596	597	598	600	601
.78	603	604	605	607	608	610	611	612	614	615
.79	617	618	619	621	622	624	625	627	628	630
.80	631	632	634	635	637	638	640	641	643	644
.81	646	647	649	650	652	653	655	656	658	659
.82	661	662	664	665	667	668	670	671	673	675
.83	676	678	679	681	682	684	685	687	689	690
.84	692	693	695	697	698	700	701	703	705	706
.85	708	710	711	713	714	716	718	719	721	723
.86	724	726	728	729	731	733	735	736	738	740
.87	741	743	745	746	748	750	752	753	755	757
.88	759	760	762	764	766	767	769	771	773	774
.89	776	778	780	782	783	785	787	789	791	793
.90	794	796	798	800	802	804	805	807	809	811
.91	813	815	817	818	820	822	824	826	828	830
.92	832	834	836	838	839	841	843	845	847	849
.93	851	853	855	857	859	861	863	865	867	869
.94	871	873	875	877	879	881	883	885	887	889
.95	891	893	895	897	899	902	904	906	908	910
.96	912	914	916	918	920	923	925	927	929	931
.97	933	935	938	940	942	944	946	948	951	953
.98	955	957	959	962	964	966	968	971	973	975
.99	977	979	982	984	986	989	991	993	995	998

Antilogarithms

x	0	1	2	3	4	5	6	7	8	9
.00	100	100	100	101	101	101	101	102	102	102
.01	102	103	103	103	103	104	104	104	104	104
.02	105	105	105	105	106	106	106	106	107	107
.03	107	107	108	108	108	108	109	109	109	109
.04	110	110	110	110	111	111	111	111	112	112
.05	112	112	113	113	113	114	114	114	114	115
.06	115	115	115	116	116	116	116	117	117	117
.07	117	118	118	118	119	119	119	119	120	120
.08	120	121	121	121	121	122	122	122	122	123
.09	123	123	124	124	124	124	125	125	125	126
.10	126	126	126	127	127	127	127	128	128	129
.11	129	129	129	130	130	130	131	131	131	132
.12	132	132	132	133	133	133	134	134	134	135
.13	135	135	136	136	136	136	137	137	137	138
.14	138	138	139	139	139	140	140	140	141	141
.15	141	142	142	142	143	143	143	144	144	144
.16	145	145	145	146	146	146	147	147	147	148
.17	148	148	149	149	149	150	150	150	151	151
.18	151	152	152	152	153	153	153	154	154	155
.19	155	155	156	156	156	157	157	157	158	158
.20	158	159	159	160	160	160	161	161	161	162
.21	162	163	163	163	164	164	164	165	165	166
.22	166	166	167	167	167	168	168	168	169	169
.23	170	170	171	171	171	172	172	172	173	173
.24	174	174	175	175	175	176	176	177	177	177
.25	178	178	179	179	179	180	180	181	181	182
.26	182	182	183	183	184	184	185	185	185	186
.27	186	187	187	188	188	188	189	189	190	190
.28	191	191	191	192	192	193	193	194	194	195
.29	195	195	196	196	197	197	198	198	199	199
.30	200	200	200	201	201	202	202	203	203	204
.31	204	204	205	206	206	207	207	207	208	208
.32	209	209	210	210	211	211	212	212	213	213
.33	214	214	215	215	216	216	217	217	218	218
.34	219	219	220	220	221	221	222	222	223	223
.35	224	224	225	225	226	226	226	227	228	229
.36	229	230	230	231	231	232	232	233	233	234
.37	234	235	236	236	237	237	238	238	239	239
.38	240	240	241	242	242	243	243	244	244	245
.39	245	246	247	247	248	248	249	249	250	251
.40	251	252	252	253	254	254	255	255	256	256
.41	257	258	258	259	259	260	261	261	262	262
.42	263	264	264	265	265	266	267	267	268	269
.43	269	270	270	271	272	272	273	274	274	275
.44	275	276	277	278	278	279	279	280	281	281
.45	282	282	283	284	284	285	286	286	287	288
.46	288	289	290	290	291	292	292	293	294	294
.47	295	296	296	297	298	299	299	300	301	301
.48	302	303	303	304	305	305	306	307	308	308
.49	309	310	310	311	312	313	313	314	315	316

$\theta \rightarrow \sin \theta$

θ	.0	.1	.2	.3	.4	.5	.6	.7	.8	.9
45	0.707	708	710	711	712	713	714	716	717	718
46	.719	721	722	723	724	725	727	728	729	730
47	.731	733	734	735	736	737	738	740	741	742
48	.743	744	745	747	748	749	750	751	752	754
49	.755	756	757	758	759	760	762	763	764	765
50	0.766	767	768	769	771	772	773	774	775	776
51	.777	778	779	780	782	783	784	785	786	787
52	.788	789	790	791	792	793	794	795	797	798
53	.799	800	801	802	803	804	805	806	807	808
54	.809	810	811	812	813	814	815	816	817	818
55	0.819	820	821	822	823	824	825	826	827	828
56	.829	830	831	832	833	834	835	836	837	838
57	.839	840	841	842	842	843	844	845	846	847
58	.848	849	850	851	852	853	854	854	855	856
59	.857	858	859	860	861	862	863	863	864	865
60	0.866	867	868	869	869	870	871	872	873	874
61	.875	875	876	877	878	879	880	880	881	882
62	.883	884	885	885	886	887	888	889	889	890
63	.891	892	893	893	894	895	896	896	897	898
64	.899	900	900	901	902	903	903	904	905	906
65	0.906	907	908	909	909	910	911	911	912	913
66	.914	914	915	916	916	917	918	918	919	920
67	.921	921	922	923	923	924	925	925	926	927
68	.927	928	928	929	930	930	931	932	932	933
69	.934	934	935	935	936	937	937	938	938	939
70	0.940	940	941	941	942	943	943	944	944	945
71	.946	946	947	947	948	948	949	949	950	951
72	.951	952	952	953	953	954	954	955	955	956
73	.956	957	957	958	958	959	959	960	960	961
74	.961	962	962	963	963	964	964	965	965	965
75	0.966	966	967	967	968	968	969	969	969	970
76	.970	971	971	972	972	972	973	973	974	974
77	.974	975	975	976	976	976	977	977	977	978
78	.978	979	979	979	980	980	980	981	981	981
79	.982	982	982	983	983	983	984	984	984	985
80	0.985	985	985	986	986	986	987	987	987	987
81	.988	988	988	988	989	989	989	990	990	990
82	.990	991	991	991	991	991	992	992	992	992
83	.993	993	993	993	993	994	994	994	994	994
84	.995	995	995	995	995	995	996	996	996	996
85	0.996	996	996	997	997	997	997	997	997	997
86	.998	998	998	998	998	998	998	998	998	999
87	.999	999	999	999	999	999	999	999	999	999
88	.999	999	1.000	1.000	1.000	1.000	1.000	1.000	1.000	1.000
89	1.000	1.000	1.000	1.000	1.000	1.000	1.000	1.000	1.000	1.000
90	1.000	1.000	1.000	1.000	1.000	1.000	1.000	1.000	1.000	1.000

Sines of angles

θ	.0	.1	.2	.3	.4	.5	.6	.7	.8	.9
0	0.000	002	003	005	007	009	010	012	014	016
1	.017	019	021	023	024	026	028	030	031	033
2	.035	037	038	040	042	044	045	047	049	051
3	.052	054	056	058	059	061	063	065	066	068
4	.070	071	073	075	077	078	080	082	084	085
5	0.087	089	091	092	094	096	098	099	101	103
6	.105	106	108	110	111	113	115	117	118	120
7	.122	124	125	127	129	131	132	134	136	137
8	.139	141	143	144	146	148	150	151	153	155
9	.156	158	160	162	163	165	167	168	170	172
10	0.174	175	177	179	181	182	184	186	187	189
11	.191	193	194	196	198	199	201	203	204	206
12	.208	210	211	213	215	216	218	220	222	223
13	.225	227	228	230	232	233	235	237	239	240
14	.242	244	245	247	249	250	252	254	255	257
15	0.259	261	262	264	266	267	269	271	272	274
16	.276	277	279	281	282	284	286	287	289	291
17	.292	294	296	297	299	301	302	304	306	307
18	.309	311	312	314	316	317	319	321	322	324
19	.326	327	329	331	332	334	335	337	339	340
20	0.342	344	345	347	349	350	352	353	355	357
21	.358	360	362	363	365	367	368	370	371	373
22	.375	376	378	379	381	383	384	386	388	389
23	.391	392	394	396	397	399	400	402	404	405
24	.407	408	410	412	413	415	416	418	419	421
25	0.423	424	426	427	429	431	432	434	435	437
26	.438	440	442	443	445	446	448	449	451	452
27	.454	456	457	459	460	462	463	465	466	468
28	.469	471	473	474	476	477	479	480	482	483
29	.485	486	488	489	491	492	494	495	497	498
30	0.500	502	503	505	506	508	509	511	512	514
31	.515	517	518	520	521	522	524	525	527	528
32	.530	531	533	534	536	537	539	540	542	543
33	.545	546	548	549	550	552	553	555	556	558
34	.559	561	562	564	565	566	568	569	571	572
35	0.574	575	576	578	579	581	582	584	585	586
36	.588	589	591	592	593	595	596	598	599	600
37	.602	603	605	606	607	609	610	612	613	614
38	.616	617	618	620	621	623	624	625	627	628
39	.629	631	632	633	635	636	637	639	640	641
40	0.643	644	645	647	648	649	651	652	653	655
41	.656	657	659	660	661	663	664	665	667	668
42	.669	670	672	673	674	676	677	678	679	681
43	.682	683	685	686	687	688	690	691	692	693
44	.695	696	697	698	700	701	702	703	705	706

θ → cos θ

θ	.0	.1	.2	.3	.4	.5	.6	.7	.8	.9
45	0.707	706	705	703	702	701	700	698	697	696
46	.695	693	692	691	690	688	687	686	685	683
47	.682	681	679	678	677	676	674	673	672	670
48	.669	668	667	665	664	663	661	660	659	657
49	.656	655	653	652	651	649	648	647	645	644
50	0.643	641	640	639	637	636	635	633	632	631
51	.629	628	627	625	624	623	621	620	618	617
52	.616	614	613	612	610	609	607	606	605	603
53	.602	600	599	598	596	595	593	592	591	589
54	.588	586	585	584	582	581	579	578	576	575
55	0.574	572	571	569	568	566	565	564	562	561
56	.559	558	556	555	553	552	550	549	548	546
57	.545	543	542	540	539	537	536	534	533	531
58	.530	528	527	525	524	522	521	520	518	517
59	.515	514	512	511	509	508	506	505	503	502
60	0.500	498	497	495	494	492	491	489	488	486
61	.485	483	482	480	479	477	476	474	473	471
62	.469	468	466	465	463	462	460	459	457	456
63	.454	452	451	449	448	446	445	443	442	440
64	.438	437	435	434	432	431	429	427	426	424
65	0.423	421	419	418	416	415	413	412	410	408
66	.407	405	404	402	400	399	397	396	394	392
67	.391	389	388	386	384	383	381	379	378	376
68	.375	373	371	370	368	367	365	363	362	360
69	.358	357	355	353	352	350	349	347	345	344
70	0.342	340	339	337	335	334	332	331	329	327
71	.326	324	322	321	319	317	316	314	312	311
72	.309	307	306	304	302	301	299	297	296	294
73	.292	291	289	287	286	284	282	281	279	277
74	.276	274	272	271	269	267	266	264	262	261
75	0.259	257	255	254	252	250	249	247	245	244
76	.242	240	239	237	235	233	232	230	228	227
77	.225	223	222	220	218	216	215	213	211	210
78	.208	206	204	203	201	199	198	196	194	193
79	.191	189	187	186	184	182	181	179	177	175
80	0.174	172	170	168	167	165	163	162	160	158
81	.156	155	153	151	150	148	146	144	143	141
82	.139	137	136	134	132	131	129	127	125	124
83	.122	120	118	117	115	113	111	110	108	106
84	.105	103	101	099	098	096	094	092	091	089
85	0.087	085	084	082	080	078	077	075	073	071
86	.070	068	066	065	063	061	059	058	056	054
87	.052	051	049	047	045	044	042	040	038	037
88	.035	033	031	030	028	026	024	023	021	019
89	.017	016	014	012	010	009	007	005	003	002
90	0.000									

Cosines of angles

θ	.0	.1	.2	.3	.4	.5	.6	.7	.8	.9
0	1.000	000	000	000	000	000	000	000	000	000
1	1.000	000	000	000	000	000	000	000	000	0.999
2	0.999	999	999	999	999	999	999	999	999	998
3	.999	999	998	998	998	998	998	998	998	998
4	.998	997	997	997	997	997	997	997	996	996
5	0.996	996	996	996	996	995	995	995	995	995
6	.995	994	994	994	994	994	993	993	993	993
7	.993	992	992	992	992	991	991	991	991	991
8	.990	990	990	990	989	989	989	988	988	988
9	.988	987	987	987	987	986	986	986	985	985
10	0.985	985	984	984	984	983	983	983	982	982
11	.982	981	981	981	980	980	980	979	979	979
12	.978	978	977	977	977	976	976	976	975	975
13	.974	974	974	973	973	972	972	972	971	971
14	.970	970	969	969	969	968	968	967	967	966
15	0.966	965	965	965	964	964	963	963	962	962
16	.961	961	960	960	959	959	958	958	957	957
17	.956	956	955	955	954	954	953	953	952	952
18	.951	951	950	949	949	948	948	947	947	946
19	.946	945	944	944	943	943	942	941	941	940
20	0.940	939	938	938	937	937	936	935	935	934
21	.934	933	932	932	931	930	930	929	928	928
22	.927	927	926	925	925	924	923	923	922	921
23	.921	920	919	918	918	917	916	916	915	914
24	.914	913	912	911	911	910	909	909	908	907
25	0.906	906	905	904	903	903	902	901	900	900
26	.899	898	897	896	896	895	894	893	893	892
27	.891	890	889	888	888	887	886	885	885	884
28	.883	882	881	880	880	879	878	877	876	875
29	.875	874	873	872	871	870	869	869	868	867
30	0.866	865	864	863	863	862	861	860	859	858
31	.857	856	855	854	854	853	852	851	850	849
32	.848	847	846	845	844	843	842	842	841	840
33	.839	838	837	836	835	834	833	832	831	830
34	.829	828	827	826	825	824	823	822	821	820
35	0.819	818	817	816	815	814	813	812	811	810
36	.809	808	807	806	805	804	803	802	801	800
37	.799	798	797	795	794	793	792	791	790	789
38	.788	787	786	785	784	783	782	780	779	778
39	.777	776	775	774	773	772	771	769	768	767
40	0.766	765	764	763	762	760	759	758	757	756
41	.755	754	752	751	750	749	748	747	745	744
42	.743	742	741	740	738	737	736	735	734	733
43	.731	730	729	728	727	725	724	723	722	721
44	.719	718	717	716	714	713	712	711	710	708

θ → tan θ

θ	.0	.1	.2	.3	.4	.5	.6	.7	.8	.9
45	1.000	003	007	011	014	018	021	025	028	032
46	.036	039	043	046	050	054	057	061	065	069
47	.072	076	080	084	087	091	095	099	103	107
48	.111	115	118	122	126	130	134	138	142	146
49	.150	154	159	163	167	171	175	179	183	188
50	1.192	196	200	205	209	213	217	222	226	230
51	.235	239	244	248	253	257	262	266	271	275
52	.280	285	289	294	299	303	308	313	317	322
53	.327	332	337	342	347	351	356	361	366	371
54	.376	381	387	392	397	402	407	412	418	423
55	1.428	433	439	444	450	455	460	466	471	477
56	.483	488	494	499	505	511	517	522	528	534
57	.540	546	552	558	564	570	576	582	588	594
58	.600	607	613	619	625	632	638	645	651	658
59	.664	671	678	684	691	698	704	711	718	725
60	1.732	739	746	753	760	767	775	782	789	797
61	.804	811	819	827	834	842	849	857	865	873
62	.881	889	897	905	913	921	929	937	946	954
63	1.963	971	980	988	997	2.006	2.014	2.023	2.032	2.041
64	2.050	059	069	078	087	097	106	116	125	135
65	2.145	154	164	174	184	194	204	215	225	236
66	.246	257	267	278	289	300	311	322	333	344
67	.356	367	379	391	402	414	426	438	450	463
68	.475	488	500	513	526	539	552	565	578	592
69	.605	619	633	646	660	675	689	703	718	733
70	2.747	762	778	793	808	824	840	856	872	888
71	2.904	921	937	954	971	989	3.006	3.024	3.042	3.060
72	3.078	096	115	133	152	172	191	211	230	251
73	.271	291	312	333	354	376	398	420	442	465
74	.487	511	534	558	582	606	630	655	681	706
75	3.732	758	785	812	839	867	895	923	952	981
76	4.011	041	071	102	134	165	198	230	264	297
77	.331	366	402	437	474	511	548	586	625	665
78	.705	745	787	829	872	915	959	5.005	5.050	5.097
79	5.145	193	242	292	343	396	449	503	558	614
80	5.671	730	789	850	912	976	6.041	6.107	6.174	6.243
81	6.314	386	460	535	612	691	772	855	940	7.026
82	7.115	207	300	396	495	596	700	806	916	8.028
83	8.144	264	386	513	643	777	915	9.058	9.205	9.357
84	9.514	9.677	9.845	10.02	10.20	10.39	10.58	10.78	10.99	11.20
85	11.43	11.66	11.91	12.16	12.43	12.71	13.00	13.30	13.62	13.95
86	14.30	14.67	15.06	15.46	15.89	16.35	16.83	17.34	17.89	18.46
87	19.08	19.74	20.45	21.20	22.02	22.90	23.86	24.90	26.03	27.27
88	28.64	30.14	31.82	33.69	35.80	38.19	40.92	44.07	47.74	52.08
89	57.29	63.66	71.62	81.85	95.49	114.6	143.2	191.0	286.5	573.0

Tangents of angles

θ	.0	.1	.2	.3	.4	.5	.6	.7	.8	.9
0	0.000	002	003	005	007	009	010	012	014	016
1	.017	019	021	023	024	026	028	030	031	033
2	.035	037	038	040	042	044	045	047	049	051
3	.052	054	056	058	059	061	063	065	066	068
4	.070	072	073	075	077	079	080	082	084	086
5	0.087	089	091	093	095	096	098	100	102	103
6	.105	107	109	110	112	114	116	117	119	121
7	.123	125	126	128	130	132	133	135	137	139
8	.141	142	144	146	148	149	151	153	155	157
9	.158	160	162	164	166	167	169	171	173	175
10	0.176	178	180	182	184	185	187	189	191	193
11	.194	196	198	200	202	203	205	207	209	211
12	.213	214	216	218	220	222	224	225	227	229
13	.231	233	235	236	238	240	242	244	246	247
14	.249	251	253	255	257	259	260	262	264	266
15	0.268	270	272	274	275	277	279	281	283	285
16	.287	289	291	292	294	296	298	300	302	304
17	.306	308	310	311	313	315	317	319	321	323
18	.325	327	329	331	333	335	337	338	340	342
19	.344	346	348	350	352	354	356	358	360	362
20	0.364	366	368	370	372	374	376	378	380	382
21	.384	386	388	390	392	394	396	398	400	402
22	.404	406	408	410	412	414	416	418	420	422
23	.424	427	429	431	433	435	437	439	441	443
24	.445	447	449	452	454	456	458	460	462	464
25	0.466	468	471	473	475	477	479	481	483	486
26	.488	490	492	494	496	499	501	503	505	507
27	.510	512	514	516	518	521	523	525	527	529
28	.532	534	536	538	541	543	545	547	550	552
29	.554	557	559	561	563	566	568	570	573	575
30	0.577	580	582	584	587	589	591	594	596	598
31	.601	603	606	608	610	613	615	618	620	622
32	.625	627	630	632	635	637	640	642	644	647
33	.649	652	654	657	659	662	664	667	669	672
34	.675	677	680	682	685	687	690	692	695	698
35	0.700	703	705	708	711	713	716	719	721	724
36	.727	729	732	735	737	740	743	745	748	751
37	.754	756	759	762	765	767	770	773	776	778
38	.781	784	787	790	793	795	798	801	804	807
39	.810	813	816	818	821	824	827	830	833	836
40	0.839	842	845	848	851	854	857	860	863	866
41	.869	872	875	879	882	885	888	891	894	897
42	.900	904	907	910	913	916	920	923	926	929
43	.933	936	939	942	946	949	952	956	959	962
44	.966	969	972	976	979	983	986	990	993	997

θ → log sin θ

θ	.0	.1	.2	.3	.4	.5	.6	.7	.8	.9
45	1̄.849	850	851	852	852	853	854	855	855	856
46	.857	858	858	859	860	861	861	862	863	863
47	.864	865	866	866	867	868	868	869	870	870
48	.871	872	872	873	874	874	875	876	876	877
49	.878	878	879	880	880	881	882	882	883	884
50	1̄.884	885	886	886	887	887	888	889	889	890
51	.891	891	892	892	893	894	894	895	895	896
52	.897	897	898	898	899	899	900	901	901	902
53	.902	903	903	904	905	905	906	906	907	907
54	.908	909	909	910	910	911	911	912	912	913
55	1̄.913	914	914	915	915	916	917	917	918	918
56	.919	919	920	920	921	921	922	922	923	923
57	.924	924	925	925	926	926	927	927	927	928
58	.928	929	929	930	930	931	931	932	932	933
59	.933	934	934	934	935	935	936	936	937	937
60	1̄.938	938	938	939	939	940	940	941	941	941
61	.942	942	943	943	943	944	944	945	945	946
62	.946	946	947	947	948	948	948	949	949	949
63	.950	950	951	951	951	952	952	953	953	953
64	.954	954	954	955	955	955	956	956	957	957
65	1̄.957	958	958	958	959	959	959	960	960	960
66	.961	961	961	962	962	962	963	963	963	964
67	.964	964	965	965	965	966	966	966	967	967
68	.967	967	968	968	968	969	969	969	970	970
69	.970	970	971	971	971	972	972	972	972	973
70	1̄.973	973	974	974	974	974	975	975	975	975
71	.976	976	976	976	977	977	977	977	978	978
72	.978	978	979	979	979	979	980	980	980	980
73	.981	981	981	981	982	982	982	982	982	983
74	.983	983	983	983	984	984	984	984	985	985
75	1̄.985	985	985	986	986	986	986	986	987	987
76	.987	987	987	987	988	988	988	988	988	989
77	.989	989	989	989	989	990	990	990	990	990
78	.990	991	991	991	991	991	991	991	992	992
79	.992	992	992	992	993	993	993	993	993	993
80	1̄.993	993	994	994	994	994	994	994	994	994
81	.995	995	995	995	995	995	995	995	996	996
82	.996	996	996	996	996	996	996	996	997	997
83	.997	997	997	997	997	997	997	997	997	998
84	.998	998	998	998	998	998	998	998	998	998
85	1̄.998	998	998	999	999	999	999	999	999	999
86	1̄.999	999	999	999	999	999	999	999	999	999
87	1̄.999	999	999	0.000	0.000	0.000	0.000	0.000	0.000	0.000
88	0.000	0.000	0.000	0.000	0.000	0.000	0.000	0.000	0.000	0.000
89	0.000	0.000	0.000	0.000	0.000	0.000	0.000	0.000	0.000	0.000
90	0.000	0.000	0.000	0.000	0.000	0.000	0.000	0.000	0.000	0.000

Logarithms of sines

θ	.0	.1	.2	.3	.4	.5	.6	.7	.8	.9
0		3̄.242	3̄.543	3̄.719	3̄.844	3̄.941	2̄.020	2̄.087	2̄.145	2̄.196
1	2̄.242	283	321	356	388	418	446	472	497	521
2	.543	564	584	603	622	640	657	673	689	704
3	.719	733	747	760	773	786	798	810	821	833
4	.844	854	865	875	885	895	904	913	923	932
5	2̄.940	949	957	966	974	982	989	997	1̄.005	1̄.012
6	1̄.019	026	033	040	047	054	060	067	073	080
7	.086	092	098	104	110	116	121	127	133	138
8	.144	149	154	159	165	170	175	180	185	190
9	.194	199	204	208	213	218	222	227	231	235
10	1̄.240	244	248	252	257	261	265	269	273	277
11	.281	284	288	292	296	300	303	307	311	314
12	.318	321	325	328	332	335	339	342	345	349
13	.352	355	359	362	365	368	371	374	378	381
14	.384	387	390	393	396	399	402	404	407	410
15	1̄.413	416	419	421	424	427	430	432	435	438
16	.440	443	446	448	451	453	456	458	461	463
17	.466	468	471	473	476	478	481	483	485	488
18	.490	492	495	497	499	501	504	506	508	510
19	.513	515	517	519	521	523	526	528	530	532
20	1̄.534	536	538	540	542	544	546	548	550	552
21	.554	556	558	560	562	564	566	568	570	572
22	.574	575	577	579	581	583	585	586	588	590
23	.592	594	595	597	599	601	602	604	606	608
24	.609	611	613	614	616	618	619	621	623	624
25	1̄.626	628	629	631	632	634	636	637	639	640
26	.642	643	645	646	648	650	651	653	654	656
27	.657	659	660	661	663	664	666	667	669	670
28	.672	673	674	676	677	679	680	681	683	684
29	.686	687	688	690	691	692	694	695	696	698
30	1̄.699	700	702	703	704	705	707	708	709	711
31	.712	713	714	716	717	718	719	721	722	723
32	.724	725	727	728	729	730	731	733	734	735
33	.736	737	738	740	741	742	743	744	745	746
34	.748	749	750	751	752	753	754	755	756	758
35	1̄.759	760	761	762	763	764	765	766	767	768
36	.769	770	771	772	773	774	775	776	777	778
37	.779	780	781	782	783	784	785	786	787	788
38	.789	790	791	792	793	794	795	796	797	798
39	.799	800	801	802	803	804	804	805	806	807
40	1̄.808	809	810	811	812	813	813	814	815	816
41	.817	818	819	820	820	821	822	823	824	825
42	.826	826	827	828	829	830	831	831	832	833
43	.834	835	835	836	837	838	839	839	840	841
44	.842	843	843	844	845	846	846	847	848	849

θ → log cos θ

θ	.0	.1	.2	.3	.4	.5	.6	.7	.8	.9
45	1̄.849	849	848	847	846	846	845	844	843	843
46	.842	841	840	839	839	838	837	836	835	835
47	.834	833	832	831	831	830	829	828	827	826
48	.826	825	824	823	822	821	820	820	819	818
49	.817	816	815	814	813	813	812	811	810	809
50	1̄.808	807	806	805	804	804	803	802	801	800
51	.799	798	797	796	795	794	793	792	791	790
52	.789	788	787	786	785	784	783	782	781	780
53	.779	778	777	776	775	774	773	772	771	770
54	.769	768	767	766	765	764	763	762	761	760
55	1̄.759	758	756	755	754	753	752	751	750	749
56	.748	746	745	744	743	742	741	740	738	737
57	.736	735	734	733	731	730	729	728	727	725
58	.724	723	722	721	719	718	717	716	714	713
59	.712	711	709	708	707	705	704	703	702	700
60	1̄.699	698	696	695	694	692	691	690	688	687
61	.686	684	683	681	680	679	677	676	674	673
62	.672	670	669	667	666	664	663	661	660	659
63	.657	656	654	653	651	650	648	646	645	643
64	.642	640	639	637	636	634	632	631	629	628
65	1̄.626	624	623	621	619	618	616	614	613	611
66	.609	608	606	604	602	601	599	597	595	594
67	.592	590	588	586	585	583	581	579	577	575
68	.574	572	570	568	566	564	562	560	558	556
69	.554	552	550	548	546	544	542	540	538	536
70	1̄.534	532	530	528	526	523	521	519	517	515
71	.513	510	508	506	504	501	499	497	495	492
72	.490	488	485	483	481	478	476	473	471	468
73	.466	463	461	458	456	453	451	448	446	443
74	.440	437	435	432	430	427	424	421	419	416
75	1̄.413	410	407	404	402	399	396	393	390	387
76	.384	381	378	374	371	368	365	362	359	355
77	.352	349	345	342	339	335	332	328	325	321
78	.318	314	311	307	303	300	296	292	288	284
79	.281	277	273	269	265	261	257	252	248	244
80	1̄.240	235	231	227	222	218	213	208	204	199
81	.194	190	185	180	175	170	165	159	154	149
82	.144	138	133	127	121	116	110	104	098	092
83	.086	080	073	067	060	054	047	040	033	026
84	1̄.019	012	005	2̄.997	2̄.989	2̄.982	2̄.974	2̄.966	2̄.957	2̄.949
85	2̄.940	932	923	913	904	895	885	875	865	854
86	.844	833	821	810	798	786	773	760	747	733
87	.719	704	689	673	657	640	622	603	584	564
88	.543	521	497	472	446	418	388	356	321	283
89	2̄.242	196	145	087	020	3̄.941	3̄.844	3̄.719	3̄.543	3̄.242

Logarithms of cosines

θ	.0	.1	.2	.3	.4	.5	.6	.7	.8	.9
0	.0000	000	000	000	000	000	000	000	000	000
1	.000	000	000	000	000	000	000	000	000	000
2	0.000	000	000	000	000	000	000	000	1̄.999	1̄.999
3	1̄.999	999	999	999	999	999	999	999	999	999
4	.999	999	999	999	999	999	999	999	998	998
5	1̄.998	998	998	998	998	998	998	998	998	998
6	.998	998	997	997	997	997	997	997	997	997
7	.997	997	997	996	996	996	996	996	996	996
8	.996	996	996	995	995	995	995	995	995	995
9	.995	994	994	994	994	994	994	994	994	993
10	1̄.993	993	993	993	993	993	993	992	992	992
11	.992	992	992	991	991	991	991	991	991	991
12	.990	990	990	990	990	990	989	989	989	989
13	.989	989	988	988	988	988	988	987	987	987
14	.987	987	987	986	986	986	986	986	985	985
15	1̄.985	985	985	984	984	984	984	983	983	983
16	.983	983	982	982	982	982	982	981	981	981
17	.981	980	980	980	980	979	979	979	979	978
18	.978	978	978	977	977	977	977	976	976	976
19	.976	975	975	975	975	974	974	974	974	973
20	1̄.973	973	972	972	972	972	971	971	971	970
21	.970	970	970	969	969	969	968	968	968	967
22	.967	967	967	966	966	966	965	965	965	964
23	.964	964	963	963	963	962	962	962	961	961
24	.961	960	960	960	959	959	959	958	958	958
25	1̄.957	957	957	956	956	955	955	955	954	954
26	.954	953	953	953	952	952	951	951	951	950
27	.950	949	949	949	948	948	948	947	947	946
28	.946	946	945	945	944	944	943	943	943	942
29	.942	941	941	941	940	940	939	939	938	938
30	1̄.938	937	937	936	936	935	935	934	934	934
31	.933	933	932	932	931	931	930	930	929	929
32	.928	928	927	927	927	926	926	925	925	924
33	.924	923	923	922	922	921	921	920	920	919
34	.919	918	918	917	917	916	915	915	914	914
35	1̄.913	913	912	912	911	911	910	910	909	909
36	.908	907	907	906	906	905	905	904	903	903
37	.902	902	901	901	900	899	899	898	898	897
38	.897	896	895	895	894	894	893	892	892	891
39	.891	890	889	889	888	887	887	886	886	885
40	1̄.884	884	883	882	882	881	880	880	879	878
41	.878	877	876	876	875	874	874	873	872	872
42	.871	870	870	869	868	868	867	866	866	865
43	.864	863	863	862	861	861	860	859	858	858
44	.857	856	855	855	854	853	852	852	851	850

$\theta \rightarrow \log \tan \theta$

θ	.0	.1	.2	.3	.4	.5	.6	.7	.8	.9
45	0.000	002	003	005	006	008	009	011	012	014
46	.015	017	018	020	021	023	024	026	027	029
47	.030	032	033	035	036	038	039	041	043	044
48	.046	047	049	050	052	053	055	056	058	059
49	.061	062	064	065	067	069	070	072	073	075
50	0.076	078	079	081	082	084	085	087	089	090
51	.092	093	095	096	098	099	101	103	104	106
52	.107	109	110	112	113	115	117	118	120	121
53	.123	124	126	128	129	131	132	134	136	137
54	.139	140	142	144	145	147	148	150	152	153
55	0.155	156	158	160	161	163	164	166	168	169
56	.171	173	174	176	178	179	181	183	184	186
57	.187	189	191	192	194	196	197	199	201	203
58	.204	206	208	209	211	213	214	216	218	220
59	.221	223	225	226	228	230	232	233	235	237
60	0.239	240	242	244	246	247	249	251	253	254
61	.256	258	260	262	263	265	267	269	271	272
62	.274	276	278	280	282	284	285	287	289	291
63	.293	295	297	298	300	302	304	306	308	310
64	.312	314	316	318	320	322	323	325	327	329
65	0.331	333	335	337	339	341	343	345	347	349
66	.351	353	356	358	360	362	364	366	368	370
67	.372	374	376	379	381	383	385	387	389	391
68	.394	396	398	400	402	405	407	409	411	414
69	.416	418	420	423	425	427	430	432	434	437
70	0.439	441	444	446	448	451	453	456	458	461
71	.463	465	468	470	473	475	478	481	483	486
72	.488	491	493	496	499	501	504	507	509	512
73	.515	517	520	523	526	528	531	534	537	540
74	.543	545	548	551	554	557	560	563	566	569
75	0.572	575	578	581	584	587	590	594	597	600
76	.603	606	610	613	616	620	623	626	630	633
77	.637	640	644	647	651	654	658	661	665	669
78	.673	676	680	684	688	692	695	699	703	707
79	.711	715	720	724	728	732	736	741	745	749
80	0.754	758	763	767	772	776	781	786	791	795
81	.800	805	810	815	820	826	831	836	841	847
82	.852	858	863	869	875	881	886	892	898	905
83	.911	917	924	930	937	943	950	957	964	971
84	0.978	986	993	1.001	1.009	1.016	1.024	1.033	1.041	1.049
85	1.058	067	076	085	094	104	114	124	134	145
86	.155	166	178	189	201	214	226	239	253	266
87	.281	295	311	326	343	360	378	396	415	436
88	.457	479	503	528	554	582	612	644	679	717
89	1.758	804	855	913	980	2.059	2.156	2.281	2.457	2.758

Logarithms of tangents

θ	.0	.1	.2	.3	.4	.5	.6	.7	.8	.9
0		$\overline{3}.242$	$\overline{3}.543$	$\overline{3}.719$	$\overline{3}.844$	$\overline{3}.941$	$\overline{2}.020$	$\overline{2}.087$	$\overline{2}.145$	$\overline{2}.196$
1	$\overline{2}.242$	283	321	356	388	418	446	472	497	521
2	.543	564	585	604	622	640	657	674	689	705
3	.719	734	747	761	774	786	799	811	822	834
4	.845	855	866	876	886	896	906	915	924	933
5	$\overline{2}.942$	951	959	967	976	984	991	999	007	014
6	$\overline{1}.022$	029	036	043	050	057	063	070	076	083
7	.089	095	102	108	114	119	125	131	137	142
8	.148	153	159	164	169	174	180	185	190	195
9	.200	205	209	214	219	224	228	233	237	242
10	$\overline{1}.246$	251	255	259	264	268	272	276	280	285
11	.289	293	297	301	305	308	312	316	320	324
12	.327	331	335	339	342	346	349	353	356	360
13	.363	367	370	374	377	380	384	387	390	394
14	.397	400	403	406	410	413	416	419	422	425
15	$\overline{1}.428$	431	434	437	440	443	446	449	452	455
16	.457	460	463	466	469	472	474	477	480	483
17	.485	488	491	493	496	499	501	504	507	509
18	.512	514	517	519	522	525	527	530	532	535
19	.537	539	542	544	547	549	552	554	556	559
20	$\overline{1}.561$	563	566	568	570	573	575	577	580	582
21	.584	586	589	591	593	595	598	600	602	604
22	.606	609	611	613	615	617	619	621	624	626
23	.628	630	632	634	636	638	640	642	644	647
24	.649	651	653	655	657	659	661	663	665	667
25	$\overline{1}.669$	671	673	675	677	678	680	682	684	686
26	.688	690	692	694	696	698	700	702	703	705
27	.707	709	711	713	715	716	718	720	722	724
28	.726	728	729	731	733	735	737	738	740	742
29	.744	746	747	749	751	753	754	756	758	760
30	$\overline{1}.761$	763	765	767	768	770	772	774	775	777
31	.779	780	782	784	786	787	789	791	792	794
32	.796	797	799	801	803	804	806	808	809	811
33	.813	814	816	817	819	821	822	824	826	827
34	.829	831	832	834	836	837	839	840	842	844
35	$\overline{1}.845$	847	848	850	852	853	855	856	858	860
36	.861	863	864	866	868	869	871	872	874	876
37	.877	879	880	882	883	885	887	888	890	891
38	.893	894	896	897	899	901	902	904	905	907
39	.908	910	911	913	915	916	918	919	921	922
40	$\overline{1}.924$	925	927	928	930	931	933	935	936	938
41	.939	941	942	944	945	947	948	950	951	953
42	.954	956	957	959	961	962	964	965	967	968
43	.970	971	973	974	976	977	979	980	982	983
44	.985	986	988	989	991	992	994	995	997	998

$x \rightarrow x^2$

x	0	1	2	3	4	5	6	7	8	9
5.5	30.25	30.36	30.47	30.58	30.69	30.80	30.91	31.02	31.14	31.25
5.6	31.36	31.47	31.58	31.70	31.81	31.92	32.04	32.15	32.26	32.38
5.7	32.49	32.60	32.72	32.83	32.95	33.06	33.18	33.29	33.41	33.52
5.8	33.64	33.76	33.87	33.99	34.11	34.22	34.34	34.46	34.57	34.69
5.9	34.81	34.93	35.05	35.16	35.28	35.40	35.52	35.64	35.76	35.88
6.0	36.00	36.12	36.24	36.36	36.48	36.60	36.72	36.84	36.97	37.09
6.1	37.21	37.33	37.45	37.58	37.70	37.82	37.95	38.07	38.19	38.32
6.2	38.44	38.56	38.69	38.81	38.94	39.06	39.19	39.31	39.44	39.56
6.3	39.69	39.82	39.94	40.07	40.20	40.32	40.45	40.58	40.70	40.83
6.4	40.96	41.09	41.22	41.34	41.47	41.60	41.73	41.86	41.99	42.12
6.5	42.25	42.38	42.51	42.64	42.77	42.90	43.03	43.16	43.30	43.43
6.6	43.56	43.69	43.82	43.96	44.09	44.22	44.36	44.49	44.62	44.76
6.7	44.89	45.02	45.16	45.29	45.43	45.56	45.70	45.83	45.97	46.10
6.8	46.24	46.38	46.51	46.65	46.79	46.92	47.06	47.20	47.33	47.47
6.9	47.61	47.75	47.89	48.02	48.16	48.30	48.44	48.58	48.72	48.86
7.0	49.00	49.14	49.28	49.42	49.56	49.70	49.84	49.98	50.13	50.27
7.1	50.41	50.55	50.69	50.84	50.98	51.12	51.27	51.41	51.55	51.70
7.2	51.84	51.98	52.13	52.27	52.42	52.56	52.71	52.85	53.00	53.14
7.3	53.29	53.44	53.58	53.73	53.88	54.02	54.17	54.32	54.46	54.61
7.4	54.76	54.91	55.06	55.20	55.35	55.50	55.65	55.80	55.95	56.10
7.5	56.25	56.40	56.55	56.70	56.85	57.00	57.15	57.30	57.46	57.61
7.6	57.76	57.91	58.06	58.22	58.37	58.52	58.68	58.83	58.98	59.14
7.7	59.29	59.44	59.60	59.75	59.91	60.06	60.22	60.37	60.53	60.68
7.8	60.84	61.00	61.15	61.31	61.47	61.62	61.78	61.94	62.09	62.25
7.9	62.41	62.57	62.73	62.88	63.04	63.20	63.36	63.52	63.68	63.84
8.0	64.00	64.16	64.32	64.48	64.64	64.80	64.96	65.12	65.29	65.45
8.1	65.61	65.77	65.93	66.10	66.26	66.42	66.59	66.75	66.91	67.08
8.2	67.24	67.40	67.57	67.73	67.90	68.06	68.23	68.39	68.56	68.72
8.3	68.89	69.06	69.22	69.39	69.56	69.72	69.89	70.06	70.22	70.39
8.4	70.56	70.73	70.90	71.06	71.23	71.40	71.57	71.74	71.91	72.08
8.5	72.25	72.42	72.59	72.76	72.93	73.10	73.27	73.44	73.62	73.79
8.6	73.96	74.13	74.30	74.48	74.65	74.82	75.00	75.17	75.34	75.52
8.7	75.69	75.86	76.04	76.21	76.39	76.56	76.74	76.91	77.09	77.26
8.8	77.44	77.62	77.79	77.97	78.15	78.32	78.50	78.68	78.85	79.03
8.9	79.21	79.39	79.57	79.74	79.92	80.10	80.28	80.46	80.64	80.82
9.0	81.00	81.18	81.36	81.54	81.72	81.90	82.08	82.26	82.45	82.63
9.1	82.81	82.99	83.17	83.36	83.54	83.72	83.91	84.09	84.27	84.46
9.2	84.64	84.82	85.01	85.19	85.38	85.56	85.75	85.93	86.12	86.30
9.3	86.49	86.68	86.86	87.05	87.24	87.42	87.61	87.80	87.98	88.17
9.4	88.36	88.55	88.74	88.92	89.11	89.30	89.49	89.68	89.87	90.06
9.5	90.25	90.44	90.63	90.82	91.01	91.20	91.39	91.58	91.78	91.97
9.6	92.16	92.35	92.54	92.74	92.93	93.12	93.32	93.51	93.70	93.90
9.7	94.09	94.28	94.48	94.67	94.87	95.06	95.26	95.45	95.65	95.84
9.8	96.04	96.24	96.43	96.63	96.83	97.02	97.22	97.42	97.61	97.81
9.9	98.01	98.21	98.41	98.60	98.80	99.00	99.20	99.40	99.60	99.80

Squares

x	0	1	2	3	4	5	6	7	8	9
1.0	1.00	1.02	1.04	1.06	1.08	1.10	1.12	1.14	1.17	1.19
1.1	1.21	1.23	1.25	1.28	1.30	1.32	1.35	1.37	1.39	1.42
1.2	1.44	1.46	1.49	1.51	1.54	1.56	1.59	1.61	1.64	1.66
1.3	1.69	1.72	1.74	1.77	1.80	1.82	1.85	1.88	1.90	1.93
1.4	1.96	1.99	2.02	2.04	2.07	2.10	2.13	2.16	2.19	2.22
1.5	2.25	2.28	2.31	2.34	2.37	2.40	2.43	2.46	2.50	2.53
1.6	2.56	2.59	2.62	2.66	2.69	2.72	2.76	2.79	2.82	2.86
1.7	2.89	2.92	2.96	2.99	3.03	3.06	3.10	3.13	3.17	3.20
1.8	3.24	3.28	3.31	3.35	3.39	3.42	3.46	3.50	3.53	3.57
1.9	3.61	3.65	3.69	3.72	3.76	3.80	3.84	3.88	3.92	3.96
2.0	4.00	4.04	4.08	4.12	4.16	4.20	4.24	4.28	4.33	4.37
2.1	4.41	4.45	4.49	4.54	4.58	4.62	4.67	4.71	4.75	4.80
2.2	4.84	4.88	4.93	4.97	5.02	5.06	5.11	5.15	5.20	5.24
2.3	5.29	5.34	5.38	5.43	5.48	5.52	5.57	5.62	5.66	5.71
2.4	5.76	5.81	5.86	5.90	5.95	6.00	6.05	6.10	6.15	6.20
2.5	6.25	6.30	6.35	6.40	6.45	6.50	6.55	6.60	6.66	6.71
2.6	6.76	6.81	6.86	6.92	6.97	7.02	7.08	7.13	7.18	7.24
2.7	7.29	7.34	7.40	7.45	7.51	7.56	7.62	7.67	7.73	7.78
2.8	7.84	7.90	7.95	8.01	8.07	8.12	8.18	8.24	8.29	8.35
2.9	8.41	8.47	8.53	8.58	8.64	8.70	8.76	8.82	8.88	8.94
3.0	9.00	9.06	9.12	9.18	9.24	9.30	9.36	9.42	9.49	9.55
3.1	9.61	9.67	9.73	9.80	9.86	9.92	9.99	10.05	10.11	10.18
3.2	10.24	10.30	10.37	10.43	10.50	10.56	10.63	10.69	10.76	10.82
3.3	10.89	10.96	11.02	11.09	11.16	11.22	11.29	11.36	11.42	11.49
3.4	11.56	11.63	11.70	11.76	11.83	11.90	11.97	12.04	12.11	12.18
3.5	12.25	12.32	12.39	12.46	12.53	12.60	12.67	12.74	12.82	12.89
3.6	12.96	13.03	13.10	13.18	13.25	13.32	13.40	13.47	13.54	13.62
3.7	13.69	13.76	13.84	13.91	13.99	14.06	14.14	14.21	14.29	14.36
3.8	14.44	14.52	14.59	14.67	14.75	14.82	14.90	14.98	15.05	15.13
3.9	15.21	15.29	15.37	15.44	15.52	15.60	15.68	15.76	15.84	15.92
4.0	16.00	16.08	16.16	16.24	16.32	16.40	16.48	16.56	16.65	16.73
4.1	16.81	16.89	16.97	17.06	17.14	17.22	17.31	17.39	17.47	17.56
4.2	17.64	17.72	17.81	17.89	17.98	18.06	18.15	18.23	18.32	18.40
4.3	18.49	18.58	18.66	18.75	18.84	18.92	19.01	19.10	19.18	19.27
4.4	19.36	19.45	19.54	19.62	19.71	19.80	19.89	19.98	20.07	20.16
4.5	20.25	20.34	20.43	20.52	20.61	20.70	20.79	20.88	20.98	21.07
4.6	21.16	21.25	21.34	21.44	21.53	21.62	21.72	21.81	21.90	22.00
4.7	22.09	22.18	22.28	22.37	22.47	22.56	22.66	22.75	22.85	22.94
4.8	23.04	23.14	23.23	23.33	23.43	23.52	23.62	23.72	23.81	23.91
4.9	24.01	24.11	24.21	24.30	24.40	24.50	24.60	24.70	24.80	24.90
5.0	25.00	25.10	25.20	25.30	25.40	25.50	25.60	25.70	25.81	25.91
5.1	26.01	26.11	26.21	26.32	26.42	26.52	26.63	26.73	26.83	26.94
5.2	27.04	27.14	27.25	27.35	27.46	27.56	27.67	27.77	27.88	27.98
5.3	28.09	28.20	28.30	28.41	28.52	28.62	28.73	28.84	28.94	29.05
5.4	29.16	29.27	29.38	29.48	29.59	29.70	29.81	29.92	30.03	30.14

$x \rightarrow \sqrt{x}$

x	0	1	2	3	4	5	6	7	8	9
5.5	2.35	2.35	2.35	2.35	2.35	2.36	2.36	2.36	2.36	2.36
5.6	2.37	2.37	2.37	2.37	2.37	2.38	2.38	2.38	2.38	2.39
5.7	2.39	2.39	2.39	2.39	2.40	2.40	2.40	2.40	2.40	2.41
5.8	2.41	2.41	2.41	2.41	2.42	2.42	2.42	2.42	2.42	2.43
5.9	2.43	2.43	2.43	2.44	2.44	2.44	2.44	2.44	2.45	2.45
6.0	2.45	2.45	2.45	2.46	2.46	2.46	2.46	2.46	2.47	2.47
6.1	2.47	2.47	2.47	2.48	2.48	2.48	2.48	2.48	2.49	2.49
6.2	2.49	2.49	2.49	2.50	2.50	2.50	2.50	2.50	2.51	2.51
6.3	2.51	2.51	2.51	2.52	2.52	2.52	2.52	2.52	2.53	2.53
6.4	2.53	2.53	2.53	2.54	2.54	2.54	2.54	2.54	2.55	2.55
6.5	2.55	2.55	2.55	2.56	2.56	2.56	2.56	2.56	2.57	2.57
6.6	2.57	2.57	2.57	2.57	2.58	2.58	2.58	2.58	2.58	2.59
6.7	2.59	2.59	2.59	2.59	2.60	2.60	2.60	2.60	2.60	2.61
6.8	2.61	2.61	2.61	2.61	2.62	2.62	2.62	2.62	2.62	2.62
6.9	2.63	2.63	2.63	2.63	2.63	2.64	2.64	2.64	2.64	2.64
7.0	2.65	2.65	2.65	2.65	2.65	2.66	2.66	2.66	2.66	2.66
7.1	2.66	2.67	2.67	2.67	2.67	2.67	2.68	2.68	2.68	2.68
7.2	2.68	2.69	2.69	2.69	2.69	2.69	2.69	2.70	2.70	2.70
7.3	2.70	2.70	2.71	2.71	2.71	2.71	2.71	2.71	2.72	2.72
7.4	2.72	2.72	2.72	2.73	2.73	2.73	2.73	2.73	2.73	2.74
7.5	2.74	2.74	2.74	2.74	2.75	2.75	2.75	2.75	2.75	2.75
7.6	2.76	2.76	2.76	2.76	2.76	2.77	2.77	2.77	2.77	2.77
7.7	2.77	2.78	2.78	2.78	2.78	2.78	2.79	2.79	2.79	2.79
7.8	2.79	2.79	2.80	2.80	2.80	2.80	2.80	2.81	2.81	2.81
7.9	2.81	2.81	2.81	2.82	2.82	2.82	2.82	2.82	2.82	2.83
8.0	2.83	2.83	2.83	2.83	2.84	2.84	2.84	2.84	2.84	2.84
8.1	2.85	2.85	2.85	2.85	2.85	2.85	2.86	2.86	2.86	2.86
8.2	2.86	2.87	2.87	2.87	2.87	2.87	2.87	2.88	2.88	2.88
8.3	2.88	2.88	2.88	2.89	2.89	2.89	2.89	2.89	2.89	2.90
8.4	2.90	2.90	2.90	2.90	2.91	2.91	2.91	2.91	2.91	2.91
8.5	2.92	2.92	2.92	2.92	2.92	2.92	2.93	2.93	2.93	2.93
8.6	2.93	2.93	2.94	2.94	2.94	2.94	2.94	2.94	2.95	2.95
8.7	2.95	2.95	2.95	2.95	2.96	2.96	2.96	2.96	2.96	2.96
8.8	2.97	2.97	2.97	2.97	2.97	2.97	2.98	2.98	2.98	2.98
8.9	2.98	2.98	2.99	2.99	2.99	2.99	2.99	2.99	3.00	3.00
9.0	3.00	3.00	3.00	3.00	3.01	3.01	3.01	3.01	3.01	3.01
9.1	3.02	3.02	3.02	3.02	3.02	3.02	3.03	3.03	3.03	3.03
9.2	3.03	3.03	3.04	3.04	3.04	3.04	3.04	3.04	3.05	3.05
9.3	3.05	3.05	3.05	3.05	3.06	3.06	3.06	3.06	3.06	3.06
9.4	3.07	3.07	3.07	3.07	3.07	3.07	3.08	3.08	3.08	3.08
9.5	3.08	3.08	3.09	3.09	3.09	3.09	3.09	3.09	3.10	3.10
9.6	3.10	3.10	3.10	3.10	3.10	3.11	3.11	3.11	3.11	3.11
9.7	3.11	3.12	3.12	3.12	3.12	3.12	3.12	3.13	3.13	3.13
9.8	3.13	3.13	3.13	3.14	3.14	3.14	3.14	3.14	3.14	3.14
9.9	3.15	3.15	3.15	3.15	3.15	3.15	3.16	3.16	3.16	3.16

Square roots from 1 to 10

x	0	1	2	3	4	5	6	7	8	9
1.0	1.00	1.00	1.01	1.01	1.02	1.02	1.03	1.03	1.04	1.04
1.1	1.05	1.05	1.06	1.06	1.07	1.07	1.08	1.08	1.09	1.09
1.2	1.10	1.10	1.10	1.11	1.11	1.12	1.12	1.13	1.13	1.14
1.3	1.14	1.14	1.15	1.15	1.16	1.16	1.17	1.17	1.17	1.18
1.4	1.18	1.19	1.19	1.20	1.20	1.20	1.21	1.21	1.22	1.22
1.5	1.22	1.23	1.23	1.24	1.24	1.24	1.25	1.25	1.26	1.26
1.6	1.26	1.27	1.27	1.28	1.28	1.28	1.29	1.29	1.30	1.30
1.7	1.30	1.31	1.31	1.32	1.32	1.32	1.33	1.33	1.33	1.34
1.8	1.34	1.35	1.35	1.35	1.36	1.36	1.36	1.37	1.37	1.37
1.9	1.38	1.38	1.39	1.39	1.39	1.40	1.40	1.40	1.41	1.41
2.0	1.41	1.42	1.42	1.42	1.43	1.43	1.44	1.44	1.44	1.45
2.1	1.45	1.45	1.46	1.46	1.46	1.47	1.47	1.47	1.48	1.48
2.2	1.48	1.49	1.49	1.49	1.50	1.50	1.50	1.51	1.51	1.51
2.3	1.52	1.52	1.52	1.53	1.53	1.53	1.54	1.54	1.54	1.55
2.4	1.55	1.55	1.56	1.56	1.56	1.57	1.57	1.57	1.57	1.58
2.5	1.58	1.58	1.59	1.59	1.59	1.60	1.60	1.60	1.61	1.61
2.6	1.61	1.62	1.62	1.62	1.62	1.63	1.63	1.63	1.64	1.64
2.7	1.64	1.65	1.65	1.65	1.66	1.66	1.66	1.66	1.67	1.67
2.8	1.67	1.68	1.68	1.68	1.69	1.69	1.69	1.69	1.70	1.70
2.9	1.70	1.71	1.71	1.71	1.71	1.72	1.72	1.72	1.73	1.73
3.0	1.73	1.73	1.74	1.74	1.74	1.75	1.75	1.75	1.75	1.76
3.1	1.76	1.76	1.77	1.77	1.77	1.77	1.78	1.78	1.78	1.79
3.2	1.79	1.79	1.79	1.80	1.80	1.80	1.81	1.81	1.81	1.81
3.3	1.82	1.82	1.82	1.82	1.83	1.83	1.83	1.84	1.84	1.84
3.4	1.84	1.85	1.85	1.85	1.85	1.86	1.86	1.86	1.87	1.87
3.5	1.87	1.87	1.88	1.88	1.88	1.88	1.89	1.89	1.89	1.89
3.6	1.90	1.90	1.90	1.91	1.91	1.91	1.91	1.92	1.92	1.92
3.7	1.92	1.93	1.93	1.93	1.93	1.94	1.94	1.94	1.94	1.95
3.8	1.95	1.95	1.95	1.96	1.96	1.96	1.96	1.97	1.97	1.97
3.9	1.97	1.98	1.98	1.98	1.98	1.99	1.99	1.99	1.99	2.00
4.0	2.00	2.00	2.00	2.01	2.01	2.01	2.01	2.02	2.02	2.02
4.1	2.02	2.03	2.03	2.03	2.03	2.04	2.04	2.04	2.04	2.05
4.2	2.05	2.05	2.05	2.06	2.06	2.06	2.06	2.07	2.07	2.07
4.3	2.07	2.08	2.08	2.08	2.08	2.09	2.09	2.09	2.09	2.10
4.4	2.10	2.10	2.10	2.10	2.11	2.11	2.11	2.11	2.12	2.12
4.5	2.12	2.12	2.13	2.13	2.13	2.13	2.14	2.14	2.14	2.14
4.6	2.14	2.15	2.15	2.15	2.15	2.16	2.16	2.16	2.16	2.17
4.7	2.17	2.17	2.17	2.17	2.18	2.18	2.18	2.18	2.19	2.19
4.8	2.19	2.19	2.20	2.20	2.20	2.20	2.20	2.21	2.21	2.21
4.9	2.21	2.22	2.22	2.22	2.22	2.22	2.23	2.23	2.23	2.23
5.0	2.24	2.24	2.24	2.24	2.24	2.25	2.25	2.25	2.25	2.26
5.1	2.26	2.26	2.26	2.26	2.27	2.27	2.27	2.27	2.28	2.28
5.2	2.28	2.28	2.28	2.29	2.29	2.29	2.29	2.30	2.30	2.30
5.3	2.30	2.30	2.31	2.31	2.31	2.31	2.32	2.32	2.32	2.32
5.4	2.32	2.33	2.33	2.33	2.33	2.33	2.34	2.34	2.34	2.34

$x \rightarrow \sqrt{x}$

x	.0	.1	.2	.3	.4	.5	.6	.7	.8	.9
55	7.42	7.42	7.42	7.44	7.44	7.45	7.46	7.46	7.47	7.48
56	7.48	7.49	7.50	7.50	7.51	7.52	7.52	7.53	7.54	7.54
57	7.55	7.56	7.56	7.57	7.58	7.58	7.59	7.60	7.60	7.61
58	7.62	7.62	7.63	7.64	7.64	7.65	7.66	7.66	7.67	7.67
59	7.68	7.69	7.69	7.70	7.71	7.71	7.72	7.73	7.73	7.74
60	7.75	7.75	7.76	7.77	7.77	7.78	7.78	7.79	7.80	7.80
61	7.81	7.82	7.82	7.83	7.84	7.84	7.85	7.85	7.86	7.87
62	7.87	7.88	7.89	7.89	7.90	7.91	7.91	7.92	7.92	7.93
63	7.94	7.94	7.95	7.96	7.96	7.97	7.97	7.98	7.99	7.99
64	8.00	8.01	8.01	8.02	8.02	8.03	8.04	8.04	8.05	8.06
65	8.06	8.07	8.07	8.08	8.09	8.09	8.10	8.11	8.11	8.12
66	8.12	8.13	8.14	8.14	8.15	8.15	8.16	8.17	8.17	8.18
67	8.19	8.19	8.20	8.20	8.21	8.22	8.22	8.23	8.23	8.24
68	8.25	8.25	8.26	8.26	8.27	8.28	8.28	8.29	8.29	8.30
69	8.31	8.31	8.32	8.32	8.33	8.34	8.34	8.35	8.35	8.36
70	8.37	8.37	8.38	8.38	8.39	8.40	8.40	8.41	8.41	8.42
71	8.43	8.43	8.44	8.44	8.45	8.46	8.46	8.47	8.47	8.48
72	8.49	8.49	8.50	8.50	8.51	8.51	8.52	8.53	8.53	8.54
73	8.54	8.55	8.56	8.56	8.57	8.57	8.58	8.58	8.59	8.60
74	8.60	8.61	8.61	8.62	8.63	8.63	8.64	8.64	8.65	8.65
75	8.66	8.67	8.67	8.68	8.68	8.69	8.69	8.70	8.71	8.71
76	8.72	8.72	8.73	8.73	8.74	8.75	8.75	8.76	8.76	8.77
77	8.77	8.78	8.79	8.79	8.80	8.80	8.81	8.81	8.82	8.83
78	8.83	8.84	8.84	8.85	8.85	8.86	8.87	8.87	8.88	8.88
79	8.89	8.89	8.90	8.91	8.91	8.92	8.92	8.93	8.93	8.94
80	8.94	8.95	8.96	8.96	8.97	8.97	8.98	8.98	8.99	8.99
81	9.00	9.01	9.01	9.02	9.02	9.03	9.03	9.04	9.04	9.05
82	9.06	9.06	9.07	9.07	9.08	9.08	9.09	9.09	9.10	9.10
83	9.11	9.12	9.12	9.13	9.13	9.14	9.14	9.15	9.15	9.16
84	9.17	9.17	9.18	9.18	9.19	9.19	9.20	9.20	9.21	9.21
85	9.22	9.22	9.23	9.24	9.24	9.25	9.25	9.26	9.26	9.27
86	9.27	9.28	9.28	9.29	9.30	9.30	9.31	9.31	9.32	9.32
87	9.33	9.33	9.34	9.34	9.35	9.35	9.36	9.36	9.37	9.38
88	9.38	9.39	9.39	9.40	9.40	9.41	9.41	9.42	9.42	9.43
89	9.43	9.44	9.44	9.45	9.46	9.46	9.47	9.47	9.48	9.48
90	9.49	9.49	9.50	9.50	9.51	9.51	9.52	9.52	9.53	9.53
91	9.54	9.54	9.55	9.56	9.56	9.57	9.57	9.58	9.58	9.59
92	9.59	9.60	9.60	9.61	9.61	9.62	9.62	9.63	9.63	9.64
93	9.64	9.65	9.65	9.66	9.66	9.67	9.67	9.68	9.69	9.69
94	9.70	9.70	9.71	9.71	9.72	9.72	9.73	9.73	9.74	9.74
95	9.75	9.75	9.76	9.76	9.77	9.77	9.78	9.78	9.79	9.79
96	9.80	9.80	9.81	9.81	9.82	9.82	9.83	9.83	9.84	9.84
97	9.85	9.85	9.86	9.86	9.87	9.87	9.88	9.88	9.89	9.89
98	9.90	9.90	9.91	9.91	9.92	9.92	9.93	9.93	9.94	9.94
99	9.95	9.95	9.96	9.96	9.97	9.97	9.98.	9.98	9.99	9.99

Square roots from 10 to 100

x	.0	.1	.2	.3	.4	.5	.6	.7	.8	.9
10	3.16	3.18	3.19	3.21	3.22	3.24	3.26	3.27	3.29	3.30
11	3.32	3.33	3.35	3.36	3.38	3.39	3.41	3.42	3.44	3.45
12	3.46	3.48	3.49	3.51	3.52	3.54	3.55	3.56	3.58	3.59
13	3.61	3.62	3.63	3.65	3.66	3.67	3.69	3.70	3.71	3.73
14	3.74	3.75	3.77	3.78	3.79	3.81	3.82	3.83	3.85	3.86
15	3.87	3.89	3.90	3.91	3.92	3.94	3.95	3.96	3.97	3.99
16	4.00	4.01	4.02	4.04	4.05	4.06	4.07	4.09	4.10	4.11
17	4.12	4.14	4.15	4.16	4.17	4.18	4.20	4.21	4.22	4.23
18	4.24	4.25	4.27	4.28	4.29	4.30	4.31	4.32	4.34	4.35
19	4.36	4.37	4.38	4.39	4.40	4.42	4.43	4.44	4.45	4.46
20	4.47	4.48	4.49	4.51	4.52	4.53	4.54	4.55	4.56	4.57
21	4.58	4.59	4.60	4.62	4.63	4.64	4.65	4.66	4.67	4.68
22	4.69	4.70	4.71	4.72	4.73	4.74	4.75	4.76	4.77	4.79
23	4.80	4.81	4.82	4.83	4.84	4.85	4.86	4.87	4.88	4.89
24	4.90	4.91	4.92	4.93	4.94	4.95	4.96	4.97	4.98	4.99
25	5.00	5.01	5.02	5.03	5.04	5.05	5.06	5.07	5.08	5.09
26	5.10	5.11	5.12	5.13	5.14	5.15	5.16	5.17	5.18	5.19
27	5.20	5.21	5.22	5.22	5.23	5.24	5.25	5.26	5.27	5.28
28	5.29	5.30	5.31	5.32	5.33	5.34	5.35	5.36	5.37	5.38
29	5.39	5.39	5.40	5.41	5.42	5.43	5.44	5.45	5.46	5.47
30	5.48	5.49	5.50	5.50	5.51	5.52	5.53	5.54	5.55	5.56
31	5.57	5.58	5.59	5.59	5.60	5.61	5.62	5.63	5.64	5.65
32	5.66	5.67	5.67	5.68	5.69	5.70	5.71	5.72	5.73	5.74
33	5.74	5.75	5.76	5.77	5.78	5.79	5.80	5.81	5.81	5.82
34	5.83	5.84	5.85	5.86	5.87	5.87	5.88	5.89	5.90	5.91
35	5.92	5.92	5.93	5.94	5.95	5.96	5.97	5.97	5.98	5.99
36	6.00	6.01	6.02	6.02	6.03	6.04	6.05	6.06	6.07	6.07
37	6.08	6.09	6.10	6.11	6.12	6.12	6.13	6.14	6.15	6.16
38	6.16	6.17	6.18	6.19	6.20	6.20	6.21	6.22	6.23	6.24
39	6.24	6.25	6.26	6.27	6.28	6.28	6.29	6.30	6.31	6.32
40	6.32	6.33	6.34	6.35	6.36	6.36	6.37	6.38	6.39	6.40
41	6.40	6.41	6.42	6.43	6.43	6.44	6.45	6.46	6.47	6.47
42	6.48	6.49	6.50	6.50	6.51	6.52	6.53	6.53	6.54	6.55
43	6.56	6.57	6.57	6.58	6.59	6.60	6.60	6.61	6.62	6.63
44	6.63	6.64	6.65	6.66	6.66	6.67	6.68	6.69	6.69	6.70
45	6.71	6.72	6.72	6.73	6.74	6.75	6.75	6.76	6.77	6.77
46	6.78	6.79	6.80	6.80	6.81	6.82	6.83	6.83	6.84	6.85
47	6.86	6.86	6.87	6.88	6.88	6.89	6.90	6.91	6.91	6.92
48	6.93	6.94	6.94	6.95	6.96	6.96	6.97	6.98	6.99	6.99
49	7.00	7.01	7.01	7.02	7.03	7.04	7.04	7.05	7.06	7.06
50	7.07	7.08	7.09	7.09	7.10	7.11	7.11	7.12	7.13	7.13
51	7.14	7.15	7.16	7.16	7.17	7.18	7.18	7.19	7.20	7.20
52	7.21	7.22	7.22	7.23	7.24	7.25	7.25	7.26	7.27	7.27
53	7.28	7.29	7.29	7.30	7.31	7.31	7.32	7.33	7.33	7.34
54	7.35	7.36	7.36	7.37	7.38	7.38	7.39	7.40	7.40	7.41

$x \rightarrow \dfrac{1}{x}$

x	0	1	2	3	4	5	6	7	8	9
5.5	0.182	181	181	181	181	180	180	180	179	179
5.6	.179	178	178	178	177	177	177	176	176	176
5.7	.175	175	175	175	174	174	174	173	173	173
5.8	.172	172	172	172	171	171	171	170	170	170
5.9	.169	169	169	169	168	168	168	168	167	167
6.0	0.167	166	166	166	166	165	165	165	164	164
6.1	.164	164	163	163	163	163	162	162	162	162
6.2	.161	161	161	161	160	160	160	159	159	159
6.3	.159	158	158	158	158	157	157	157	157	156
6.4	.156	156	156	156	155	155	155	155	154	154
6.5	0.154	154	153	153	153	153	152	152	152	152
6.6	.152	151	151	151	151	150	150	150	150	149
6.7	.149	149	149	149	148	148	148	148	147	147
6.8	.147	147	147	146	146	146	146	146	145	145
6.9	.145	145	145	144	144	144	144	143	143	143
7.0	0.143	143	142	142	142	142	142	141	141	141
7.1	.141	141	140	140	140	140	140	139	139	139
7.2	.139	139	139	138	138	138	138	138	137	137
7.3	.137	137	137	136	136	136	136	136	136	135
7.4	.135	135	135	135	134	134	134	134	134	134
7.5	0.133	133	133	133	133	132	132	132	132	132
7.6	.132	131	131	131	131	131	131	130	130	130
7.7	.130	130	130	129	129	129	129	129	129	128
7.8	.128	128	128	128	128	127	127	127	127	127
7.9	.127	126	126	126	126	126	126	125	125	125
8.0	0.125	125	125	125	124	124	124	124	124	124
8.1	.123	123	123	123	123	123	123	122	122	122
8.2	.122	122	122	122	121	121	121	121	121	121
8.3	.120	120	120	120	120	120	120	119	119	119
8.4	.119	119	119	119	118	118	118	118	118	118
8.5	0.118	118	117	117	117	117	117	117	117	116
8.6	.116	116	116	116	116	116	115	115	115	115
8.7	.115	115	115	115	114	114	114	114	114	114
8.8	.114	114	113	113	113	113	113	113	113	112
8.9	.112	112	112	112	112	112	112	111	111	111
9.0	0.111	111	111	111	111	110	110	110	110	110
9.1	.110	110	110	110	109	109	109	109	109	109
9.2	.109	109	108	108	108	108	108	108	108	108
9.3	.108	107	107	107	107	107	107	107	107	106
9.4	.106	106	106	106	106	106	106	106	105	105
9.5	0.105	105	105	105	105	105	105	104	104	104
9.6	.104	104	104	104	104	104	104	103	103	103
9.7	.103	103	103	103	103	103	102	102	102	102
9.8	.102	102	102	102	102	102	101	101	101	101
9.9	.101	101	101	101	101	101	100	100	100	100

Reciprocals

x	0	1	2	3	4	5	6	7	8	9
1.0	1.000	0.990	980	971	962	952	943	935	926	917
1.1	0.909	901	893	885	877	870	862	855	847	840
1.2	.833	826	820	813	806	800	794	787	781	775
1.3	.769	763	758	752	746	741	735	730	725	719
1.4	.714	709	704	699	694	690	685	680	676	671
1.5	0.667	662	658	654	649	645	641	637	633	629
1.6	.625	621	617	613	610	606	602	599	595	592
1.7	.588	585	581	578	575	571	568	565	562	559
1.8	.556	552	549	546	543	541	538	535	532	529
1.9	.526	524	521	518	515	513	510	508	505	503
2.0	0.500	498	495	493	490	488	485	483	481	478
2.1	.476	474	472	469	467	465	463	461	459	457
2.2	.455	452	450	448	446	444	442	441	439	437
2.3	.435	433	431	429	427	426	424	422	420	418
2.4	.417	415	413	412	410	408	407	405	403	402
2.5	0.400	398	397	395	394	392	391	389	388	386
2.6	.385	383	382	380	379	377	376	375	373	372
2.7	.370	369	368	366	365	364	362	361	360	358
2.8	.357	356	355	353	352	351	350	348	347	346
2.9	.345	344	342	341	340	339	338	337	336	334
3.0	0.333	332	331	330	329	328	327	326	325	324
3.1	.323	322	321	319	318	317	316	315	314	313
3.2	.313	312	311	310	309	308	307	306	305	304
3.3	.303	302	301	300	299	299	298	297	296	295
3.4	.294	293	292	292	291	290	289	288	287	287
3.5	0.286	285	284	283	282	282	281	280	279	279
3.6	.278	277	276	275	275	274	273	272	272	271
3.7	.270	270	269	268	267	267	266	265	265	264
3.8	.263	262	262	261	260	260	259	258	258	257
3.9	.256	256	255	254	254	253	253	252	251	251
4.0	0.250	249	249	248	248	247	246	246	245	244
4.1	.244	243	243	242	242	241	240	240	239	239
4.2	.238	238	237	236	236	235	235	234	234	233
4.3	.233	232	231	231	230	230	229	229	228	228
4.4	.227	227	226	226	225	225	224	224	223	223
4.5	0.222	222	221	221	220	220	219	219	218	218
4.6	.217	217	216	216	216	215	215	214	214	213
4.7	.213	212	212	211	211	211	210	210	209	209
4.8	.208	208	207	207	207	206	206	205	205	204
4.9	.204	204	203	203	202	202	202	201	201	200
5.0	0.200	200	199	199	198	198	198	197	197	196
5.1	.196	196	195	195	195	194	194	193	193	193
5.2	.192	192	192	191	191	190	190	190	189	189
5.3	.189	188	188	188	187	187	187	186	186	186
5.4	.185	185	185	184	184	183	183	183	182	182

Answers

Exercise 1a (p. 15)

1. 30°
2. 36°, 72°, 72°
3. 90°, 55°, 35°
4. 30°, 150°, 120°, 60°; a trapezium
5. 120°, 144°, 162°
6. 129°
7. 12
8. 36°, 72°, 72
9. 120° each
10. answer varies
11. 145°
12. 124°

Exercise 1d (p. 17)

1. (a) congruent (b) congruent
   (c) congruent (d) congruent
   (e) congruent (f) congruent
   (g) not congruent

Exercise 1e (p. 18)

1. (a) $a = 40°$ (b) $d = 65°$ (c) $x = 124°$
   (d) $m = 51°$ (e) $n = 72°$ (f) $c = 140°$
   (g) $u = 28°$ (h) $a = 63°$
       $v = 152°$ $b = 117°$
   (i) $m = 70°$ (j) $x = 76°$
       $n = 40°$ $w = 38°$

2. 70°
3. 61°
4. $31\frac{1}{2}°$
5. 111°
6. $127\frac{1}{2}°$
7. 78°
8. 62°
9. 106°
10. 90°
13. 30°
14. 20°, 80°, 80°
15. 60°
16. 32°
17. 30°
18. 115°

Exercise 1f (p. 21)

1. 9
2. (a) ABCD, QAPC, AYCX
   (b) $\triangle QAX \equiv \triangle PCY$, $\triangle ABY \equiv \triangle CDX$,
       $\triangle QBC \equiv \triangle PDA$
3. 8 cm
4. 4 cm
5. 132°, 48°, 132°, 48°
6. rhombus, rectangle
7. $22\frac{1}{2}°$
10. 32.5 cm

Exercise P1.1 (p. 22)

3. 36°
4. 9

5. (a) $x + \frac{x}{3} + \frac{x}{6} = 180$
   (b) 120°, 40°, 20°
6. (a) 6 sides, (b) 120°
8. (a) $837 + 3x = 1080°$
   (b) 81°

Exercise 2a (p. 24)

1. (a) $\frac{5x}{9}$ (b) $\frac{11y}{6}$ (c) $\frac{3b - 2a}{ab}$
   (d) $\frac{9}{14x}$ (e) $\frac{9b - 16a}{12ab}$ (f) $\frac{9c + 8a}{6abc}$

2. $\frac{2bc + 7ac - 3ab}{abc}$
3. $\frac{6 - x^2 + 10x}{2x}$

4. $\frac{15e + 8c}{6cde}$

5. (a) $\frac{3x - 4}{3}$ (b) $\frac{2x - 1}{5}$ (c) $\frac{2a + 2}{15}$
   (d) $\frac{11x + 7}{6}$ (e) $\frac{41x + 1}{30}$ (f) $\frac{7a - 31}{15}$

6. $\frac{9x + 6}{12} = \frac{3x + 2}{4}$

7. (a) $\frac{5x + 3}{2a}$ (b) $\frac{5}{6}$
   (c) $\frac{7ab + b - 2a}{ab}$ (d) $\frac{5}{2a}$
   (e) $\frac{3a - 2b}{10ab}$ (f) $\frac{y - 2x}{xy}$

8. $\frac{3a + 23b - 9}{6b}$

9. (a) $\frac{3a - b}{a - b}$ (b) $\frac{x + 4y}{x + 2y}$
   (c) $\frac{3a - 18}{(a + 4)(a - 2)}$ (d) $\frac{5t + 7}{(t + 1)(t + 2)}$
   (e) $\frac{2x^2 + 7x}{(x - 1)(x + 2)}$ (f) $\frac{24 - 2x - x^2}{x(x - 3)}$
   (g) $\frac{2}{(n - 6)(n - 4)(n - 5)}$
   (h) $\frac{-4}{(x + 2)(x + 3)}$

10. $\frac{5 - a}{7a + 4}$

Mathematics for Caribbean Schools

Answers

Exercise 2b (p. 25)

1. $2(m + 4n)$
2. $3(a - 5b)$
3. $5(2x - 1)$
4. $-3(h + 4k)$
5. $-2(x - 9)$
6. $a(5 - 8b)$
7. $3x(3 + z)$
8. $4m(2c + 3d - 4e)$
9. $3x(x^2 - 4x - 3)$
10. $5ab(2ab - 3a + 4b)$
11. $4(a - 2b)$
12. $3(3x + 4y)$
13. $3a(b - 2c + d)$
14. $4x(2p - q + 2r)$
15. $m(3m^2 - 2m + 1)$
16. $2n^2(3n^2 - n + 2)$
17. $ab(5 + 4a - 6b)$
18. $ax(2b + 7c - 3a)$
19. $2a^2(2a^2 + ab - 5b^2)$
20. $6a^2c(4bc + 5cx - 3ax^2)$

Exercise 2c (p. 25)

1. $(x + 3b)(3 + 5a)$
2. $(e - f)(2c + d)$
3. $(u + v)(4 + x)$
4. $(n - 3y)(m - 3x)$
5. $(2a - 1)(3a + 2)$
6. $(d - e)(c - d)$
7. $2(3g - h)(2e - f)$
8. $(2a + 3n)(2b - 1)$
9. $(3x - 2)(a - 2b)$
10. $(x - 1)(a + 1)$
11. $(c + d)(a - b)$
12. $(4a - 7b)(1 - 4x)$
13. $(p + q)(1 + 5a)$
14. $(c - 4m)(2c - 3m)$
15. $(x + 2)(x - 5)$
16. $(p + q)(2r - s)$
17. $(y - 5)(y + 4)$
18. $(2a - b)(3c - d)$
19. $(2a - 5c)(b + 2d)$
20. $(3x - 2)(2y + z)$

Exercise 2d (p. 26)

1. 6
2. 11
3. 16
4. 4
5. 27
6. 11
7. 0
8. 6
9. 8
10. 15
11. 14
12. 15
13. 6
14. 2
15. 3
16. $2\frac{1}{2}$
17. $\frac{1}{2}$
18. 12
19. ± 5
20. ± 6

Exercise 2e (p. 26)

1. (a) -3 (b) 0 (c) 4 (d) 3
   (e) -2 (f) 2 (g) -2 (h) 1
   (i) 6 (j) ± 4
2. (a) -3 (b) 7 (c) 11 (d) 1
   (e) 40 (f) 2 (g) 17 (h) 65
   (i) -2 (j) 4
3. -12
4. (a) 6 (b) 2 (c) 0 (d) 0
5. (a) 5 (b) 0 (c) -4 (d) -3
   (e) 0 (f) 5
6. (a) 13 (b) 5 (c) 21 (d) 61
7. (a) 6 (b) 0 (c) 0 (d) 0
   (e) -6
8. ± 30
9. (a) -2 (b) 4 (c) $\frac{4}{9}$
10. (a) $\frac{4}{25}$ (b) 231 (c) $\frac{49}{4}$ (or $12\frac{1}{4}$)

Exercise 2f (p. 27)

1. 10^9
2. $20y^5$
3. m^3
4. $3x^2$
5. 1
6. $\frac{1}{64}$
7. 16
8. x^8
9. $\frac{1}{27x^3}$
10. $\frac{40}{x^4}$

Exercise 2g (p. 28)

1. a^6
2. b^8
3. c^{15}
4. d^{12}
5. e^9
6. 1
7. g^{-10}
8. h^{-20}
9. 5^{-2}
10. 3^4
11. 10^{14}
12. 2^{-6}

Exercise 2h (p. 28)

1. $9m^8$
2. $8n^{15}$
3. $4v^6$
4. $-2a^6$
5. c^6
6. $-e^{12}$
7. $-c^{20}$
8. d^{20}
9. m^4n^8
10. a^6b^3
11. $u^{12}v^8$
12. $125m^3n^9$

Exercise 2i (p. 29)

1. (a) 3 (b) $4\frac{1}{2}$ (c) $4\frac{1}{2}$ (d) $5\frac{1}{4}$
2. (b) 3
3. (a) $\frac{19}{2}$ (b) 1 (c) 4

Exercise 2j (p. 30)

1. (a) 4 (b) -11 (c) no
2. (a) 27 (b) 48 (c) $\frac{(52)^2}{3}$
   (d) $\frac{(56)^2}{3}$ (e) (i) no (ii) no
3.
4. (a) 4794 (b) 317 (c) 2700

Exercise P2.1 (p. 31)

1. $3(1 - a)$
2. $\frac{1}{4}f + 2$
3. $4k - 9$
4. $3c + 4$
5. $\frac{y}{6} - \frac{2}{3}$
6. $\frac{5}{4}(d + 2)$
7. $\frac{11x - 9}{12}$
8. $\frac{7b + 3}{6}$
9. $\frac{3h + 1}{2h}$
10. $\frac{7 - 3n}{6}$

Exercise P2.2 (p. 31)

1. $k + 3$
2. $m - 2$
3. $7 + 5y$
4. $\frac{7}{6}b + \frac{1}{2}$
5. $-2(c + 2)$
6. $-5(t + 1) + 4t^2$
7. $\frac{(11x - 9)}{12}$
8. $\frac{(3x + 2)}{12}$
9. $\frac{x}{6(2x + 1)}$

Answers

Exercise P2.3 (p. 31)

1. $3(a - 2b)$
2. $4d(3c - 25d)$
3. $3(a^2 - ab^2 + 3ab)$
4. $2(3 + v)$
5. $\dfrac{(2 - x)}{2}$
5. $\dfrac{2(x - 8)}{5(x + 2)}$
7. $2c^2 - 8c + 7$
8. $6x^2 + 3xy - 2y^2$

Exercise P2.4 (p. 31)

1. $(f + g)(a + b)$
2. $(b - c)(a + g)$
3. $(k - h)(2a - c)$
4. $(a - 2b)(3a - 2c)$
5. $(a - 1)(2a + xy)$
6. $2(m - n)(j + k)$
7. $(h - 2k)(m - 2n)$
8. $(t + 3s)(x - 2v)$

Exercise P2.5 (p. 32)

1. 20
2. 16
3. 48
4. 6
5. 9
6. 33

Exercise P2.6 (p. 32)

1. (a) 64   (b) 1024   (c) -64
   (d) 0.027   (e) 1.331
2. (a) m^{10} (b) k^7 (c) m^3p^3
   (d) m^3 (e) $\dfrac{1}{k^6}$ (f) mp
3. (a) 3^7 (b) 5^6 (c) 2^8
   (d) 0.4^5 (e) $(-2)^8$ (f) 10^9
   (g) 3^5 (h) $\frac{12}{2}$ (j) 2^5
   (k) 0.9^2 (m) $(-5)^3$ (n) 10^{-3}
4. (a) (i) $\frac{1}{2^4}$ (ii) $\frac{1}{16}$ (iii) 0.0625
   (b) (i) $\frac{1}{4^3}$ (ii) $\frac{1}{64}$ (iii) 0.016
   (c) (i) $\frac{1}{5^2}$ (ii) $\frac{1}{25}$ (iii) 0.04
   (d) (i) $\frac{1}{3^5}$ (ii) $\frac{1}{243}$ (iii) 0.004
   (e) (i) $\frac{1}{6^2}$ (ii) $\frac{1}{36}$ (iii) 0.028
   (f) (i) $\frac{1}{10^4}$ (ii) $\frac{1}{10\,000}$ (iii) 0
   (g) (i) $\frac{1}{3^4}$ (ii) $\frac{1}{81}$ (iii) 0.012
   (h) (i) $\frac{1}{11}$ (ii) $\frac{1}{11}$ (iii) 0.091
5. (a) 5^{-1} (b) 7^{-2} (c) 2^{-3} (d) 10^{-1}
   (e) 5^{-3} (f) 3^{-3} (g) 10^{-3} (h) 2^{-7}
6. $\frac{5}{12}$
7. (a) (i) 3^5 (ii) 243
   (b) (i) 5^4 (ii) 625
   (c) (i) 2^0 (ii) 1
   (d) (i) 0.4^6 (ii) $\frac{64}{15\,625}$
   (e) (i) 3^4 (ii) 81
   (f) (i) 7^0 (ii) 1
   (g) (i) $\frac{12}{2}$ (ii) $\frac{1}{4}$
   (h) (i) 0.9^1 (ii) $\frac{9}{10}$
   (i) (i) $\frac{1}{5^4}$ (ii) $\frac{1}{625}$
   (j) (i) 10^4 (ii) 10 000

Exercise P2.7 (p. 32)

1. (a) (i) 18 (ii) $27\sqrt{2}$
   (iii) $18\sqrt{6}$ (iv) 108
   (b) does not obey commutative law:
   (i) and (ii); does not obey associative
   law: (iii) and (iv)
2. (a) -7 (b) 7 (c) 8
   (d) -55 (e) 55
3. (b) 5
4. (a) (i) 0 (ii) 6 (iii) 2 (iv) -4
   (b) (i) no: see (a) (i) and (ii);
   (ii) no: see (a) (iii) and (iv)
5. (a) (i) 9 (ii) 25 (iii) 8 (iv) 81
   (b) (i) no; see (a) (i) and (ii);
   (ii) no; since $(5 \oslash 2) \oslash 3 \neq$
   $5 \oslash (2 \oslash 3)$

Exercise 3a (p. 33)

1. $x = 5$
2. $m = 4$
3. $a = 3$
4. $d = 12$
5. $y = 3$
6. $n = 4\frac{1}{2}$
7. $x = 4\frac{1}{3}$
8. $a = 3$
9. $m = 2\frac{1}{3}$
10. $t = 12$
11. $x = 4$
12. $a = 2\frac{2}{5}$
13. $y = 6$
14. $b = 3\frac{1}{2}$
15. $t = 6$
16. $x = 6$
17. $a = 4\frac{4}{5}$
18. $y = 7$
19. $e = 2\frac{1}{5}$
20. $x = 6$

Exercise 3b (p. 34)

1. $a = 4$
2. $b = 3$
3. $m = -11$
4. $n = 6\frac{2}{3}$
5. $t = 6$
6. $x = 11$
7. $y = 3$
8. $d = \frac{1}{5}$
9. $a = -1$
10. $x = 6$
11. $c = 3\frac{2}{3}$
12. $m = \frac{1}{3}$
13. $x = -8$
14. $a = 6$
15. $n = 1\frac{3}{5}$
16. $x = 4$
17. $v = 3$
18. $y = 2$
19. $z = 4$
20. $x = 3$

Exercise 3c (p. 35)

1. $x = 3$
2. $a = -5$
3. $d = 1\frac{1}{3}$
4. $y = 1\frac{2}{3}$
5. $x = 8\frac{1}{6}$
6. $y = \frac{1}{5}$
7. $d = \frac{3}{4}$
8. $a = -4$
9. $x = 1\frac{3}{4}$

⑩ $x = -54$ ⑪ $x = 2$ ⑫ $n = 2$
⑬ $a = 1\frac{1}{2}$ ⑭ $y = -2$ ⑮ $x = 4$
⑯ $x = -8$ ⑰ $z = 4$ ⑱ $m = 1\frac{2}{5}$
⑲ $x = -\frac{2}{3}$ ⑳ $x = -6$

Exercise 3d (p. 36)

① $x = 17$ ② 7 cm
③ Nina: $64, Indra: $83
④ 12 years   ⑤ $1\frac{1}{2}$ hours
⑥ 24 5-c coins   ⑦ 13 days
⑧ 11 ml, 48 ml
⑨ (a) $4(x - 3) = 5x - 3$ (b) $x = -9$
⑩ (a) $6 + \frac{1}{3}n = 2n + 1$ (b) $n = 3$
⑪ 17 years   ⑫ 50 goats
⑬ 60 km/h   ⑭ 40 litres, 200 litres
⑮ 69 eggs
⑯ (a) $\dfrac{x + 15}{4} = 80 - x$ (b) $x = 61$
⑰ 45 cm   ⑱ 90 km
⑲ (a) $\dfrac{m}{30}$ (b) 1440
⑳ 480 km

Exercise 3e (p. 39)

① (a) 309 K   (b) 127°C
② (a) 12.8 cm   (b) 3.5 cm
③ (a) $y = 7$ (b) $x = \frac{7}{9}$
④ (a) 15 m²   (b) 8 cm
   (c) 18 cm   (d) 2.5 m
⑤ (a) $300   (b) $460
⑥ (a) 3 amps   (b) 3 volts   (c) 90 ohms
⑦ (a) 11 m   (b) 7 m
⑧ (a) 60 000 kg   (b) 2 m   (c) $\frac{2}{3}$ m
⑨ (a) 44 cm   (b) 3.5 m
   (c) $6\frac{2}{7}$ m   (d) 0.437 5 m
⑩ (a) 60 km/h   (b) $6\frac{1}{4}$ s

Exercise 3f (p. 40)

① $y = 4, 3, 2, 1, 0$
② $d = 1, 2, 3, 4, 5$
③ $y = 1, 3, 5, 7, 9$
④ $y = 3x + 2$

x	−1	0	1	2	3
$3x$	−3	0	3	6	9
+2	+2	+2	+2	+2	+2
y	−1	+2	+5	+8	+11

⑤ $y = 17 - 6x$

x	0	1	2	3	4	5
17	17	17	17	17	17	17
$-6x$	0	−6	−12	−18	−24	−30
y	17	11	5	−1	−7	−13

⑥ (a) $295   (b) 120 km
⑦ (a) 1 h 13 min   (b) 2.6 kg
⑧ (a) (i) $150   (ii) $1195   (b) $2500
⑨ (a) 38 litres   (b) $4\frac{1}{2}$ h   (c) 6.3 h
⑩ (a) 440 cm²   (b) 16 cm

Exercise 3g (p. 41)

① (a) $y = 0, 40, 160, 360, 640, 1000$
   (b) $x = \pm\frac{1}{2}, \pm3, \pm5, \pm10$
② (a) $y = 0, 12, 16, 12, 0$
   (b) $x = \pm4, \pm3, \pm2, \pm1$
③ (a) 4   (b) −3   (c) −6
   (d) −5   (e) 0
④ $993.75
⑤ (a) $m = 100, 4, 1, \frac{1}{4}$
   (b) $n = \pm10, \pm5, \pm3\frac{1}{3}, \pm2$
⑥ (a) 154 m²   (b) 14 cm   (c) 3.5 m
⑦ (a) 308 cm³   (b) 14 cm   (c) $2\frac{1}{3}$ cm
⑧ (a) 44.1 m   (b) 6.4 m   (c) 10 s
   (d) 5 s   (e) 44.1 m
⑨ (a) 95 m   (b) 5
⑩ (a) 12.5 m   (b) 5.5 m   (c) 50 km
⑪ (a) $6\frac{1}{2}$ cm   (b) 14
⑫ (a) 1275   (b) 13   (c) 10

Exercise P3.1 (p. 42)

① $\frac{1}{5}$ ② $\frac{1}{2}$ ③ $-1\frac{2}{5}$ ④ 6
⑤ 2   ⑥ 8   ⑦ 6   ⑧ 2

Exercise P3.2 (p. 43)

① 7   ② 6

Exercise P3.3 (p. 43)

① (a) $4w + 6 = 70$
   (b) 16
② (a) $4x - 40 = 500$
   (b) 115
③ (a) $110(m - 12) - 75m = 2880$
   (b) 120

Answers

④ (a) $n + 3n + 2n + (n + 2) + 4 = 34$
   (b) 4
⑤ (a) $8 \times \frac{1}{2} + 6t = 13$ (b) $\frac{3}{2}$

Exercise P3.4 (p. 43)

① (a) x notebooks with ruled pages
   (b) $9x + 7(10 - x) = 82$
   (c) ruled pages: 6, plain pages: 4
② (a) new price $= p$ cents
   (b) $p - 15 = \frac{90}{100} \times p$
   (c) New price $=$ \$1.50; old price $=$ \$1.35
③ (a) salary $=$ \$s
   (b) $s - \left(\frac{s}{5} + \frac{s}{10}\right) = 4060$
   (c) \$5800
④ (a) length of short side $= x$ cm
   (b) $x + 2(2x) = 15$
   (c) short side $= 3$ cm; long side $= 6$ cm
⑤ (a) smaller odd number $= n$
   (b) $7n - 6(n + 2) = 17$
   (c) numbers are 29 and 31
⑥ (a) speed of car B $= v$ km/h
   (b) $(v + 2v)\frac{5}{2} = 240$
   (c) speed of car B: 32 km/h;
       of car A: 64 km/h
⑦ (a) height of shorter column $= h$ m
   (b) $h + 30 = \frac{5h}{4}$
   (c) shorter column $= 120$ m;
       taller $= 150$ m
⑧ (a) width of short wall $= w$ m
   (b) $2w + 2(w + 1.5) - (2 \times 1.3) = 27.2$
   (c) short wall $= 6.7$ m; long wall $= 8.2$ m

Exercise P3.5 (p. 43)

① $w = 7h$ ② $s = c + 45$
③ $l = c - 5$ ④ $m = 10l$
⑤ $n = \frac{l}{5}$ ⑥ $t = 90c$
⑦ $l = 20 - p$ ⑧ $c = 5n$
⑨ $p = u - 35$ ⑩ $l = c + t$

Exercise P3.6 (p. 44)

① weight of 1 book $= e$ gm;
   total weight $= w$ gm; $w = 250e$
② \$p $=$ prize; \$a $=$ amount each person
   receives; $\frac{p}{10} = a$

③ w kg $=$ old weight; n kg $=$ new weight;
   $n = w + 8$
④ l m $=$ length of train; c m $=$ length of
   1 carriage; $l = 8c$
⑤ \$s $=$ sale price; \$u $=$ usual price;
   $s = u - 20$

Exercise P3.7 (p. 44)

① (a) -13 (b) $-\frac{2}{5}$ (c) $\frac{35}{88}$
   (d) $5\frac{1}{4}$ (e) 3
② (a) $A = 2\pi r(r + h)$ (b) 283 cm²
③ (a) 34 litres   (b) 4.5 hours
   (c) 1.9 hours
④ 7

Exercise 4a (p. 46)

① (a) $q = 82°$ (b) $r = 53°$
   (c) $s = 252°$ (d) $t = 270°, u = 90°$
   (e) $v = 64°, w = 32°$
② 105°   ③ 54°
④ (a) $i = 108°, j = 54°$
   (b) $e = 46°, f = 92°, q = h = 44°$
   (c) $a = b = 43°, c = 94°, d = 47°$
   (d) $t = 46°, u = 134°, v = w = 23°$
⑤ $2(a + b)$

Exercise 4b (p. 48)

① $a = b = 40°, c = 32°$
② $d = 115°, e = 90°$
③ $f = 90°, q = 34°, h = 56°$
④ $i = 65°, j = 25°$
⑤ $k = 37°, l = 53°$
⑥ $m = p = 20°, n = 70°$
⑦ $q = 80°, r = s = 70°, t = 10°, u = 60°, v = 40°$
⑧ $w = 252°, x = 108°, y = 54°, z = 126°$
⑨ ADC $= 34°$

Exercise P4.1 (p. 49)

① (a) $x = 34°$ (b) $y = 80°$
   (c) $z = 90°$
② (a) $a = x$ (b) $b = 90° - x$
   (c) $c = x, d = 2x$
③ (a) $x = 90°$, angle in semicircle
   (b) They are equal.
   (c) $a = y + z$

Exercise 5a (p. 53)

1 $125.80

2 (a) $202.50   (b) 12.5 cents

3 $187.40   **4** $207.41   **5** $2 854.42

6 (a) 335 kWh
   (b) (i) $10.95   (ii) $79.58

Exercise 5b (p. 55)

1 $1.477 cents   **2** $1.76

3 (a) 28 cu metres   (b) 6160 gals
   (c) 205 gals

4 (a) 30 cu metres   (b) 90 cu metres
   (c) $162

5 (a) $630   (b) $871

6 (a) $68 000   (b) $2720
   (c) $81 328

Exercise 5c (p. 57)

1 (a) $2.05   (b) $1.70
   (c) $4.00   (d) $1.05

2 (a) $36.72   (b) $281.50

3 (a) 76 cents per min (b) 38 cents per min
   (c) 50%

4 (a) $100   (b) $467.60
   (c) $1 532.80

5 (a) (i) 77   (ii) $9.40
     (iii) $60.00   (iv) $85.50
   (b) 95

Exercise 5d (p. 60)

1 (a) (i) $160   (ii) 9 months
   (b) $378

2 $1.80

3 (a) $11 820   (b) $11 583.60

4 $375   **5** 108

6 (a) 7.5%   (b) $20 812.50   (c) 9.2%

Exercise P5.1 (p. 61)

1 (a) 122 kWh, $89.06
   (b) 271 kWh, $149.05

2 (a) $47.93   (b) $204.39

3 $252.73

Exercise P5.2 (p. 62)

1 (a) *A1 Savings*   (b) $2.36

2 $15 633.17

3 (a) $765   (b) $592.42   (c) 30.3%

4 (a) $36 501   (b) $240.10

5 (a) $1200   (b) $200
   (c) 10 months

Exercise 6a (p. 64)

1 (a) {f, c, s, e}   (b) {f, a, c, o, r, i}
   (c) {f, a, c, o, r, i, s, e}   (d) {f, c}

2 $P' \cup Q' = (P \cap Q)' = $ {f, a, c, o, r, i, s, e}
   $P' \cap Q' = (P \cup Q)' = $ {f, c}

3 (a) {e, n}   (b) {n, i, c, e, h}
   (c) {k}   (d) {c, h, i, k}
   (e) {h}   (f) {n, i, c, e, k}

4 (a) {9, 8, 7, 6, ...}
   (b) {−3, −2, −1, 0, 1, ...}
   (c) {6, 7, 8, 9, ...}
   (d) {−1, −2, −3, −4, ...}
   (e) {4}   (f) {6}
   (g) {−6, −5, −4, −3, −2, −1, 0, 1}
   (h) {−8, −7, −6, −5, −4, −3, −2}

5 (a) 7   (b) 4
   (c) {3, 6, 7, 8, 9}   (d) {9}

6 (a) {1, 4, 9, 16}
   (b) {14, 15, 16, ..., 19, 20}
   (c) {1, 2, 4, 5, 8, 10, 20}
   (d) {(1, 4), (2, 7), (3, 10), (4, 13), (5, 16), (6, 19)}
   (e) {(1, 1), (2, 7), (3, 17)}
   (f) {(1, 19), (1, 20), (2, 20)}

7 6 countries   **8** 7 boys

9
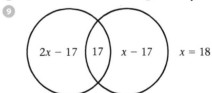

10 (a) $q = 43 - r$ (b) $s = 32 - r$
    (c) 32   (d) 184 (when $r = 0$)
    (e) $r = 26, q = 17, p = 210$

Exercise 6b (p. 67)

1

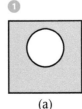

   (a)   (b)   (c)

Answers

2

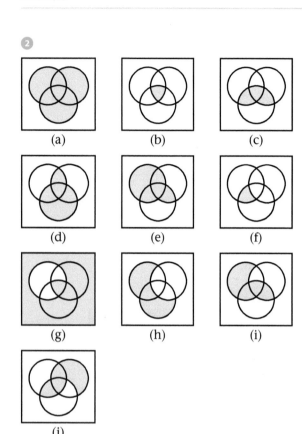

(a)   (b)   (c)

(d)   (e)   (f)

(g)   (h)   (i)

(j)

3 (a) and (b)

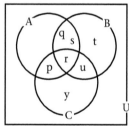

(c) (i) {p, r, u}   (ii) {p, q, r, s, u, y}

4 (a) 55   (b) 95   (c) 140   (d) 0

5 (a) 8   (b) 14   (c) 27
     (d) 0   (e) 11   (f) 30

6 (a) 6   (b) 25   (c) 40
     (d) 48   (e) 10   (f) 23

7 22

8 21%

9

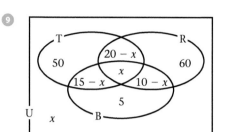

$x = 10$, 20 students read a book

10

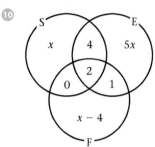

$x = 5$, 32 students speak English

Exercise 6c (p. 69)

1 (a) All houses have staircases
   (b) Not all houses have staircases
   (c) No house has a staircase

2

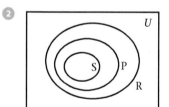

3 (a)

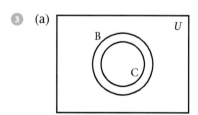

(b)

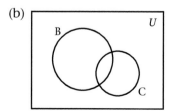

4

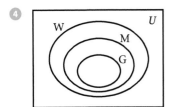

5

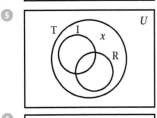

6

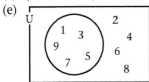

Exercise P6.1 (p. 70)

1 true   **2** true

3 false, *gram* is not a unit of time

4 false, ∈ is not the symbol for a subset

5 false, *fraction* and *decimal* are not *integers*

Exercise P6.2 (p. 70)

(note that answers may vary)

1 (a) {whole numbers from 1 to 20}

(b) {even numbers}

(c) {square numbers from 4 to 36}

(d) {the three smallest prime numbers}

2 (i) (a) W

(b) N = {odd numbers less than 10}

(c) {1, 3, 5, 7, 9}   (d) N' = {2, 4, 6, 8}

(e)

(ii) and (iii) *(similar approach)*

Exercise P6.3 (p. 70)

1 (a) (i) {four-legged animals}

(ii) {house pets}   (iii) {wild animals}

(b) {house pets} ⊂ {four-legged animals}

2 and **3** *(similar approach)*

Exercise P6.4 (p. 71)

1 (a) I   (b) F   (c) F   (d) F   (e) I   (f) { }

2 (a) (i) 4   (ii) { }, {p}, {q}, {p, q}

(b) (i) 8

(ii) { }, {n}, {i}, {e}, {n, i}, {n, e}, {i, e}, {n, i, e}

Exercise P6.5 (p. 71)

1 equal   **2** not equivalent

3 not equivalent   **4** not equivalent

5 not equivalent   **6** equal

7 equivalent   **8** not equivalent

Exercise P6.6 (p. 71)

1

(a) 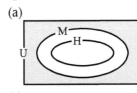 (b)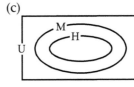

(c) 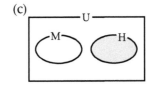 (d)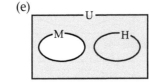

(e) 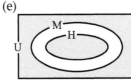 (f)

2

(a) 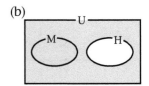 (b)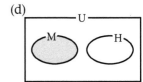

(c) 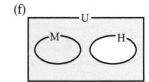 (d)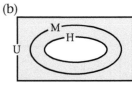

(e) 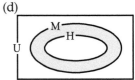 (f)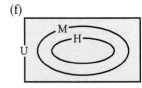

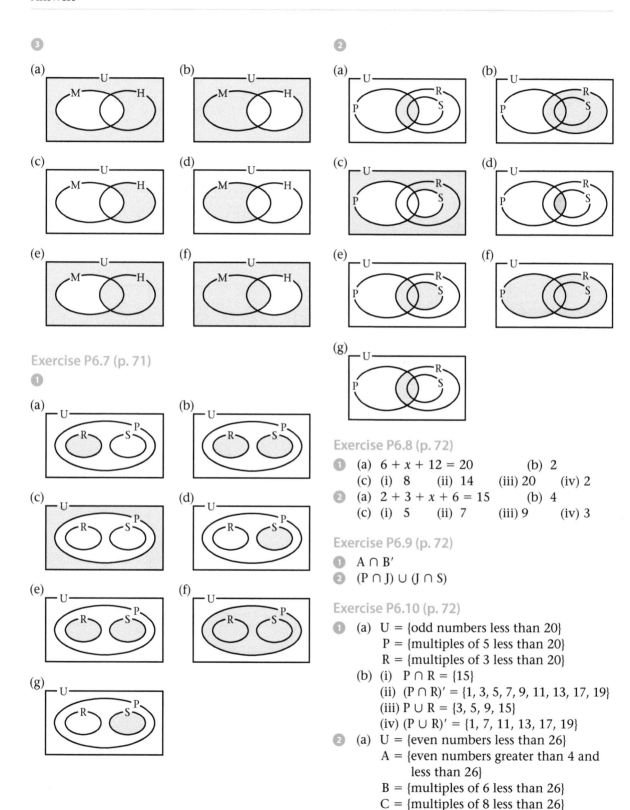

3

(a)

(b)

(c)

(d)

(e)

(f)

2

(a)

(b)

(c)

(d)

(e)

(f)

(g)

Exercise P6.7 (p. 71)

1

(a)

(b)

(c)

(d)

(e)

(f)

(g)

Exercise P6.8 (p. 72)

1 (a) $6 + x + 12 = 20$ (b) 2
   (c) (i) 8   (ii) 14   (iii) 20   (iv) 2
2 (a) $2 + 3 + x + 6 = 15$ (b) 4
   (c) (i) 5   (ii) 7   (iii) 9   (iv) 3

Exercise P6.9 (p. 72)

1 $A \cap B'$
2 $(P \cap J) \cup (J \cap S)$

Exercise P6.10 (p. 72)

1 (a) U = {odd numbers less than 20}
       P = {multiples of 5 less than 20}
       R = {multiples of 3 less than 20}
   (b) (i) $P \cap R$ = {15}
       (ii) $(P \cap R)'$ = {1, 3, 5, 7, 9, 11, 13, 17, 19}
       (iii) $P \cup R$ = {3, 5, 9, 15}
       (iv) $(P \cup R)'$ = {1, 7, 11, 13, 17, 19}
2 (a) U = {even numbers less than 26}
       A = {even numbers greater than 4 and
           less than 26}
       B = {multiples of 6 less than 26}
       C = {multiples of 8 less than 26}

(b) (i) A ∩ B = {6, 12, 18, 24}
A ∩ C = {8, 16, 24}
B ∩ C = {24}

(ii) (A ∩ B)′ = {2, 4, 8, 10, 14, 16, 20, 22}
(A ∩ C)′ = {2, 4, 6, 10, 12, 14, 18, 20, 22}
(B ∩ C)′ = {2, 4, 6, 8, 10, 12, 14, 16, 18, 20, 22}

(iii) A ∪ B = {6, 8, 10, 12, 14, 16, 18, 20, 22, 24}
A ∪ C = {6, 8, 10, 12, 14, 16, 18, 20, 22, 24}
B ∪ C = {6, 8, 12, 16, 18, 24}

(iv) (A ∪ B)′ = {2, 4}
(A ∪ C)′ = {2, 4}
(B ∪ C)′ = {2, 4, 10, 14, 20, 22}

Exercise P6.11 (p. 72)

 (i) (a) L = {x: x is a prime number greater than 10}
G = {x: x is a prime number less than 50}

(b) L = {11, 13, 17, ...}
G = {47, 43, 41, ..., 3, 2}

(c) {11, 13, 17, 19, 23, 29, 31, 37, 41, 43, 47}

(ii) (a) F = {x: x is a factor of 24}
T = {x: x is a factor of 30}

(b) F = {1, 2, 3, 4, 6, 8, 12, 24}
T = {1, 2, 3, 5, 6, 10, 15, 30}
T ∪ F = {2, 3, 6}

(c) HCF = 6

2 (a)

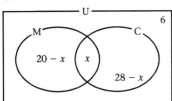

(b) 6 + 20 − x + x + 28 − x = 36
(c) 18

Exercise 7a (p. 74)

1

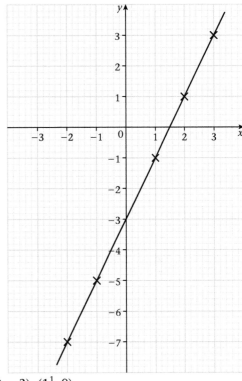

$(0, -3)$, $(1\frac{1}{2}, 0)$

2 (a)

x	−3	0	3
y	14	5	−4

(b)

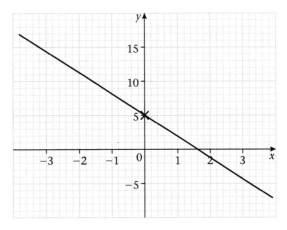

Answers

3 (a)

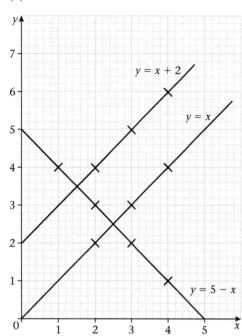

(b) The first two lines are parallel to each other. The third line is perpendicular to the other two.

4 (a)

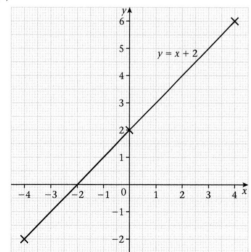

(b) 5

(c) −1

(d) (−2, 0), (0, 2)

5 (a)

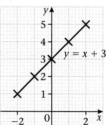

(b)

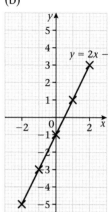

(c)

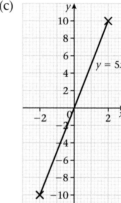

(d)

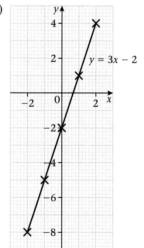

6　(a), (b)

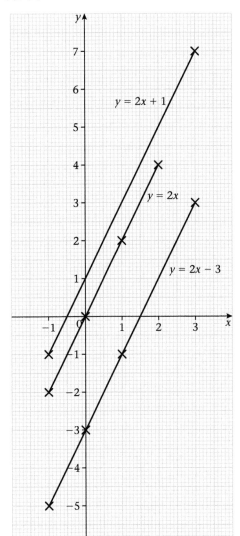

Exercise 7b (p. 76)

1　$\frac{2}{7}$　　2　$\frac{3}{2}$　　3　$\frac{3}{5}$　　4　$-\frac{4}{5}$　　5　$-\frac{4}{3}$

6　$\frac{2}{7}$　　7　-2　　8　$-\frac{3}{4}$　　9　3　　10　$-\frac{1}{2}$

Exercise 7c (p. 78)

1　(a)　$y = 3x - 3$　　　(b)　$y = 3x$
　　(c)　$x + y = 6$　　　(d)　$3x + 4y = 18$
　　(e)　$4x + y + 5 = 0$　(f)　$2y = 5x + 9$

2　(a)　$7x - 3y = 0$　　(b)　$7x + 3y = 0$
　　(c)　$x + y = 3$　　　(d)　$3x - 10y = 42$
　　(e)　$5x + 16y = 67$　(f)　$5x + 2y + 12 = 0$

3　(a)　$-\frac{1}{3}$　　　　　　(b)　$x + 3y = 7$

4　(a)　$y = 1 - x$　　　　(d)　$y = 1$
　　(b)　$x = 1$　　　　　　(e)　$2y = x - 2$
　　(c)　$y = x + 1$

Exercise 7d (p. 80)

1　$x = 1, y = 2$　　　　2　$x = 3, y = 2$
3　$x = 1, y = 0$　　　　4　$x = 0, y = 2$
5　$x = -1, y = 2$　　　6　$x = 3, y = -1$
7　$x = -1, y = -1$　　8　$x = 2\frac{1}{2}, y = 1\frac{1}{2}$
9　$x = 1.6, y = 1.4$　　10　$x = 1.3, y = -1.2$

Exercise P7.1 (p. 80)

1　(a)　function
　　(b)　not a function; two x-values are each
　　　　mapped to more than one y-value
　　(c)　function

2　(a)　each and every object of the first set is
　　　　mapped onto one and only one object
　　　　in the second set
　　(b)　the x-values are all connected to two
　　　　corresponding y-values

Exercise P7.2 (p. 81)

1　(a)　answers will vary　　(b)　undefined
　　(c)　does not cut y-axis　(d)　$x = -\frac{5}{2}$

2　(a)　answers will vary　　(b)　0
　　(c)　8　　　　　　　　　　(d)　$y = 8$

3　(a)　answers will vary　　(b)　-1
　　(c)　2　　　　　　　　　　(d)　$y + x = 2$

4　(a)　answers will vary　　(b)　$\frac{1}{2}$
　　(c)　0　　　　　　　　　　(d)　$2y = x$

5　(a)　answers will vary　　(b)　2
　　(c)　8　　　　　　　　　　(d)　$y = 2x + 8$

6　(a)　answers will vary　　(b)　$-\frac{1}{2}$
　　(c)　8　　　　　　　　　　(d)　$2y + x = 16$

Exercise P7.3 (p. 81)

1　(a)　0　　　　　　　　(b)　$y = 3$
2　(a)　$\frac{1}{3}$　　　　　　(b)　$3y = x - 13$
3　(a)　$\frac{1}{3}$　　　　　　(b)　$3y = x + 13$
4　(a)　-3　　　　　　(b)　$y + 3x = 11$
5　(a)　2　　　　　　　(b)　$y - 2x + 7 = 0$
6　(a)　$-\frac{1}{3}$　　　　　(b)　$3y + x = 12$
7　(a)　$-\frac{2}{7}$　　　　　(b)　$7y + 2x = 8$
8　(a)　1　　　　　　　(b)　$y = x + 3$

Answers

⑨ (a) $-\frac{1}{2}$ (b) $2y + x = 16$

⑩ (a) $-\frac{7}{6}$ (b) $6y + 7x + 52 = 0$

⑪ (a) 0 (b) $y = 0$

⑫ (a) 3 (b) $y = 3x + 3$

Exercise P7.4 (p. 82)

① $y = 2x + 4$ ② $y + 4x = 3$

③ $y = 5$ ④ $y + x + 5 = 0$

⑤ $y + 3x + 16 = 0$

Exercise P7.5 (p. 82)

①

x	−2	0	1	2
3	3	3	3	3
$x + 3$	1	3	4	5

②

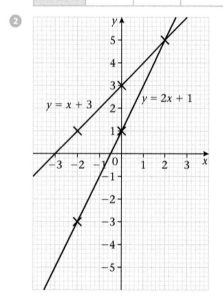

③

x	−2	0	1	2
$2x$	−4	0	2	4
1	1	1	1	1
$2x + 1$	−3	1	3	5

④ (2, 5)

Exercise P7.6 (p. 82)

① (a)

x	−3	−2	−1	0	1	2	3	4
$3x$	−9	−6	−3	0	3	6	9	12
−2	−2	−2	−2	−2	−2	−2	−2	−2
$3x − 2$	−11	−8	−5	−2	1	4	7	10

(b)–(d)

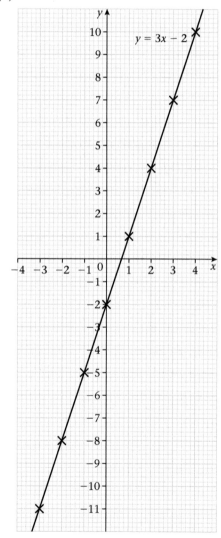

2 (a)

x	−4	−3	−2	−1	0	1	2	3	4
−2	−2	−2	−2	−2	−2	−2	−2	−2	−2
$x-2$	−6	−5	−4	−3	−2	−1	0	1	2

(b)–(d)

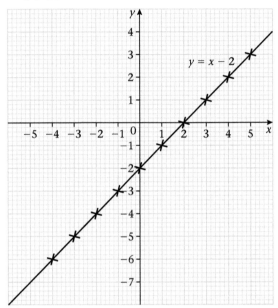

2 (a)

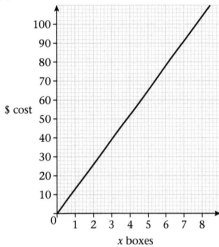

(b) (i) $91   (ii) 6 boxes

3 (a)

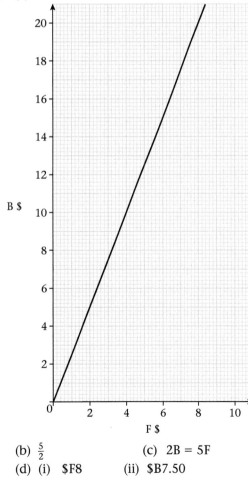

(b) $\frac{5}{2}$ (c) 2B = 5F

(d) (i) $F8   (ii) $B7.50

Exercise P7.7 (p. 82)

1 (a)

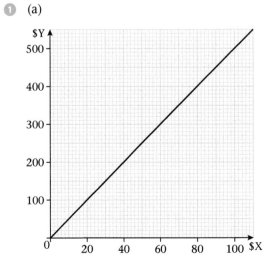

(b) (i) $Y250   (ii) $X90

Answers

④ (a) $45
(b) 350 units
(c) $D = 0.15m + 30$

⑤ (a)

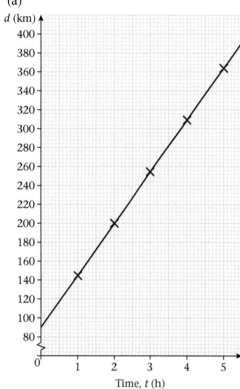

d (km)

Time, t (h)

(c) $d = 55t + 90$
(d) 1 h 24 min

Exercise 8a (p. 84)

① $x = 1, y = 2$　② $x = 3, y = 2$
③ $a = -1, b = 3$　④ $m = -1, n = 2$
⑤ $x = 3, y = 1$　⑥ $x = -1, y = -1$
⑦ $a = 5, b = -2$　⑧ $x = 2, y = 2$
⑨ $x = 1, y = 3$　⑩ $a = 7, b = 5$
⑪ $x = -2, y = -3$　⑫ $x = 1, y = -3$

Exercise 8b (p. 85)

① $a = 4, b = 3$　② $p = 3, q = -1$
③ $x = 7, y = -2$　④ $x = 2, y = -4$
⑤ $x = 2, y = -3$　⑥ $x = 0, y = 3$
⑦ $a = 2\frac{1}{2}, b = 1$　⑧ $x = 5, y = -2$
⑨ $x = 0, y = -2$　⑩ $h = 2, k = 1\frac{1}{2}$
⑪ $p = -1, q = -2$　⑫ $r = -2, s = 11$
⑬ $x = -3, y = 0$　⑭ $x = 12, y = -5$

⑮ $a = 3, b = 4$　⑯ $u = -2, v = 1$
⑰ $d = -2, e = 2$　⑱ $x = 1\frac{1}{3}, y = 3$
⑲ $f = 2\frac{1}{3}, g = 1\frac{1}{2}$　⑳ $y = 2\frac{1}{2}, z = -3\frac{1}{4}$

Exercise 8c (p. 86)

① 12, 7　② 39 yr, 14 yr
③ 11, 6　④ 42c, 18c
⑤ 10c, 45c　⑥ 15c, 12c
⑦ 5 five cent coins;　⑧ 14 yr, 11 yr
　3 ten cent coins
⑨ $x = 3, y = 2$; 150 cm²
⑩ $x = 3, y = 2$

Exercise 8d (p. 87)

① mug: $5, plate: $3.50
② (a) $3.75　(b) $4
③ $1.50, $1.25　④ 45, 30
⑤ $x = 2, y = 3$
⑥ $a = 1\frac{4}{5}, b = 2\frac{2}{5}$; 7.2 cm
⑦ $x = 3, y = 2\frac{1}{2}$, 10 cm, 10 cm, 8 cm
⑧ $m = 25, n = 30, \widehat{A} = 80°, \widehat{B} = 50°$
⑨ 56
⑩ 71 or 17
⑪ 78
⑫ 25 years
⑬ Charles: 17 years, Tom: 12 years
⑭ 16 km/h, 6 km/h
⑮ $\frac{6}{13}$

Exercise P8.1 (p. 88)

① $x = 4, y = 2$　② $x = 1, y = -1$
③ $x = 1, y = -1$　④ $x = 2, y = -3$
⑤ $x = -2, y = -1$　⑥ $x = 3, y = 3$
⑦ $x = \frac{3}{2}, y = \frac{3}{2}$　⑧ $x = 1, y = \frac{5}{4}$
⑨ $x = 4, y = \frac{1}{3}$

Exercise P8.2 (p. 89)

① $x = 2, y = -2$　② $x = 3, y = 2$
③ $x = 2, y = -2$　④ $x = \frac{1}{2}, y = -\frac{1}{2}$
⑤ $x = 1, y = -\frac{1}{2}$　⑥ $x = 1, y = 2$
⑦ $x = \frac{3}{2}, y = -\frac{2}{3}$　⑧ $x = 3, y = 8$
⑨ $x = 4, y = \frac{2}{3}$

Exercise P8.3 (p. 89)

① $\frac{23}{43}$　　　　② Dani 44, Fred 27

③ (a) $3x + y - 3 = 5y - 4$; $2y - 1 = x + 2$
(b) $x = 5, y = 4$
(c) width $= 16\,$cm, height $= 7\,$cm

④ 57

⑤ 1 barrel: 1.2 kl; 1 tank: 8 kl

⑥ 1 television: 8 kg; 1 radio: 2.5 kg

⑦ unskilled worker \$10/h, skilled worker \$20/h

⑧ 8 buses, 36 cars　　⑨ $u = 6\frac{2}{3}, v = 10\frac{2}{3}$

⑩ father 52, son 24　⑪ 57 and 41

⑫ $s = 2, t = 3$

⑬ Rob \$50, Dan \$22.50

⑭ (a) $2y - 1 = x + 2 = 2x + y - 1$
(b) $x = 1, y = 2$　　(c) 9 cm

⑮ Ann has 8 25¢-coins and 12 10¢-coins.
John has 11 25¢-coins and 9 10¢-coins.¢

Revision exercise 1 (p. 91)

① 20 sides

② $B\widehat{C}P = 131°$, $B\widehat{P}C = 34°$

④ 93°

⑤ $x = 73, 34$ or $53\frac{1}{2}$; $y = 34, 73$ or $53\frac{1}{2}$
respectively

⑥ 123°

⑦ (a) 102°　　(b) 78°　　(c) 78°
(d) 12°　　(e) 44°　　(f) 46°

⑧ (a) 27°　　(b) 63°

⑨ (a) 63°　　(b) 126°

⑩ RP∥SQ

Revision test 1 (p. 92)

① A　　② B　　③ B　　④ C　　⑤ A

⑦ $a = 71, b = 54\frac{1}{2}$

⑨ $180 - (50 + x) = x + 28$ (i.e. $F\widehat{A}B = B\widehat{D}E$),
$x = 51$

⑩ (a) 115°　　　　(b) 119°

Revision exercise 2 (p. 93)

① (a) 8　　　　　(b) -22

② $\frac{15}{4}$

③ (a) $\dfrac{-(7x + 1)}{x}$　　(b) $\dfrac{29t + 17}{30t}$
(c) $12d - 13d^2 - 4$

④ (a) 8　　　　　(b) 8

⑤ (a) $7(2a + 3)$　　(b) $9(x + z)$
(c) $4(2 + 2m + n)$　(d) $3x(5x + 8 + 3y)$
(e) $5ab(7b - 5a)$

⑥ (a) $x = 9$　　　(b) $x = \frac{11}{3}$

⑦
d	3	4	5	6	7	8
A	-24	-10	8	30	56	86

⑧ (a) $(m + n)(a - 3)$　(b) $(a - 7)(a + 3)$
(c) $(y - 2z)(3x - 5a)$

⑨ (a) 65　　　　(b) (i) 95　　(ii) 23

⑩ $66\frac{2}{3}\,$km h^{-1}, $133\frac{1}{3}\,$km h^{-1}

Revision test 2 (p. 93)

① A　② C　③ C　④ A　⑤ A

⑥ (a) (i) 12　(ii) ± 7　　(b) 3

⑦ 11

⑧ \$2500

⑨ (a) $4x(13 - 2x)$　　(b) $2a(a + b)$
(c) $(x + 5)(x - 9)$　(d) $(2a - b)(3x + 2y)$

⑩ (a) \$116　　　　(b) 15 hours

Revision exercise 3 (p. 94)

① (a) P $= \{1, 2, 3, 4, 6, 9, 12, 18\}$
(b) Q $= \{2, 4, 6, …, 18, 20\}$
(c) 12　　　　　　(d) 1

②
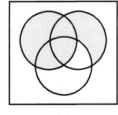
(a)

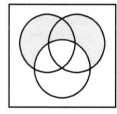
(b)

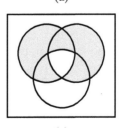
(c)

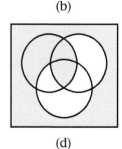
(d)

③ (a) \$2000　　　　(b) \$9000

④ (a) \$2973.75　　　(b) \$3419.81

⑤ (a) \$231.60　　　(b) \$278.28

⑥ (a) 6　　(b) 33　　(c) 34　　(d) 25

Answers

7

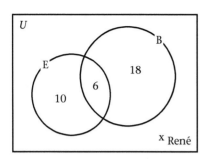

8 (a) (i) $1620 (ii) $1498.36
   (b) Investment at simple interest
   (c) $121.64

9 (a) $2023
   (b) $72\frac{1}{4}$%

10 13

Revision test 3 (p. 95)

1 B
2 B
3 D
4 C
5 A
6 (a) (i) $48 000   (ii) $36 000
      (iii) $27 000
   (b) (i) $20 000   (ii) $33\frac{1}{3}$%
7 $47.03
8 (a)

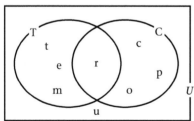

   (b) (i) {c, o, u, p}   (ii) {m, u, t, e}
      (iii) {u}   (iv) {c, r, o, u, p}

9

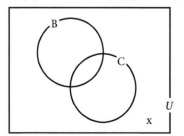

10 $n = 13$; there are 82 pages with maps

Revision exercise 4 (p. 96)

1 (a)

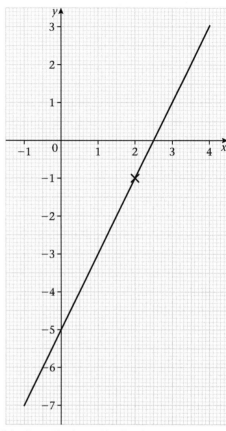

   (b) (0, −5), (1, −3), (3, 1)
   (c) $(2\frac{1}{2}, 0)$, (0, −5)
2 (a) 1
   (b) −2
   (c) 2
   (d) $\frac{1}{2}$

③

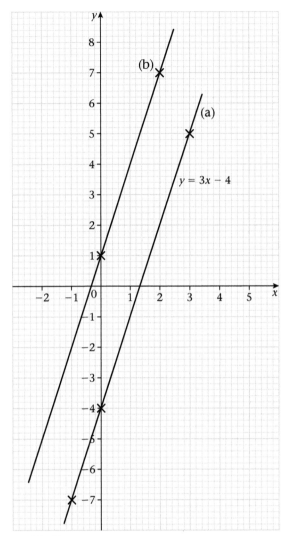

(c) 3
(d) They are parallel

④ (a)

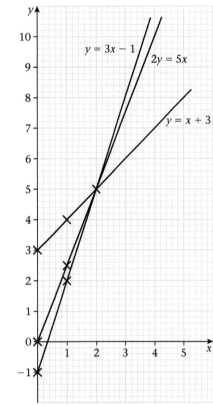

(b) All lines meet at (2, 5)
(c) $y = x + 3$, $2y = 5x$
   $y = x + 3$, $y = 3x - 1$
   $2y = 5x$, $y = 3x - 1$

⑤ $x = -3$, $y = -5$

⑥ (a) $y = \frac{5}{2}$, $z = -\frac{13}{4}$
   (b) $x = -\frac{1}{3}$, $y = \frac{1}{2}$

⑦ (a) $x = -4$, $y = -2$
   (b) $p = 21$, $q = -15$

⑧ (a) $2y + x + 4 = 0$
   (b) $y = 2x + 17$
   (c) $y + 4x + 9 = 0$

⑨ $1.20

⑩ (a) $x + 4y = 292$
   $2x + 5y = 482$
   (b) $156, $34

Revision test 4 (p. 97)
❶ D   ❷ B   ❸ C   ❹ D   ❺ A
❻ (a) $x = 2$, $y = 3$ (b) $a = 2$, $b = -1$

Answers

7 $AB = 6\,\text{cm}$, $BC = 7\,\text{cm}$, $AC = 12\,\text{cm}$

8 (a) (i) 2   (ii) $\frac{1}{2}$ (iii) -1

    (b) (i) 3   (ii) $\frac{-5}{3}$ (iii) 5

    (c) (i) -2 (ii) $\frac{3}{2}$ (iii) 3

    (d) (i) -3 (ii) 0   (iii) 0

9 $x = 3$, $y = 5\frac{1}{2}$

10 $x = 1$, $y = 2$

General revision test A (p. 98)

1 D   **2** C   **3** C   **4** B   **5** A

6 B   **7** D   **8** B   **9** A   **10** A

11 (a) $3(x^2 - 3y)$

    (b) $2(mn + 2m - 3n^2)$

    (c) $(3 - y)(x + 2)$

12 (a) $2(x - 3)$ (b) $-2b(a - b)$

13 (a) $x = 4$ (b) $y = 6$

    (c) $x = -2$ (d) $x = \frac{-19}{4}$

14 (a) 15   (b) 2   (c) -43 (d) $\frac{1}{2}$

15 (a) 6   (b) $\dfrac{8b^2}{9apq}$

    (c) $\dfrac{9x}{14}$ (d) $\dfrac{2x - 5}{(x + 2)(2x + 1)}$

16 (a) 7.5%

    (b) $20\,812.50

    (c) 8.4%

17 $x = 66$

18 $n = 11$

19 $A\widehat{D}B$ is angle in a semicircle

20 $x = -2$, $y = 5$

Exercise 9a (p. 103)

1 (a) 5.7   (b) 0.8, 4.2

2 (b) -4.2 (c) 4.2, -1.2

3 (b) $-1 < x < +3$ (c) 1.4

    (d) 2.4, -0.4

4 (b) 0, 3   (c) -1.25

    (d) -0.8, 3.8   (e) $0 < x < 3$

5 (a) $0 < x < +1$ (b) 0.5

    (c) 0.25   (d) -2.6

    (e) 2.8, -1.8

6 (a)

x	-2	-1	0	1	2	3	4
y	7	2	-1	-2	-1	2	7

    (c) -1.9 (d) 3.8, -1.8 (e) -2

7 (a)

x	-1	0	1	2	3	4
y	0	4	6	6	4	0

    (c) $-1 < x < +4$ (d) 1.8

    (e) 3.3, -0.3 (f) 6.25

8 (a)

x	-3	-2	-1	0	1	2	3	4
y	6	0	-4	-6	-6	-4	0	6

    (b) 3, -2 (c) -6.25

9 (a) 1.4, -4.4 (b) 6.25

Exercise P9.1 (p. 105)

1 (a) -11 (b) 38   (c) $\frac{5}{4}$

2 (a) $\{20, 8, 0, -4, -4, 0, 8\}$

    (b) $\{9, -5, -3, 15, 49\}$

Exercise P9.2 (p. 105)

1 (a) $y = x^2 + 3$

x	-2	-1	0	1	2	3
x^2	4	1	0	1	4	9
$+3$	3	3	3	3	3	3
$x^2 + 3$	7	4	3	4	7	12

    (b)–(d)

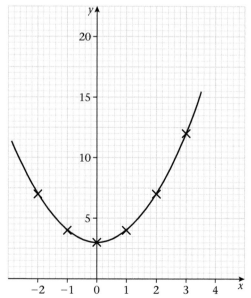

2 (a) $y = \frac{1}{2}x^2 - 4$

x	-3	-2	-1	0	1	2	3
$\frac{1}{2}x^2$	$\frac{9}{2}$	2	$\frac{1}{2}$	0	$\frac{1}{2}$	2	$\frac{9}{2}$
-4	-4	-4	-4	-4	-4	-4	-4
$\frac{1}{2}x^2 - 4$	$\frac{1}{2}$	-2	$-\frac{7}{2}$	-4	$-\frac{7}{2}$	-2	$\frac{1}{2}$

(b)–(d)

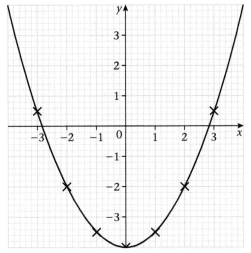

3 (a) $y = 2 - x^2$

$-x^2$	-4	-1	0	-1	-4
$2 - x^2$	-2	1	2	1	-2

(b)–(d)

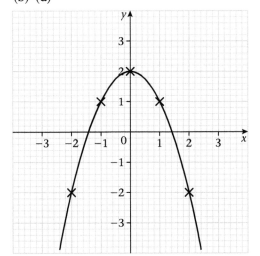

Exercise P9.3 (p. 105)

1 (a)

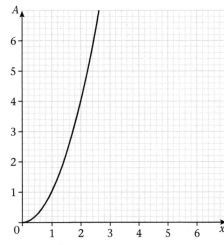

(b) $A = x^2$
(c) $5.3\,\text{cm}^2$

Exercise 10a (p. 107)

1 (a) A (b) $\frac{4}{3}$
2 (a) 0 (b) $\frac{5}{3}$ (c) $3\frac{1}{3}\,\text{cm}$
3 (a) A′(2, 2), B′(6, 2), C′(2, 8)
4 (b) 2, (4, 3)
5 P′(2, 3), Q′(4, 5), R′(6, 2.2)

Exercise 10b (p. 108)

2 (a) $1\frac{1}{4}$ (b) $15\,\text{cm}$
3 (b) $-\frac{1}{3}$ $(2\frac{1}{2}, \frac{1}{2})$
5 No. Corresponding sides are not enlarged by the same scale factor.
6 $6\,\text{cm}$, $8\,\text{cm}$, $12\,\text{cm}$
8 (i) 2 (ii) (2, 3)

Answers

1

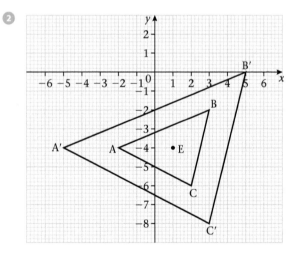

2

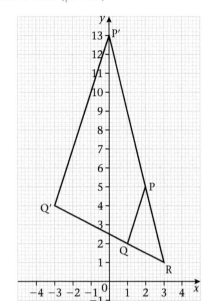

3

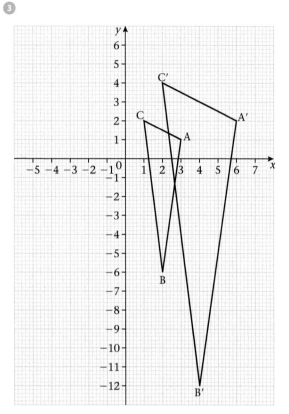

(c) ABC and A′B′C′ are similar triangles. Corresponding angles are $A\widehat{B}C$ and $A'\widehat{B}'C'$, $B\widehat{C}A$ and $B'\widehat{C}'A'$, and $C\widehat{A}B$ and $C'\widehat{A}'B'$.

4

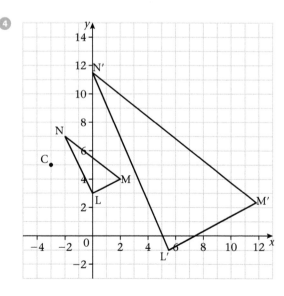

5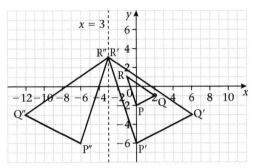

6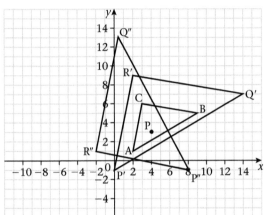

7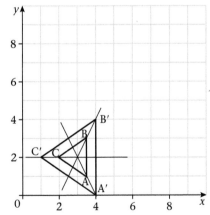

(b) scale factor 2

(c) centre of enlargement (3, 2)

8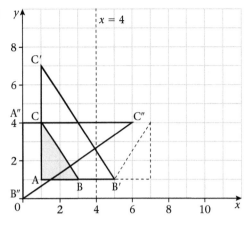

9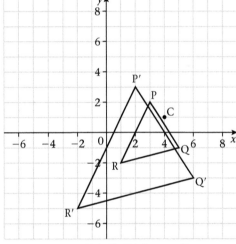

(b) P′(2, 3), Q′(6, 3), R′ (−5, −2)

Exercise 11a (p. 111)

1 (a) $l = 2\pi$; $m = 4\pi$; $n = 6\pi$

(b) $p = 5\pi$; $q = 7\pi$

(c) $x = \dfrac{9\pi}{2}$; $y = \dfrac{14\pi}{3}$; $z = \dfrac{17\pi}{6}$

2 (a) 11 cm   (b) 44 m

(c) 8.8 cm   (d) 13.2 cm

(e) 45°   (f) 60°

(g) 126 m   (h) 70 cm

3 $\dfrac{7\pi}{12}$ cm   **4** $\dfrac{11}{21}$ m   **5** 90°

6 (a) 36°   (b) 108°   (c) 180°

(d) 144°   (e) 54°   (f) 36°

7 50 mm   **8** 64 mm

9 315°   **10** 22 cm

11 22 cm   **12** 6.3 cm

Answers

⑬ (a) 120°   (b) $12\frac{4}{7}$ cm

⑭ 41 mm

⑮ (a) $327\frac{3}{11}°$ (b) 10   (c) 26.4 m

Exercise 11b (p. 113)

① (a) 4π   (b) 12π   (c) $\frac{63\pi}{2}$

② (a) $38\frac{1}{2}$ cm²   (b) 22 cm²   (c) 1540 cm²

    (d) 270°   (e) 63°   (f) 9 m

③ 77 cm²   ④ 419 cm²

⑤ 8 cm²   ⑥ 126°

⑦ (a) 42 cm²   (b) 56 cm²   (c) 112 cm²

⑧ 16.8 cm² (to 3 s.f.)

⑨ 27 cm²

⑩ (a) 253.7°   (b) 2.69 m²

Exercise 11c (p. 116)

① 100 : 9   ② 16 : 49

③ (a) 4 : 25   (b) 24 cm²

④ (a) $\frac{2}{3}$ (b) 18 cm

⑤ 200 ha   ⑥ 24 cm²

⑦ $30.00   ⑧ $1.25

⑨ 72 cm   ⑩ 220 cm by 55 cm

Exercise 11d (p. 117)

① 8 : 27   ② 1 : 8   ③ 8 : 1

④ (a) 1 : 27   (b) 324 g

⑤ (a) 3 : 2   (b) 24 cm

⑥ 3.072 litres

⑦ 6.4 litres

⑧ $40.50   ⑨ $\frac{1}{24}$ m³   ⑩ 12 cm

Exercise 11e (p. 118)

① $20   ② $1   ③ 405 ml

④ $54   ⑤ 2.7 t   ⑥ 4.608 kg

⑦ 12 500 litres

⑧ (a) 1 : 1375   (b) 38.5 m

⑨ (a) 1 : 80   (b) 1296 m²

⑩ 675 kg

Exercise P11.1 (p. 119)

① 134 m²

② (a) 1 : 2.25   (b) 1 : 1.5

③ (a) 100.57 cm²   (b) 804.57 cm²

④ (a) 3465 litres   (b) 3465 m³

    (c) 5.39 m   (d) 128.3 m³

⑤ 12 cm

⑥ (a) (i) 241 cm²   (ii) 2406 cm³

    (b) 520 cm³

⑦ (a) 3.53 m³   (b) 0.75 m

⑧ (a) 1 m³   (b) 1.0 m

Exercise 12a (p. 121)

① (a) bicycles   (b) motorbikes

    (c) 85   (d) 25

② (a) 5900 m   (b) 6200 m

    (c) Blanc   (d) 1800 m

③ (a) $\frac{1}{3}$ (b) 25%   (c) 25 min

④ (a) 37.5°C   (b) 40.5°C

    (c) 5   (d) 0500

⑥ (a) $\frac{1}{3}$ (b) $\frac{1}{4}$

    (c)

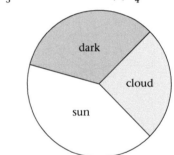

    (d) Period of sunshine: 150°

⑦

age	12	13	14	15	16
no. of students	5	8	6	8	9

(a)

(b)

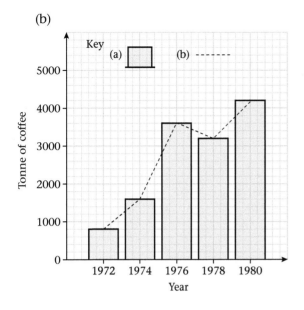

Exercise 12b (p. 124)

②

Sticks of matches per box	31–43	44–46	47–49
Frequency	17	18	15

④

Marks	Frequency
0–9	3
10–19	1
20–29	2
30–39	2
40–49	3
50–59	6
60–69	4
70–79	2
80–89	4

Exercise 12c (p. 126)

① (a) continuous
   (b) discrete
   (c) discrete
   (d) continuous

② (a) 10, 11, 12, 13, 14, 15
   (b) 59.5–65.5 kg
   (c) 15.5–19.5 kg
   (d) 19.5–24.5 cm

③

1–100	101–200	201–300	301–400	401–500
5	17	19	5	4

There is very little change in the pattern of the distribution.

④

0–49	50–99	100–149	150–199	200–249
0	5	7	9	11

250–299	300–349	350–399	400–449	450–499
8	4	2	2	2

⑤

55–59	60–64	65–69	70–74	75–79	80–64
3	7	16	14	6	4

⑥ 50.5–55.5, 55.5–60.5, …, 80.5–85.5

Exercise 12d (p. 128)

① 11–15   ② 12.7   ③ 25.5
④ mode: 32, mean: 31.4   ⑤ 22.2 years
⑥ $27\frac{3}{4}$ absentees (28, to nearest whole person)
⑦ (a) 25 students   (b) 50.1%
⑧ Yes (mean = 199.75 = 200 to the nearest whole nail)
⑨ (a)

40–44	45–49	50–54	55–59	60–64
5	9	10	5	1

   (c) 50 kg
⑩ (a)

135–144	145–154	155–164	155–164
4	12	8	6

(b) 154.8 cm

Answers

Exercise P12.1 (p. 130)

① (a) mode = 17; mean = 16.8

(b)

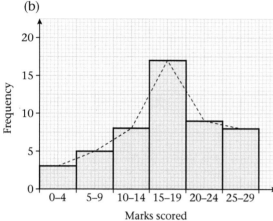

Marks scored

(c) 16

②

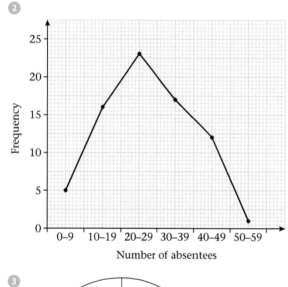

Number of absentees

③

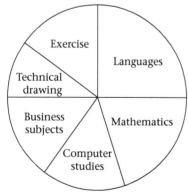

Exercise 13a (p. 132)

① (a) N   (b) S
② (a) E   (b) W
③ (a) SW   (b) NE
④ (a) SE   (b) NW
⑤ (a) NE   (b) SW
⑥ (a) W   (b) E

Exercise 13b (p. 133)

① mango tree: 040°, bridge: 066°, car: 112°, tent: 200°, well: 268°, palm tree: 308°

Exercise 13c (p. 135)

① (a) 056° (b) 240° (c) 120° (d) 270°
  (e) 327° (f) 090° (g) 133° (h) 180°

② 040°, 120°, 230°, 290°

③ 045°, 128°, 200°, 249°, 303°

④

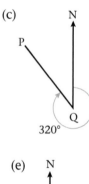

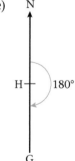

⑤ (a) (i) 250°,   (ii) 070°
  (b) (i) 150°,   (ii) 330°
  (c) (i) 057°,   (ii) 237°
  (d) (i) 270°,   (ii) 090°
  (e) (i) 319°,   (ii) 139°
  (f) (i) 015°,   (ii) 195°

Exercise 13d (p. 136)

1 $075\frac{1}{2}°$, 60 m
2 36 m
3 358°, 052°
4 061°, 40 m (30 m is an acceptable answer from a good drawing)

Exercise 13e (p. 138)

1 5 km, 037°
2 3.6 km, 236°
3 410 km, 284°
4 22 km, 109°
5 3760 m, 020°
6 (a) 145 km (b) 270 km
7 (a) 84 km (b) 15 km
8 (a) 2.4 km (b) 336°

Exercise P13.1 (p. 139)

1 488 m

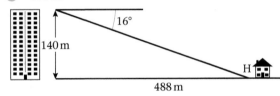

2 90°

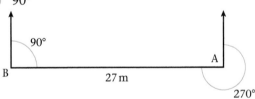

3 2665 m

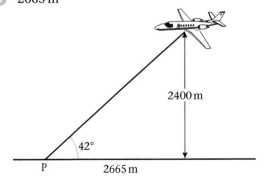

4 101 km

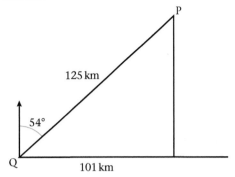

5 523 km, bearing 287°

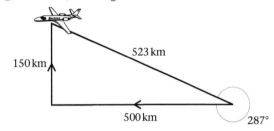

6 77 km

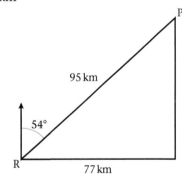

7 22 km, bearing 109°

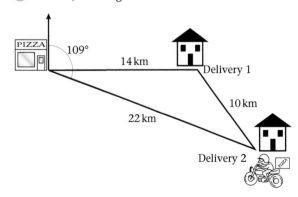

Answers

⑧ (a) 7.3 km
(b) 164°

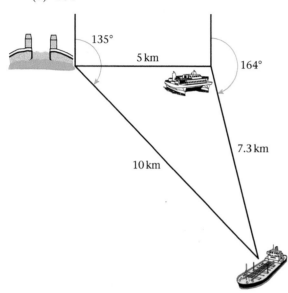

⑨ 125 km

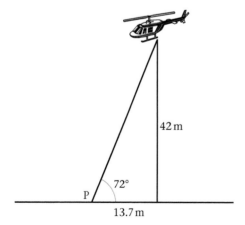

⑩ 13.7 m

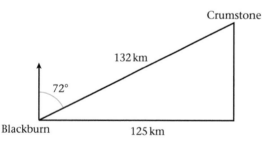

⑪ 6.3 m

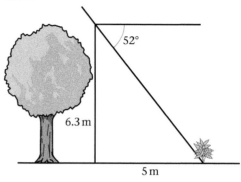

⑫ 2.2 km, bearing 244°

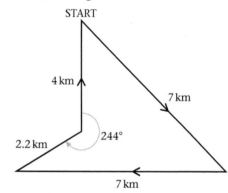

Exercise 14b (p. 142)

① 0   ② 0   ③ 0   ④ 1
⑤ 1   ⑥ 1   ⑦ 1   ⑧ 0

Exercise 14c (p. 144)

① $\frac{24}{25}$ ② $\frac{9}{10}$
③ (a) 0   (b) 0   (c) $\frac{3}{7}$ (d) $\frac{4}{7}$
④ (a) 40   (b) $\frac{5}{8}$
⑤ $\frac{3}{5}$ ⑥ 35 people
⑦ (a) $\frac{1}{5}$ (b) $\frac{1}{20}$ (c) $\frac{3}{4}$ (d) 4   (e) 1
⑧ $\frac{39}{40}$ ⑨ $\frac{9}{10}$
⑩ $\frac{1}{4}$ (however, this may not be reliable; it could be that Ebenezer has improved since the earlier games)
⑪ (a) $\frac{5}{8}$ (b) $\frac{3}{8}$ (c) 1   (d) 0
⑫ $\frac{9}{20}$
⑬ (a) 48%   (b) $\frac{12}{25}$ (approx $\frac{1}{2}$)
⑭ (a) 150   (b) 50%   (c) 26%
(d) $\frac{1}{2}$ (e) about 5

Exercise 14d (p. 146)

1
(a) $\frac{1}{6}$ (b) $\frac{1}{4}$ (c) $\frac{1}{6}$ (d) 0
(e) $\frac{1}{6}$ (f) 0   (g) 1   (h) $\frac{1}{2}$
(i) 0   (j) $\frac{1}{2}$ (k) $\frac{1}{2}$ (l) $\frac{1}{3}$
(m) $\frac{2}{3}$ (n) $\frac{5}{6}$

2
(a) $\frac{1}{52}$ (b) $\frac{1}{52}$ (c) $\frac{1}{52}$ (d) $\frac{1}{52}$
(e) $\frac{1}{13}$ (f) $\frac{1}{13}$ (g) $\frac{1}{4}$ (h) $\frac{1}{2}$
(i) $\frac{1}{26}$ (j) $\frac{1}{26}$ (k) $\frac{2}{13}$ (l) 0

3
(a) $\frac{1}{2}$ (b) $\frac{1}{4}$

4
$\frac{1}{2}$

5
(a) $\frac{1}{26}$ (b) $\frac{1}{13}$ (c) $\frac{5}{26}$
(d) $\frac{3}{26}$ (e) $\frac{4}{13}$ (f) $\frac{9}{26}$

6
(a) $\frac{2}{5}$ (b) $\frac{3}{5}$

7
$\frac{3}{8}$

8
$\frac{1}{1250}$

9
$\frac{1}{6}$

10
(a) 40
(b) (i) $\frac{9}{40}$ (ii) $\frac{31}{10}$ (iii) $\frac{7}{8}$

11
$\frac{3}{10}$

12
(b) 36   (c) 4   (d) $\frac{1}{9}$ $\left(\frac{4}{36}\right)$
(e) (i) $\frac{1}{36}$ (ii) $\frac{1}{12}$ (iii) $\frac{5}{36}$
     (iv) $\frac{1}{6}$ (v) $\frac{2}{9}$ (vi) $\frac{5}{12}$

Exercise P14.1 (p. 147)

1
$\frac{9}{32}$

2
(a) (i) $\frac{1}{5}$ (ii) $\frac{3}{10}$ (iii) $\frac{1}{2}$
(b) (i) $\frac{4}{5}$ (ii) $\frac{1}{2}$

3
(a) $\frac{2}{5}$ (b) $\frac{3}{5}$ (c) $\frac{3}{10}$ (d) $\frac{1}{2}$

4
(a) $\frac{3}{50}$ (b) $\frac{21}{100}$ (c) $\frac{1}{50}$

5
$\frac{7}{8}$

6
(a) $\frac{1}{36}$ (b) $\frac{5}{36}$ (c) $\frac{1}{12}$ (d) $\frac{5}{18}$

7
(a) $\frac{1}{10}$ (b) $\frac{9}{10}$ (c) $\frac{11}{15}$

8
(a) $\frac{5}{26}$ (b) $\frac{21}{26}$

9
(a) $\frac{2}{13}$ (b) $\frac{2}{13}$ (c) $\frac{3}{13}$ (d) $\frac{2}{13}$

10
(a) $\frac{1}{9}$ (b) $\frac{1}{3}$ (c) $\frac{2}{9}$ (d) $\frac{5}{9}$ (e) $\frac{2}{3}$

Exercise 15a (p. 149)

1
(a) $-2 < x < 4$ (b) $-2 \leqslant x \leqslant 4$
(c) $-2 < x \leqslant 4$ (d) $1 \leqslant x < 6$
(e) $-7 \leqslant x \leqslant -3$ (f) $0 < x \leqslant 3$

2

3

4

5

Exercise 15b (p. 150)

1
(a) $3 \leqslant x \leqslant 9$ (b) $-4 < x < -2$
(c) $-6 < x < 5$ (d) $-5\frac{1}{4} \leqslant x \leqslant -2\frac{2}{3}$
(e) $0 \leqslant x < 7$

2
(a) {2, 3, 4, 5, 6, 7, 8}   (b) {2, 3, 4, 5, 6, 7}
(c) {−7, −6, −5, −4, −3, −2}
(d) {−2, −1, 0, 1, 2, 3}

3 (b), (d), (e), (f)   **4** $-3\frac{1}{2} < x < \frac{2}{9}$

5 $0 < x \leqslant 1$ **6** $-4 \leqslant y \leqslant 7$

7 -2 **8** 7, 8

9 $2 < x < 6$; 4   **10** 20, 24, 28, 32

Exercise 15c (p. 151)

1 $x > 8$ **2** $y < 10.7$

3 $n \leqslant 18$ **4** 5, 6, 7

5 1, 0, −1, −2   **6** $x < 1200$

7 19   **8** $x \geqslant 12$

9 $h < 4$

10 x has a value between 3 and 6, i.e. $3 < x < 6$

⑪ b has a value from 1 to 17, i.e. $1 \leqslant b \leqslant 17$

⑫ 4

⑬ (a) 5   (b) 7

⑭ over 2 hours

⑮ over 21 km/h

Exercise P15.1 (p. 152)

① $-5 < -2$

② $h \geqslant r$

③ $(x + 2) \neq (2x - 3)$

④ $w \leqslant 5$

Exercise P15.2 (p. 152)

① $\{-1, 0, 1, 2, ...\}$

② $\{2, 1, 0, -1, ...\}$

③ $\{3, 4, 5, 6, ...\}$

④ $\{3, 2, 1, 0, ...\}$

⑤ $\{5, 4, 3, 2, ...\}$

⑥ $\{3, 4, 5, 6, ...\}$

⑦ $\{3, 4, 5, 6, ...\}$

⑧ $\{1, 2, 3, 4\}$

⑨ $\{2, 1, 0, -1, ...\}$

⑩ $\{7, 8, 9, 10, ...\}$

⑪ $\{-5, -4, -3, -2, ...\}$

⑫ $\{2, 3, 4, 5, ...\}$

⑬ $\{2, 1, 0, -1, ...\}$

⑭ $\{2, 3, 4, 5, 6\}$

⑮ $\{4, 5, 6, 7, ...\}$

⑯ $\{4, 3, 2, 1, ...\}$

⑰ $\{6, 5, 4, 3, ...\}$

⑱ $\{0, 1, 2, 3, ...\}$

Exercise P15.3 (p. 152)

① $n + 5 > 21$

② $p \geqslant 45; f < 45$

③ $65 < m \leqslant 75$

④ $70n + 50(12 - n) \leqslant 700$

⑤ $s < \sqrt{100}$

⑥ $90 < x < 180$

⑦ $3 < \dfrac{t}{6} < 4$

⑧ $\dfrac{120}{90} \leqslant v \leqslant \dfrac{120}{360}$

Exercise P15.4 (p. 153)

① $-1 < x < 3$

② $-3 < x \leqslant 2$

③ $-4 \leqslant x \leqslant 4$

Exercise P15.5 (p. 153)

① (a) $x < \frac{1}{3}$

② (a) $y \geqslant -4$

③ (a) $j > -\frac{1}{4}$

④ (a) $t \geqslant 2$

⑤ (a) $2 > n > -1$ (b) $\{0, 1\}$

⑥ $\{0, -1, -2, -3\}$

⑦ $\{-2, -1, 0, ... 3, 4\}$

⑧ $\{4, 5, 6, ... 10, 11\}$

⑨ $\{0, 1, 2, ... 7, 8\}$

⑩ $\{-6, -5, -4, ... 1, 2, 3\}$

Exercise P15.6 (p. 153)

① (a) (i) $x < 4$ (ii) $x \geqslant -3$

(b) (i)

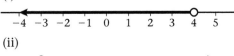

(ii)

(c) (i) $\{-3, -2, -1, 0, 1, 2, 3\}$

(ii) $\{..., -5, -4, -3, -2, ... 2, 3, 4, ...\}$

② (number line from −4 to 5)

③ (number line from −4 to 5)

Exercise P15.7 (p. 153)

① (a) If longest side is l cm, then shortest side is $(l - 5)$ cm and third side is $(l - 2)$ cm.

(b) $21 < (3l - 7) \leqslant 28$

(c) 10 cm, 8 cm, 5 cm; 11 cm, 9 cm, 6 cm

② $300

③ 9, 10, 11, 12

④ *incorrect solution*, although answer is correct; two errors of not changing the sign before 1 cancel out.

⑤ (a) $40 < v \leqslant 80$ (b) (number line 40 to 80)

Revision exercise 5 (p. 154)

① (a) (number line 0 to 8)

(b) (number line −8 to 2)

(c) (number line −6 to 2)

(d) (number line −9 to 8)

② $-7 < x < 8$ (number line −7 to 8)

③ (a) $4\frac{1}{2} < x \leqslant 7$

(b) (number line −7 to $4\frac{1}{2}$)

(c) (number line −7 to $4\frac{1}{2}$)

④ (a) (number line 0 to 8); (number line −4 to 4)

(b) (number line −4 to 8)

(c) $-2, 2, 4, 6$

⑤ (a) $x + 12 < 40$

(b) $x < 28$

6 (b) 3, −1.5
(c) 0, 1.5

7 (a)

x	−5	−4	−3	−2	−1	0	1	2
x^2	25	16	9	4	1	0	1	4
$3x$	−15	−12	−9	−6	3	0	3	6
−8	−8	−8	−8	−8	−8	−8	−8	−8
y	2	−4	−8	−10	−10	−8	−4	2

(b) 1.7, −4.7
(c) 5.75

8 (a)

x	−4	−3	−2	−1	0	1
y	9	1	−3	−3	1	9

(b) 0.2, 1.8
(c) −3.5

9 (a)

x	−1	0	1	2	3	4	5
y	10	4	0	−2	−2	0	4

(c) 1, 4   (d) $1 < x < 4$

10 (a)

x	−1	0	1	2	3	4	5	6
+4	+4	+4	+4	+4	+4	+4	+4	+4
$+5x$	−5	0	5	10	15	20	25	30
$-x^2$	−1	0	−1	−4	−9	−16	−25	−36
y	−2	4	8	10	10	8	4	−2

(c) 5.9, −0.7
(d) 10.25

Revision test 5 (p. 155)

1 A   **2** C   **3** D   **4** B   **5** D
6 $-\frac{5}{3} < x < \frac{11}{2}$ −1, 0, 1, 2, 3, 4, 5
7 (a)

x	−1	0	1	2	3	4
y	12	6	2	0	0	2

(b) 2, 3   (c) −0.25

8 (b) (i) −9.1   (ii) −0.9, 3.4
9 (a) $1\frac{1}{2}$, 2   (b) $-\frac{1}{2}$, 2
10 17, 19, 23

Revision exercise 6 (p. 156)

1 11 m   **2** $1\frac{2}{3}$ turns
3 62.5 km
4 (a) 2 : 5   (b) $27\frac{1}{2}$ cm
5 19 m²
6 (a) 9 : 16   (b) 16.5 cm
7 $240
8 (a) O(0, 0), A(7, 2), B(3, 5)
(b) O(0, 0), A′(14, 4), B′(6, 10)
9 (b) scale factor 2
10 scale factor 2, centre (0, 0)

Revision test 6 (p. 157)

1 A   **2** A   **3** C   **4** D   **5** C
6 $293\frac{1}{3}$ cm
7 (a) (i) 12.56 m   (ii) 62.8 m²
(b) 9   (c) 9.42×10^6 cm³
8 $5\frac{1}{3}$ kg
9 6 kg
10 P′(−2, 2), Q′(8, 7), R′(2, 10)

Revision exercise 7 (p. 158)

1 $41\frac{2}{3}$%
2 (a) $26 000 000   (b) $16 000 000
(c) $19 000 000
3 (a) $\frac{3}{26}$ (b) $\frac{1}{26}$ (c) $\frac{1}{4}$ (d) 0
4 (b) 18
5 (a) discrete   (b) discrete
(c) continuous   (d) discrete
6 (b) 12.3
7 (a) $\frac{x}{x + y}$ (b) 21
9 $\frac{1}{4}$
10 (a)

51–55	56–60	61–65	66–70
4	6	10	13

71–75	76–80	81–85
9	4	4

(b) 67.5 kg

Answers

1 D   **2** B   **3** B   **4** D   **5** C

8 (i) $\frac{1}{4}$ (ii) $\frac{1}{3}$ (iii) $\frac{5}{12}$ (iv) $\frac{1}{3}$

10 (c) 80–84
     (d) 68 minutes

Revision exercise 8 (p. 161)

1 (a) 24 cm
     (b) 4.8 m

2 061°, 103°, 188°, 260°, 336°

3 (a) 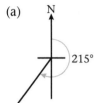   (b)

     (c)   (d)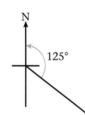

4 (a)

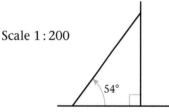

Scale 1 : 200

     (b) 5.8 m

5 (a)

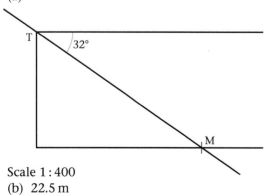

     Scale 1 : 400
     (b) 22.5 m

6 (a)

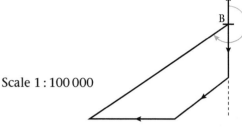

Scale 1 : 100 000

     (b) 9.8 km at 233°

7

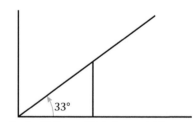

Scale 1 : 200 000

     (a) ≈ 4 km
     (b) ≈ 1.7 km

8

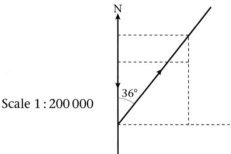

Scale 1 : 100   Height ≈ 150 cm

1 B

2 D

3 D

4 (a)

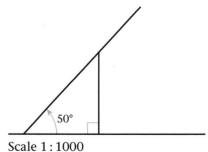

Scale 1 : 1000
     (b) 27 m

⑤ (a)

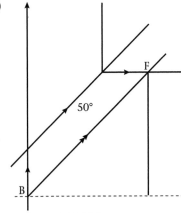

Scale 1 : 100 000

(b) (i) 4.1 km   (ii) 4.1 km

⑥ 18 cm

General revision test B (p. 162)

① C   ② B   ③ D
④ B   ⑤ D   ⑥ A
⑦ D   ⑧ C   ⑨ A
⑩ B   ⑪ 15.6
⑫ (a) $1 \leqslant x \leqslant 4$ (b) $-2 < x < 4\frac{1}{2}$

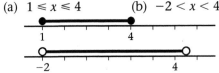

⑬ (a) taxes $33\frac{1}{3}$%, dividends $8\frac{1}{3}$%, investments
        $11\frac{1}{9}$%, materials $27\frac{7}{9}$%, payroll $19\frac{4}{9}$%
     (b) $90 000
⑭ (a) 310°, 310 km
     (b) 200 km N, 240 km W
⑮ (a) 32 cm   (b) 24.64 kg
⑯ 11.5 m
⑰ (a) scale factor $\frac{1}{2}$ (b) (0, 0)
⑱ (a)   (b) 4.5 m
                            (c) 60 m³

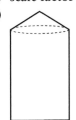

⑲ 20
⑳ (a)

x	−3	−2	−1	0	1	2
y	9	0	−5	−6	−3	4

(b)

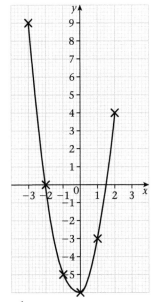

(c) $-2, 1\frac{1}{2}$ (d) -6

Exercise 16a (p. 165)

① (a) 7, 11, 13, 17, 19, 23, 29   (b) 3, 5, 7
   (c) 7, 14, 21, 28, 35, 42, 49
② (a) $12 = 2 \times 2 \times 3$
          $18 = 2 \times 3 \times 3$
          $30 = 2 \times 3 \times 5$
   (b) $28 = 2 \times 2 \times 7$
          $63 = 3 \times 3 \times 7$
          $98 = 2 \times 7 \times 7$
          $147 = 3 \times 7 \times 7$
③ (a) LCM of 12, 18, 30 = 180
          LCM of 28, 63, 98, 147 = 1764
   (b) HCF of 12, 18, 30 = 6
          HCF of 28, 63, 98, 147 = 7
④ (a) 26, 30, 34, 38   (b) 30, 35, 40, 45
   (c) 36, 49, 64, 81   (d) $\frac{3}{8}, \frac{3}{16}, \frac{3}{32}, \frac{3}{64}$
⑤ (a) 3, 7, 31, 127   (b) 3, 15, 63, 255
   (c) 511, 1023
⑥ LCM = 2520

Exercise 16b (p. 166)

① (a) 3×5^2 (b) 3×8^1
   (c) 3×10^1 (d) 3×16^0
② (a) 25_{10} (b) 18_{10} (c) 167_{10} (d) 408_{10}
③ (a) $101 000 001_2$ (b) 2241_5
   (c) 501_8

Answers

④ (a) 1110_2 (b) $11\,000_2$
(c) $110\,000_2$
⑤ (a) $11\,011_2$ (b) 305_8
(c) 117_8 (d) 1022_{16}
⑥ 16

Exercise 16c (p. 167)

① $\frac{2}{3}, \frac{7}{10}, \frac{11}{15}, \frac{4}{5}, \frac{5}{6}$ $\left(\frac{20}{30}, \frac{21}{30}, \frac{22}{30}, \frac{24}{30}, \frac{25}{30}\right)$
② (a) $4\frac{1}{6}$ (b) $2\frac{11}{24}$ (c) $2\frac{1}{8}$ (d) 1
(e) $7\frac{1}{5}$ (f) 24 (g) $3\frac{1}{3}$ (h) $6\frac{2}{3}$
③ (a) 35 (b) 26 320
④ (a) 6.966 (b) 6.966
(c) 0.069 66 (d) 0.006 966
⑤ (a) 8.626 (b) 31.6
(c) 31.9 (d) 0.185
⑥ (a) 10.065 (b) 10.1
⑦ (a) 0.004 05 km (b) 0.158 7 km
(c) 0.905 kg (d) 2400 g
(e) 7030 m (f) 1.305 litres
⑧ (a) 3.785 (b) 75.08 (c) 140
(d) 10 (e) 35 (f) 34.63
⑨ (a) 250% (b) 212.5 ha
⑩ too long by 0.3%
⑪ between 258.7 mm and 261.3 mm
⑫ $10.00
⑬ $402.50
⑭ (a) $16\frac{2}{3}\%$ (b) 3312
⑮ less, by $2\frac{1}{4}\%$
⑯ $187.50
⑰ (a) 35 min, 600 km/h, 2 h 15 min, 1530 km
(b) 665 km/h

Exercise 16d (p. 170)

① (a) 4:5 (b) 3:2 (c) 5:12 (d) 1:3
② (a) $1:1\frac{1}{2}$ (b) $1:2\frac{1}{3}$ (c) $1:\frac{3}{4}$ (d) $1:\frac{1}{5}$
③ (a) 15:6 (b) $1.70:$2
(c) 0.5 kg : 600 kg
④ (a) 7:10 (b) 30c in the $
⑤ 75 km/h
⑥ 8 for $4.80 by $6\frac{2}{3}$c per orange
⑦ (a) 1:500 000 (b) 14 km
⑧ 31 people/ha ⑨ 5 cm^3
⑩ 96 km/h ⑪ 10.8 s
⑫ 9:4 ⑬ $17\frac{1}{2}\%$ ⑭ $4\frac{1}{2}$ weeks
⑮ decreased in the ratio $n:m$

Exercise 16e (p. 171)

① 40c ② $6410 ③ $18
④ (a) $\frac{1}{3}$ (b) $\frac{2}{5}$
(c) 3:5 (d) 3:1
⑤ (a) 7:4 (b) 2:1
⑥ $40/kg
⑦ total loss = $12; loss/shirt = $1
⑧ $16.60/kg
⑨ copper 1.152 kg, sulphur 0.576 kg, oxygen 1.152 kg, water 1.620 kg
⑩ 90 km/h
⑪ Fred: $3025, Damon: $4840
⑫ $11.40 per litre

Exercise 16f (p. 172)

① 3^{11} ② $\frac{1}{256}$ ③ $5^2 = 25$ ④ $3^4 = 81$
⑤ 2^{12} ⑥ 5^{-8} ⑦ 1 ⑧ $\frac{1}{9}$
⑨ $10a^3$ ⑩ $20a^3$ ⑪ $250a^5$ ⑫ $\frac{10}{a}$
⑬ $\frac{5}{4a}$ ⑭ $250a$ ⑮ $\frac{1}{16}$

Exercise 16g (p. 172)

① (a) 9.5×10^2 (b) 9.5×10^3
(c) 9.5×10^{-1} (d) 9.5×10^{-3}
(e) 2.3×10^1 (f) 2.3×10^{-4}
(g) 2.3×10^{-2} (h) 2.3×10^4
② (a) 26 000 (b) 701 (c) 0.045 5
(d) 0.000 08 (e) 3 900 000 (f) 6020
(g) 0.001 (h) 0.87
③ (a) 0.258 (b) 2.58×10^{-1}
④ (a) 5.7×10^{-2} (b) 6.51×10^5
(c) 2.1×10^8 (d) 4.3×10^3
(e) 2.7×10^8
⑤ (a) (i) 2.6×10^2 (ii) 1.3×10^{-2}
(b) 2×10^4
⑥ (a) 1.053×10^5 (b) 8.87×10^4
⑦ (a) 1.44×10^2 (b) 4×10^4
⑧ (a) 1.296×10^7 (b) 6×10^1
⑨ (a) 1.8×10^8 (b) 3.006×10^5
(c) 5×10^2
⑩ (a) 816 (b) 8×10^{-5} m

Revision test (Ch. 16) (p. 173)

① C ② A ③ C ④ B ⑤ B
⑥ C ⑦ C ⑧ A ⑨ D ⑩ C
⑪ D ⑫ B ⑬ C ⑭ C ⑮ D
⑯ C ⑰ B ⑱ B ⑲ A ⑳ B

Exercise 17a (p. 175)

1. (a) $18   (b) $36   (c) $45
   (d) $663   (e) 18.63
2. (a) profit: $33\frac{1}{3}\%$ (b) loss: 25%
   (c) loss: $33\frac{1}{3}\%$ (d) profit: $27\frac{1}{2}\%$
   (e) loss: 40%
3. (a) $15   (b) $25   (c) $800
   (d) $289.50   (e) $290
4. $387.50   5. $10.50
6. (a) bed: $1 912.50; chest: $1 657.50;
   tables: $765
   (b) $765
7. $40 120   8. 60%   9. $81
10. (a) $32   (b) $20
    (c) 100%   (d) $1400
11. (a) $5.00   (b) $5.75
12. (a) $2550   (b) $3450   (c) 30%

Exercise 17b (p. 177)

1. (a) $315   (b) $18.75   (c) $48.93
   (d) $48.37   (e) $1 408.75
2. $52.92   3. $3\frac{1}{2}$ yr   4. $7\frac{1}{2}$ pa
5. (a) $T = 3$ yr   (b) $R = 5\%$ pa
   (c) $P = \$1000$ (d) $P = \$320$
   (e) $R = 12\frac{3}{4}\%$ pa   (f) $T = 3\frac{1}{3}$ yr
6. $1\frac{3}{4}$ yr
7. (a) (i) $2400   (ii) $600
   (iii) $1200   (iv) 8
   (b) 15%
8. (a) $2000   (b) $9000
9. $651.09   10. $6502.50
11. (a) $375   (b) $2398.93
12. $6800

Exercise 17c (p. 179)

1. (a) $189.40   (b) $359.20
2. $440.11   3. $532
4. $111.25   5. EC$10.60
6. (a) $320   (b) $18 000
7. $134.28   8. BD$564.47
9. (a) EC$4.50   (b) EC$892.50
10. (a) $42 000   (b) $28 125
11. (a) (i) $2550   (ii) $12 000
    (b) (i) 21 053   (ii) $6 315.90
    (iii) $9 473.85   (iv) $15 789.75

12. (a) $10 020   (b) $19 965   (c) 28.5%
13. $158.98
14. (a) $1560   (b) $17 490
15. (a) $36   (b) $4188
    (c) $10 212   (d) $196.01

Revision test (Ch. 17) (p. 181)

1. C   2. D   3. C   4. B   5. A
6. B   7. D   8. C   9. C   10. C
11. D   12. D   13. C   14. A   15. D
16. A   17. B   18. B   19. B   20. A

Exercise 18a (p. 184)

1. (a) {l, i, o, n, t, g, e, r}   (b) {i}
   (c) {r, e, v, t, g}   (d) {v, o, l, n}
   (e) {r, e, v, o, l, t, n, g}   (f) {v}
   (g) {v}   (h) {r, e, v, o, l, t, n, g}
   (i) {t, g, e, r}

2.

(a)   (b) 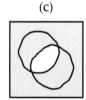   (c)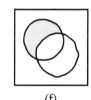

(d) 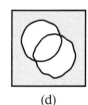   (e)   (f)

3.

(a) 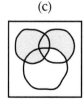   (b)   (c)

(d)   (e)   (f)

Answers

4 (a) {1, 2} (b) {5, 10}
(c) {4, 5, 6, 7} (d) {2}
(e) {1, 2, 3, 4, 5, 6, 8, 9, 10}

5 (a)

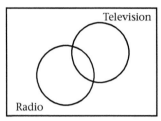

(b) 15 (c) 15 (d) 30

6 (a) 10 (b) 18 (c) 3 (d) 25

7

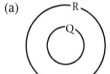

8

(a) 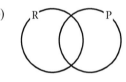 (b)

(c)

9

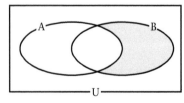

10 11 men

11 (a) $k = 3$ (b) 23 (c) 0 (d) 13

Revision test (Ch. 18) (p. 186)

1 A **2** C **3** D **4** B **5** C
6 D **7** C **8** C **9** D **10** A
11 C **12** B **13** B **14** C **15** A

Exercise 19a (p. 189)

1 (a) (i) {Gemma, Indra}
(ii) {Charles, Arthur, Ken, George}

(b) (i) {1, 8, 27, 64} (ii) {1, 2, 3, 4}
(c) (i) {10, 12, 14, 16, 18, 20}
(ii) {5, 6, 7, 8}

2 (i) A = {−5, −4, −3, −2, −1, 0, 1}
B = {−3, −2, −1, 0, 1, 2, 3}
(ii) 'is 2 less than'

3 (a) one-to-one (b) many-to-one
(c) many-to-many (d) one-to-many
(e) many-to-one

Revision test (Ch. 19) (p. 191)

1 C **2** D **3** C **4** D **5** B
6 A **7** D **8** A **9** A **10** D
11 B **12** B **13** A **14** A **15** C
16 B **17** C **18** B **19** D **20** D

Exercise 20a (p. 193)

1 36 cents, $(20 + x)$ cents
2 19 years, $(15 + y)$ years
3 $12, $(20 − a)$
4 26 years, $(23 + p)$ years
5 (a) 5 t (b) $(8 − d)$ t
6 (a) 100 (b) $5x$
7 (a) 300 (b) $100e$
8 (a) 2 (b) $\dfrac{m}{60}$

9 (a) 20 cm (b) $\dfrac{300}{b}$ cm
10 $(100x + y)$ cents
11 $\left(a + \dfrac{b}{100}\right)$ dollars

12 $\dfrac{14 − 2x}{2}$ m or $(7 − x)$ m

13 $x\left(\dfrac{20 − 2x}{2}\right)$ m² or $x(10 − x)$ m² or $(10x − x^2)$ m²
14 $(2x + 4)(x − 2)$ cm²
15 $\dfrac{d}{u}$ hours **16** $\dfrac{340}{x}$
17 vt km
18 $(ut − vt)$ km or $t(u − v)$ km

Exercise 20b (p. 194)

1 $7x$ **2** $−7a$ **3** $−33y$
4 $4y$ **5** $5x$ **6** $5z$
7 $−6a$ **8** $−7b$ **9** $−8c$
10 $5x − a$ **11** $4g − 3h$ **12** $d − 3$
13 $5a$ **14** $4x$ **15** 0
16 $4y − x$ **17** $2b − 6a$ **18** $7y − 2x$

19. $-3m - 5n$
20. $-3r + 7s + 5t$
21. $16x - 9y$
22. $2p^2 - 5q$
23. $3a - 10b$
24. $23 + 15x^2$
25. $x^2 - 10xy + 9y^2$
26. $2u^2v - 4uv^2$
27. $2d^2 - 10ad$
28. $5a^2 + 3a$
29. $7x^2 + 4x^3 + 2x^4$
30. $m^2 - mn + 4n^2$

23. $n^2 + 2n + 1$
24. $f^2 + 20f + 99$
25. $e^2 - 8e + 15$
26. $d^2 + 8d - 20$
27. $h^2 - 5h - 24$
28. $a^2 + 6a + 9$
29. $a^2 - 6a + 9$
30. $a^2 - 9$
31. $b^2 - 25$
32. $c^2 - 49$

Exercise 20c (p. 195)

1. $3a + 3b - 6$
2. $-6m + 2n - 8$
3. $5a + 9b$
4. $10y + x$
5. $u + 1$
6. $2d - 5c$
7. $12m - 2$
8. $12m - 8$
9. $3v - 3u$
10. $3d - e$
11. $5a - 3b$
12. $2n$
13. $7h + 20k$
14. $3u - v - w$
15. $-2s - 4t$
16. $7x - 11$
17. $3a + 7$
18. $13x^2 - 2x + 21$
19. $3a - 6$
20. $12x^2 - 10xy + 15y^2$

Exercise 20f (p. 197)

1. (a) $+9$ (b) $+2$ (c) -4
   (d) $+5$ (e) $+14$ (f) -10
2. (a) $+7$ (b) -7 (c) -5
   (d) -38 (e) -24
3. (a) $+7$ (b) -5 (c) $+3$
   (d) 0 (e) -6

Exercise 20g (p. 198)

1. $pr + qr + ps + qs$
2. $30 + 5y + 6x + xy$
3. $ac - bc - ad + bd$
4. $6x^2 + x - 1$
5. $4n^2 + 16n + 15$
6. $8m^2 - 30m + 27$
7. $5x^2 + 18x - 8$
8. $3a^2 + 13a - 10$
9. $x^2 + x - 72$
10. $15h^2 + 14h - 8$
11. $b^2 - 25$
12. $c^2 - 49$
13. $9t^2 - 12t + 4$
14. $4x^2 - 20xy + 25y^2$
15. $10x^2 - 17xy + 3y^2$
16. $20a^2 - 13a + 2$
17. $15a^2 - 34ad - 16d^2$
18. $6d^2 + 5de + e^2$
19. $30m^2 - mn - n^2$
20. $8k^2 - 18k - 35$

Exercise 20d (p. 196)

1. $a^2 + 5a + 6$
2. $c^2 + 5c - 6$
3. $e^2 - e - 6$
4. $d^2 - 3d - 18$
5. $x^2 - 3x + 2$
6. $a^2 + 6a + 9$
7. $b^2 - 10b + 25$
8. $m^2 - 16$
9. $n^2 + n - 20$
10. $d^2 - 4d - 21$
11. $b^2 + b - 30$
12. $p^2 - 8p + 15$
13. $q^2 - 9$
14. $u^2 - 4u - 45$
15. $v^2 - 13v + 36$
16. $2a^2 + 7a + 3$
17. $3b^2 + 14b + 8$
18. $2c^2 - 11c + 15$
19. $2d^2 - 15d - 27$
20. $4x^2 + 4x + 1$
21. $10x^2 - 11x - 6$
22. $6y^2 - 7y - 5$
23. $m^2 + 8mn + 16n^2$
24. $u^2 + 5uv + 6v^2$
25. $9d^2 - 4e^2$
26. $6b^2 + bc - 2c^2$
27. $6s^2 - 13st - 5t^2$
28. $4c^2 - 12cd + 9d^2$
29. $12m^2 - 15mn + 3n^2$
30. $8c^2 - 26ce - 45e^2$

Exercise 20h (p. 199)

1. (a) 16 (b) 5 (c) -13
   (d) 34 (e) 1 (f) 0
2. $15 - a - 2a^2$
3. (a) $2a^2 - 3a + 7$ (b) $6m - 9$
   (c) $4x^2 - 3x - 6$ (d) $4d - 1$
4. $2f^2 - 2fg + 5g^2$
5. $5b^2 - 10b$
6. $3a^2 - 10ab + 8b^2$
7. $y^2 + z^2$
8. (a) 3 (b) -9
9. $4ab$
10. $28nt - 21t^2$

Exercise 20e (p. 197)

1. $a^2 + 3a + 2$
2. $a^2 + 5a + 6$
3. $a^2 + 7a + 12$
4. $b^2 - b - 2$
5. $b^2 - b - 6$
6. $b^2 - b - 12$
7. $c^2 - 7c + 12$
8. $d^2 + 8d + 7$
9. $e^2 + 11e + 18$
10. $f^2 - 9f + 20$
11. $x^2 - 8x + 7$
12. $y^2 - 11y + 18$
13. $h^2 + 12h + 36$
14. $k^2 - 10k + 25$
15. $z^2 - 7z - 18$
16. $a^2 + 10a + 24$
17. $a^2 - 10a + 24$
18. $a^2 + 2a - 24$
19. $a^2 - 2a - 24$
20. $b^2 + 3b - 18$
21. $c^2 - 3c + 2$
22. $m^2 - 2m + 1$

Exercise 20i (p. 199)

1. $9(a - 3)$
2. $2(a + 2b - 3c + 4d)$
3. $r(3 - 8t)$
4. $a(a + c)$
5. $2m(m + 4n)$
6. $14x(3x - 2y)$
7. $3ab(14a - 17b)$
8. $x(x^2 + 3xy - 5y^2)$
9. $2x - 3$
10. $\dfrac{x(x - 5)}{5}$
11. $\dfrac{7}{5}$
12. $\dfrac{2}{3}$
13. $\dfrac{x}{2y}$
14. $(y - z)(3 + 2w)$
15. $(m + n)(5 - m)$

Answers

Exercise 20j (p. 199)

1. 1800
2. 2300
3. 2430
4. 28 000
5. 940
6. 3050

Exercise 20k (p. 200)

1. $\dfrac{5x + 1}{6}$
2. $\dfrac{12b + 11}{35}$
3. $\dfrac{19d + 48}{30}$
4. $\dfrac{13 - 22x}{21}$
5. $\dfrac{8c - 9a}{12abc}$
6. $\dfrac{6q - p}{q}$
7. $\dfrac{b + c}{b - c}$
8. $\dfrac{a}{b}$
9. $\dfrac{y - 8x}{x - 2y}$
10. $\dfrac{a}{b(a - b)}$
11. $\dfrac{x^2 - 10x - 11}{6(x - 5)}$
12. $\dfrac{5m(6 - m)}{(m - 1)(2m + 3)(m + 4)}$

Exercise 20l (p. 200)

1. 24
2. 11
3. ±15
4. −9
5. (a) 8   (b) 8   (c) 12   (d) 36
6. (a) −2   (b) −4   (c) 0   (d) 10
7. 18
8. −30
9. (a) 1   (b) 1260   (c) $1\frac{1}{3}$
10. (a) 4900   (b) $7\frac{1}{7}$

Exercise 20m (p. 201)

1. a^{12}
2. 2^6 or 64
3. c^6
4. 3×10^6
5. 6
6. $\dfrac{3}{x^3}$
7. $1\frac{1}{2}$
8. a^{-9} or $\dfrac{1}{a^9}$
9. $36a$
10. 5×10^6 or 5 000 000
11. $16v^6$
12. $-8b^6$
13. $-u^{10}$
14. $-f^{20}$
15. $x^8 y^{12}$
16. $-64u^6 v^3$
17. $a^8 m^4$
18. $-3d^4 e^{12}$
19. $-x^2$
20. $-c$

Exercise 20n (p. 201)

1. 8
2. 11
3. (a) (i) 6   (ii) −17
   (b) (i) 5   (ii) 4

4. (a) (i) 25   (ii) 626   (iii) 298
   (b) commutative but not associative
5. (a) 0   (b) 2   (c) 0   (d) 4
6. 3

Exercise 20o (p. 202)

1. −2
2. $3\frac{1}{5}$
3. 5
4. $-4\frac{1}{2}$
5. −2
6. 2
7. 5
8. 4
9. $3\frac{1}{3}$
10. $-1\frac{1}{2}$

Exercise 20p (p. 203)

1. (a) $x \geqslant 1$ (b) $x < 4$ (c) $x \leqslant 2$
   (d) $x \leqslant 4$ (e) $x > -6$ (f) $x > -5$
2. (a) $x \geqslant 2$ (b) $x \leqslant 6$ (c) $p < 47$
   (d) $x \geqslant 3$ (e) $x > 1$ (f) $x < 11$
   (g) $x \geqslant -10$ (h) $y < -47$ (i) $x \geqslant 4$
   (j) $z < 1\frac{1}{2}$
4. (a)

(b)

(c)

(d)

(e)

(f)

5. (a) $-3 \leqslant x < 1$
   (b)

6. (a) $\{-6, -5, -4\}$ (b) $\{-2, -1, 0, 1, 2\}$
7. $\{-1, 0, 1, 2, 3, 4, 5\}$
8. (a) (i) 6   (ii) −14   (iii) −9   (iv) −45
   (b) (i) 14   (ii) −6   (iii) 16   (iv) 27

Exercise 20q (p. 204)

1. 4, 2
2. 3, −5
3. −3, −4
4. 1, 2
5. −3, 4
6. −2, −5
7. 5, 2
8. −7, −2
9. $2\frac{1}{2}$, 3
10. −2, $1\frac{1}{4}$
11. 3, 0
12. −5, 1
13. −2, $1\frac{1}{2}$
14. 0, −2
15. $2\frac{1}{2}$, $-3\frac{1}{2}$
16. 4, 5
17. 2, −5
18. 8, −12
19. −2, 4
20. 1, 2

Exercise 20r (p. 205)

1. $1.65, 99c
2. 40 yr, 12 yr
3. $x = 12, y = 10$
4. $\frac{5}{8}$
5. (a) (i) $(7x + 10y)$
   (ii) $7x + 10y = 36, 3x + 20y = 39$
   (b) (i) $3   (ii) $1.50
6. 4 : 1
7. (a) length is $x + 5$ m; width is $x + 4$ m
   (b) 56 m²
8. (a) (i) $\frac{x}{2}$ (ii) $\$\left(\frac{x}{2} + 8\right)$
   (iii) $(2x + 8)$
   (b) $200 - (2x + 8) = 104$ (c) $30

Revision test (Ch. 20) (p. 206)

1	B	2	B	3	C	4	D
5	C	6	B	7	C	8	B
9	C	10	D	11	C	12	D
13	D	14	B	15	C	16	C
17	A	18	C	19	B	20	C

Exercise 21a (p. 207)

1. 62.8 cm
2. 15.84 m (1548 cm)
3. 150°, 16.5 cm
4. 50
5. 22 cm
6. 1 cm
7. 6.3 cm
8. (a) 200 m   (b) $7\frac{1}{2}$ times
9. 553.3 cm

Exercise 21b (p. 210)

1. 19.7 cm²   2. 1.113 ha
3. (a) 88 cm²   (b) 84 m²
   (c) 105 cm²   (d) 14 400 m²
4. 3850 cm², 2994 cm²
5. 5.28 kg   6. $31\frac{1}{4}\%$
7. (a) 136 cm²   (b) 10 cm
8. (a) 30.8 cm   (b) 75.5 cm²   (c) 9.4 cm²
9. 113 m
10. 14.1 cm²
11. 27.5 m
12. 15 510 cm²

Exercise 21d (p. 214)

1. 18.48 litres   2. 1.6 m
3. 1.12 m   4. 0.79 litres
5. 4.75 cm   6. 336 cm³
7. (a) 10 cm   (b) 60π cm²   (c) 96π cm³
8. 216°   9. 195π cm²
10. 20.0 cm³
11. (a) 4 cm   (b) 1.76 kg

Exercise 21e (p. 215)

1. (a) 3 cm   (b) 25 : 1
2. 9 times
3. 135 cm³
4. (a) 5 : 3   (b) 25 : 9
   (c) 125 : 27
5. (a) 6 cm   (b) 27 : 8
6. (a) 9 km²   (b) 3 km
   (c) 300 000 : 1
7. 38.5 m
8. (a) 5 : 3   (b) 25 : 9
9. 24 cm

Exercise 21f (p. 216)

1. (a) 14 min
   (b) 4.7 km h⁻¹
2. (a) 4.3 m s⁻¹
   (b) (i) 85.5 m   (ii) 84.5 m
       (iii) 20.5 s   (iv) 19.5 s
   (c) (i) 4.4 m s⁻¹   (ii) 4.1 m s⁻¹
3. (a) 4.5 m to 5.5 m
   (b) 46.75 m²
   (c) 24 m
4. (a) (i) 962.5 cm²   (ii) 19 250³
   (b) 14 cm
5. (a) (i) 8.85 ms⁻¹   (ii) 9.02 ms⁻¹
       (iii) 8.68 ms⁻¹   (iv) (8.85 ± 0.17) ms⁻¹
   (b) 48.14 cm²

Exercise 21g (p. 218)

1. 12.5 km
2. ~124 m
3. (a) 1 : 250 000
   (b) 5.5 km
4. (a) 28.2 cm²
   (b) 1 : 50
   (c) (i) 3.5 m   (ii) 7.05 m²

Answers

5 (a)

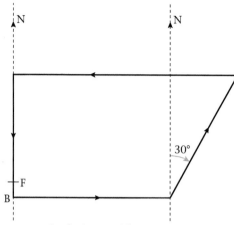

Scale 1 cm : 1 km

(b) (i) 0.5 km   (ii) due North
(c) (i) 0.5 km   (ii) due North

6 (a) 1 : 400 000
(b)
(c)

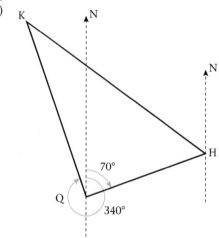

(d) 51° (50°–52°)   (e) 26 km

Revision test (Ch. 21) (p. 219)

1 A	**2** D	**3** C
4 C	**5** A	**6** B
7 C	**8** C	**9** A
10 C	**11** D	**12** B
13 C	**14** C	**15** D
16 D	**17** A	**18** D

Exercise 22a (p. 222)

1 $a = 68°$, $b = 68°$, $c = 112°$
2 $d = 70°$, $e = 95°$, $f = 15°$

3 $g = 135°$, $h = 45°$, $i = 45°$
4 $j = 48°$, $k = 48°$, $l = 132°$
5 $n = 38°$, $m = 142°$
6 $p = 55°$
7 $q = 130°$
8 $r = 130°$, $s = 80°$

Exercise 22b (p. 223)

1 $P\widehat{A}C = 105°$, $Q\widehat{B}A = 135°$, $R\widehat{C}A = 120°$
2 (a) 65°, scalene   (b) 71°, isosceles
(c) 43°, isosceles   (d) 60°, equilateral
(e) 90°, right-angled   (f) 97°, obtuse-angled
3 (a) $h = 65°$, $k = 84°$ (b) $m = n = 71°$
(c) $p = 35°$, $q = 60°$, $r = 25°$
(d) $s = 68°$, $t = 44°$, $u = 24°$
(e) $v = 47°$, $w = 29°$, $x = 151°$
(f) $y = 70°$, $z = 35°$
4 $x = 12°$
5 (a) $x = 25°$, isosceles
(b) $x = 21°$, scalene
6 (a) △XAZ   (b) △BYZ   (c) △YXC
(d) △ZDX   (e) △XEZ   (f) △GHF
7 84°
8 (a) 5 cm   (b) 6 cm   (c) 5 cm
(d) 24 cm
9 79°
10 △ABC ≡ △RPQ
△KLM ≡ △XZY

Exercise 22c (p. 226)

1

Polygon		Sum of interior angles
triangle	(3)	180°
quadrilateral	(4)	360°
pentagon	(5)	540°
hexagon	(6)	720°
heptagon	(7)	900°
octagon	(8)	1080°
decagon	(10)	1440°
dodecagon	(12)	1800°

2 (a) 100°   (b) 170°
(c) 80°   (d) 50°, 150°
3 156° each   **4** 18 sides
5 $B\widehat{C}D = 112°$, $E\widehat{A}B = 62°$, $A\widehat{E}D = 118°$
6 11°   **7** 23 cm
8 55°   **9** 11 sides
10 $x = 18°$; 72°, 90°, 108°, 126°, 144°

Exercise 22d (p. 228)

① 37° ② 67° ③ 57° ④ 50° ⑤ 64°
⑥ 284° ⑦ 56° ⑧ 59° ⑨ 75°

Exercise 22e (p. 230)

② 6.2 cm
③ AC = (4.6 ± 0.1) cm, DĈF = 50°
④ 5.9 cm
⑤ 6.9 cm
⑥ AC = 7.4 ± 0.1 cm
   CX = 5.8 ± 0.1 cm
⑦ 185 mm
⑧ 5.8 cm
⑨ 5 cm
⑩ 7.4 cm

Exercise 22f (p. 232)

① (a) D(−3, 2), E(−1, −1), F(0, 1)
   (b) P(−1, 3), Q(1, 0), R(2, 2)
   (c) $\binom{2}{1}$
② (b) A′(2, 0), B′(−1, −1), C′(1, 3)
③ (c) J(2, −2), K(−3, 4), L(−4, 0)
④ (a) (i) (2, −1), (ii) (−2, −3), (iii) (2, 0),
       (iv) (−3, 3) (v) (0, −5)
   (b) (i) (−2, 1), (ii) (2, 3), (iii) (−2, 0),
       (iv) (3, −3), (v) (0, 5)
⑤ (a), (b), (c),

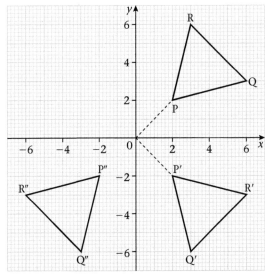

(d) a rotation of 180° about the origin or a
    reflection in the line $y = -x$

⑥ (a), (b)

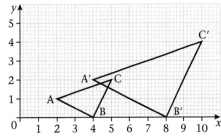

(c) A′(4, 2), B′(8, 0), C′(10, 4)
(d) $\frac{1}{4}$

Revision test (Ch. 22) (p. 233)

① D ② B ③ C ④ C ⑤ C
⑥ C ⑦ B ⑧ D ⑨ C ⑩ A
⑪ A ⑫ B ⑬ D ⑭ D ⑮ A
⑯ D ⑰ C ⑱ C ⑲ C ⑳ D

Exercise 23a (p. 237)

③ (b) 1434
⑥ 117.1
⑧ 158 cm, 159.5 cm
⑨ $47\frac{7}{15}$ kg, 46 kg
⑩ (a) 20 students   (b) 15 yr   (d) yr

⑤ (a) 4.6, 4, 4
⑦ 18 yrs 3 mo

Exercise 23b (p. 240)

① (a) (i) 20   (ii) 6
   (b) (i) 22   (ii) $3\frac{3}{4}$
   (c) (i) 9   (ii) $2\frac{3}{4}$
② (a) 10   (b) 5
③ (a) 2   (b) 1

Exercise 23c (p. 241)

① (a) 5 yr   (b) 40 children   (c) 4.7 yr
② (a) mean = 13.6 yr, median = 13.5 yr
   (b) $P_{<15} = 0.7$, $P_{\geq 15} = 0.3$
③ (b) 547.5 h
④ (b) 0–4 days   (c) 6.51 days
⑤ (c) 59.5 kg   (d) $59\frac{1}{4}$ kg
⑥ (b) (i) 51 (ii) 9.5 (c) 68%
⑦ (b) 151–160 cm   (c) 155.75 cm
⑧ (a) (i) $\frac{1}{4}$ (ii) $\frac{5}{9}$ (iii) $\frac{31}{36}$
   (b) 17 yr 4 mo
⑨ (a) $101–$125   (b) $131.50   (c) $\frac{11}{25}$
⑩ (b) 2   (c) $\frac{7}{60}$

Answers

Revision test (Ch 23) (p. 242)

① C   ② D   ③ A   ④ C   ⑤ D
⑥ B   ⑦ A   ⑧ D   ⑨ A   ⑩ B
⑪ C   ⑫ D   ⑬ B   ⑭ D   ⑮ A
⑯ D   ⑰ A   ⑱ A   ⑲ D   ⑳ B

Exercise 24a (p. 246)

① (a) 22.6° (22.7°)   (b) 28.1°   (c) 37.9°
    (d) 36.9°   (e) 10.6°
② 71.1°
③ (a) 28.1°   (b) 4 m
④ (a) 16.2° (16.3°)   (b) 7 cm
⑤ (a) 92.8°   (b) 20 cm
⑥ (a) 6   (b) 41.4°   (c) 35°   (d) 22°
⑦ (a) 18.8°   (b) 58.7 cm
⑧ (a) 32.5°   (b) 2.74 cm

Exercise 24b (p. 247)

① 180 m
② 147 m
③ (a)

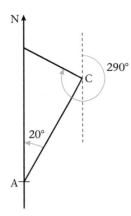

   (b) 90°
   (c) (i) 3.64 km   (ii) 10.6 km
④ $a = 11.0$ cm, $b = 22.4$ cm, $c = 15.8$ cm
⑤ (a) 5.5 cm   (b) 6.1 cm
⑥ 17.3 m
⑦ 59°
⑧ (a) 030.6°   (b) 12.8 km
⑨ 3.4 m
⑩ (a) 4.2 cm   (b) 10.9 cm   (c) 33.6 cm²

Revision test (Ch. 24) (p. 248)

① C   ② B   ③ D   ④ B   ⑤ D
⑥ A   ⑦ D   ⑧ A   ⑨ D   ⑩ C

Certificate-level practice examination
Paper 1 (p. 250)

① C   ② A   ③ D   ④ D   ⑤ B
⑥ D   ⑦ B   ⑧ D   ⑨ C   ⑩ A
⑪ D   ⑫ B   ⑬ C   ⑭ C   ⑮ B
⑯ D   ⑰ C   ⑱ A   ⑲ C   ⑳ D
⑴ D   ⑵ A   ⑶ B   ⑷ A   ⑸ B
⑹ C   ⑺ B   ⑻ B   ⑼ A   ⑽ D
⑾ A   ⑿ D   ⒀ A   ⒁ D   ⒂ C
⒃ B   ⒄ C   ⒅ A   ⒆ D   ⒇ D
⒈ C   ⒉ C   ⒊ D   ⒋ B   ⒌ C
⒍ A   ⒎ C   ⒏ B   ⒐ C   ⒑ C
⓫ D   ⓬ B   ⓭ D   ⓮ B   ⓯ C
⓰ A   ⓱ D   ⓲ B   ⓳ C   ⓴ B

Paper 2 (p. 255)

① (a) $3\frac{3}{7}$ (b) 1.05
② (a) 7.9%   (b) 5 km
③ (a) $-x + 4y$ (b) $p^2 - 3q^2$
   (c) $\dfrac{7m^2}{(m-3)(2m+1)}$
④ (a) (i) US$200   (ii) M$6
   (b) (i) 5   (ii) −15
⑤ (a) (i) 40°   (ii) 80°
   (b) 60°
   (c) (i) $\frac{1}{9}$ (ii) 3 h 20 min
⑥ (a) (i) $r = 9$ (ii) $x = 10$
   (b) (i) $x > \dfrac{-3}{2}$
   (ii)

⑦ (a)

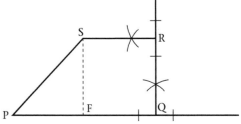

   (b) (i) 5.6 cm   (ii) 45°
   (c) △SPF is right-angled isosceles triangles
   (Pythagoras' trigonometric theorem;
   ratios)

8 (a) (i) 50 cents (ii) 70 cents
(b) $70n - 50n = 60\,000$
(c) (i) $n = 3000$ (ii) 40%

9 (a) (i) 702 (ii) $156.26
(b) (i) $23.44 (ii) $179.70

10 (a), (b)

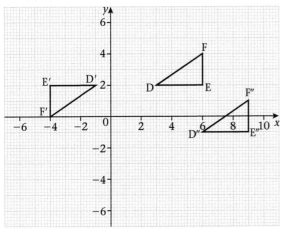

(c) (i) D″(6, −1), E″(9, −1), F″(9, 1)

(ii) translation by $\begin{pmatrix} 3 \\ -3 \end{pmatrix}$

Basic proficiency examination
Paper 1 (page 257)

1 C	**2** B	**3** B	**4** A	**5** C
6 D	**7** C	**8** B	**9** B	**10** A
11 D	**12** A	**13** A	**14** C	**15** C
16 C	**17** C	**18** D	**19** B	**20** D
21 A	**22** D	**23** C	**24** C	**25** C
26 D	**27** C	**28** B	**29** D	**30** B
31 A	**32** B	**33** D	**34** D	**35** B
36 A	**37** C	**38** C	**39** B	**40** C
41 D	**42** C	**43** D	**44** A	**45** C
46 C	**47** B	**48** B	**49** D	**50** A
51 C	**52** A	**53** B	**54** A	**55** B
56 C	**57** C	**58** D	**59** D	**60** D

Paper 2 (page 261)

1 (a) $\frac{1}{10}$ (b) 3

2 (a) $100m + n$
(b) (i) 0 (ii) 15 (iii) 25

3 (a) $343.75 (b) $41.25

4 43 kg

5 (a) $P = 5x + 5$ or $5(x + 1)$
(b) 5 cm, 12 cm, 13 cm
(c) 90°, since $13^2 = 5^2 + 12^2$
(d) 23°

6 (a) 9 cm (b) $\frac{4}{9}$

7 (a) $880
(b) $168
(c) $546
(d) $3486

8 (a) a straight line cutting the y-axis at (0, 25) and of slope 2 units
(b) (i) $25 (ii) $2 (iii) $47

9

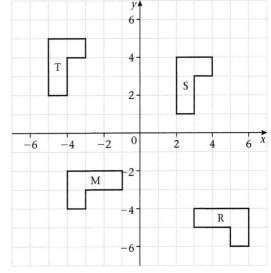

10 (a) 67
(b)

test score interval	no. of students (frequency)
91–100	1
81–90	3
71–80	3
61–70	6
51–60	7

(d) 63 (e) 19 (f) $\frac{1}{5}$

Index